POLYMER GRAFTS IN BIOCHEMISTRY

Edited by

HARRY F. HIXSON, JR.
Diagnostics Division
Abbott Laboratories
North Chicago, Illinois

and

EUGENE P. GOLDBERG
Department of Materials Science and Engineering
University of Florida
Gainesville, Florida

MARCEL DEKKER, INC. New York and Basel

CONTENTS

SYMPOSIUM ON POLYMER GRAFTS IN BIOCHEMISTRY

PREFACE

This combined issue of the Journal of Macromolecular Science-Chemistry is the outgrowth of a symposium held at the American Chemical Society meeting in April 1974 in Los Angeles. That symposium, entitled "Polymer Grafts in Biochemistry," was sponsored by the Division of Polymer Chemistry of the ACS with Dr. Eugene P. Goldberg and me as co-chairman. Dr. Goldberg, Dr. A. H. Nishikawa, and I had participated in a research program at the Xerox Research Laboratories which attempted to merge biochemistry and polymer chemistry into a new technology. During the course of that research program it became apparent to us that a large fraction of polymer chemists were not aware of the rapid developments occurring at the interface between polymer science and biochemistry, and therefore Dr. Goldberg and I proposed this symposium to the Program Committee of the Polymer Division. Obviously the proposal was accepted and this symposium resulted.

The avowed purpose of the symposium and of these journal issues is to expose the mainstream of polymer chemists and non-biochemists as well to the significant recent advances in biochemistry resulting from the developments at the interface between polymer science and biochemistry and to indicate some exciting potential new areas where progress is likely to occur. The symposium was organized into two sections according to the two largest groupings of these recent advances: Section I, Affinity or Biospecific Chromatography; and Section II, Bioactive Materials.

INTRODUCTION TO SECTION I-AFFINITY OR BIOSPECIFIC CHROMATOGRAPHY

The first group of polymer grafts which are active biologically is the affinity chromatography adsorbents. From a historical viewpoint the first application of affinity chromatography was in the purification of antibodies and antigens. These immunoadsorbents have been reviewed extensively [1] but only recently have these same principles been applied to the purification of other biological macromolecules and macroaggregates [2, 3]. The operating principle is quite simple. If a reversible equilibrium exists between two species A and B in solution,

$$A + B \rightleftarrows AB$$

then it is possible to covalently couple one member, A, of the equilibrium to a polymeric support in such a way that A, now called the ligand, may still participate in the equilibrium. If the equilibrium constant for association is large enough and sufficient time is allowed for equilibrium to be reached, then introduction of the polymer-bound ligand into a complex mixture containing B will result in the preferential adsorption of B to the polymer matrix. Washing of the matrix followed by dissociation of the A-B complex results in the isolation and purification of B. In actual practice the batch process described above and a column process in which the mixture is percolated through the polymer matrix are both used, with the latter being more common.

In such a process as that described above, the polymer-ligand adduct must possess unique properties, all of which offer significant challenge to the polymer chemist and biochemist. The polymer used for the support must permit rapid permeation and rapid approach to equilibrium for affinity chromatography to operate. Porath and co-workers [4, 5] and Cuatrecasas and Anfinsen [6] have defined the criteria for the ideal polymer support:

1) Insoluble
2) Mechanically resistant and rigid
3) Hydrophilic
4) Highly permeable to macromolecular substances

5) Resistant to chemical and microbial attack
6) Free of nonspecific adsorption sites
7) Capable of chemical modification to permit covalent coupling of ligands for specific binding to the matrix.

Polyacrylamide, cellulose, porous glass, cross-linked dextran, and agarose have all been used successfully to date; however, agarose in bead form is by far the most commonly used support material. The perfect polymeric support is not yet available as recent studies demonstrating the creation of nonspecific binding centers on agarose during cyanogen bromide derivatization have shown. Cyanogen bromide activation of agarose, dextran, and cellulose is the most frequently used method of modifying the polymer so that the ligand may be attached, but bisoxiranes [7] and divinylsulfone [8] have also been used. The chemistry of coupling ligands to porous glass and to polyacrylamide is multifaceted and varied but will not be discussed in this volume.

As stated above, the ligand (A) must be attached to the polymeric support in such a fashion that it is able to participate unhindered in the equilibrium with B. This problem takes on two aspects: (1) difunctionality and (2) distance from the polymer support. It is obvious that in addition to whatever structural feature the ligand possesses which permits it to participate in the equilibrium with B, it must also possess a functional group which enables the ligand to be coupled to the polymer support. It is also important that these two functional properties must not interact to any great extent. The design and synthesis of ligand molecules which can be permanently coupled to the support and which are fully active vis-a-vis species B and the equilibrium are critical features of any affinity chromatography system. Cuatrecasas et al. [9] demonstrated the importance of placing a spacer or extension arm between the ligand and the polymer support in the affinity purification of some enzymes. Distances of 10 to 12 A were shown to be required for satisfactory binding of these enzymes. These spacer arms are mixed blessings and in some cases actually operate to the detriment of the separation as may be seen in some of the following articles. From all of these considerations it can readily be seen that the design and synthesis of these affinity chromatography matrices is a challenging problem to the chemist and is in need of new and improved ideas and materials.

Some brief comments concerning the papers in the first section follow.

Bioaffinity Chromatography-Methodology and Application by Jerker Porath begins with a short history of affinity chromatography and then turns to the chemistry of the support matrix, especially the agar and agar derivatives which Professor Porath has been so instrumental in developing to the current level of great utility. The chemistry of such promising new derivitization methods which eliminate some of the disadvantages of the cyanogen bromide method is also discussed.

Design Parameters in Affinity Chromatography by Nishikawa, Bailon, and Ramel addresses affinity chromatography from a theoretical point of view and then tests that theory with well-designed experiments. Numerous efforts have been made to develop a theory of affinity chromatography but no one has tested his theory with direct experiment as have these authors. This is a difficult subject since all theoretical approaches require the assumption of reversible equilibrium, and the abundance of non-specific binding effects always complicates experimental testing of the theory. Nishikawa et al. have provided here a useful framework for the analysis and design of new affinity chromatography systems.

General Ligand Affinity Chromatography in Enzyme Purification by Lee and Kaplan discusses the use of ligand which is general to a class of enzymes in the isolation and purification of that class of enzymes from a complex mixture and the separation of the enzymes within that class from each other. This work represents the beginning of an attempt by numerous workers to design and synthesize a set of general ligand affinity matrices which could potentially be used to purify all enzymes. The success of such a strategy would eliminate the need to devise a new ligand-polymer combination for each new enzyme separation desired.

Application of Immunoadsorbents for Isolation of Placental Alkaline Phosphatase, Carboxypeptidase G-1, and Serum Hepatitis Antigen by Charm and Wong concerns the use of antibody-polymer combinations (immunoadsorbents) for the purification of biologically active materials. Previously, the dissociative conditions for removing the biologically active material from the immunoadsorbent frequently resulted in the destruction of its activity, but in this work suitable dissociating conditions have been worked out to retain activity for two enzymes. The system for purification of hepatitis B associated antigen from human serum represents a potential large-scale, economically useful, affinity chromatography process if the antigen is the ineffective agent for hepatitis B. Evidence at this date is that it is not, however.

Affinity Chromatography and Characterization of an Acetylcholine Receptor from Torpedo Californica by Vandlen, Schmidt, and Raftery demonstrates the diversity of uses to which affinity chromatography has been put. The purification of the membrane receptor for acetylcholine also illustrates two problems not considered in enzyme affinity chromatography. First, the ligand is a charged species, and substitution of the ligand onto the support matrix renders it an ion-exchange resin. Second, the equilibrium constants for this class of ligands are sufficiently large that the substitution level of the ligand in the matrix becomes a sensitive parameter, requiring optimization of the substitution level in the matrix.

Hydrophobic Effects in Adsorptive Protein Immobilization by Hofstee presents a systematic and thorough study of the effects of hydrophobic forces in affinity chromatography and related techniques. These effects can represent undesireable side effects in affinity chromatography or they may be used to advantage when understood and desired. This study points out that the affinity chromatography world is not an ideal world and that the practitioner must be concerned with a multitude of factors:

In summary, these articles cover the diversity of current affinity chromatography applications and also deal with recent problems and progress in the technique. Major problems which require further effort or new approaches are:

1) Better supports and coupling chemistries to minimize or eliminate nonspecific binding
2) Better dissociation methods for immunosorbents in particular so as to retain biological activity
3) Further development and especially extensive experimental testing of theoretical concepts
4) More rigorous testing of affinity methods prior to publication to determine if they are indeed affinity and not nonspecific separations

This is a new and burgeoning field, one rich in success and with even greater promise. Continued development should enable all chemists to reap the rewards of this powerful concept.

REFERENCES

[1] I. Silman and E. Katchalski, Ann. Rev. Biochem., 35, 873 (1966).

[2] W. B. Jakoby, ed., Methods Enzymol., 22, (1971).

[3] W. B. Jakoby and M. Wilchek, eds., Methods Enzymol., 34, (1974).

[4] J. Porath and L. Sundberg, The Chemistry of Biosurfaces, Vol. 2 (M. Hair, ed.), Dekker, New York, 1972.

[5] J. Porath and T. Kristiansen, in The Proteins, Vol. 1, 3rd ed. (H. Neurath and R. L. Hill, eds.), Academic, New York, 1974.

[6] P. Cuatrecasas and C. R. Anfinsen, Methods Enzymol., 22, 345 (1971)

[7] J. Porath and L. Sundberg, Protides Biol. Fluids, Proc. Colloq., 18, 401 (1970).
[8] J. Porath and L. Sundberg, J. Chromatogr., 90, 87 (1974).
[9] P. Cuatrecasas, M. Wilchek, and C. B. Anfinsen, Proc. Natl. Acad. Sci., U. S. A., 61, 636 (1968).

H. F. HIXSON, JR.

Diagnostics Division
Abbott Laboratories
February, 1976

BIOAFFINITY CHROMATOGRAPHY-METHODOLOGY AND APPLICATION

Jerker Porath

Institute of Biochemistry
University of Uppsala
Uppsala, Sweden

The term "affinity chromatography" was coined in 1968 by Cuatrecasas, Wilchek, and Anfinsen [1]. All kinds of chromatography except those depending on exclusion effects are based on affinity interactions of some kind between the solutes to be separated and the solid phase. The term is therefore too indeterminate to describe a method characterized by biological or biochemical recognition phenomena. Bioaffinity chromatography is suggested [2, 3].

Bioaffinity methods were introduced at about the same time as Tsvett made his plant pigment fractionations on columns of powdered calcium carbonate, i.e., in the first decade of this century. Hedin [4, 5] seems to have been the first one to utilize specific elution with substrate (casein) to increase the separation efficiency in batchwise purification of charcoal adsorbed trypsin. A few years later Starkenstein [6] bound liver amylase to insoluble starch. In following papers by Bockestein [7], Holmberg [8], Blom et al. [9], and others in 1920-1940, rather remarkable observations were described, e.g., the adsorption dependence on temperature and the dielectric properties of the solvent medium. Even more advanced were the experiments of

Thayer [10]. He used a substrate gradient for separation of bacterial alpha- and beta-amylase. There are interesting observations and suggestions (e.g., by Schwimmer and Balls [11, 12]) that anticipated the concept of the spacer effect.

Leloir [13] in Argentina, who made such fundamental contributions to the present knowledge of the metabolic pathways in the biosynthesis of oligosaccharides and interconversion of the monosaccharides that he received the Nobel Prize, used rather advanced bioaffinity chromatography to isolate several of the enzymes involved in these interconversion reactions.

Specific immunosorption as an analytical and diagnostic tool dates back decades. One of the earliest attempts, if not the earliest, to design an immunochromatographic procedure was made by Svensson, one of the inventors of electrofocusing [14]. He adsorbed diazotized serum albumin on charcoal and tried with some success to isolate the corresponding antibodies. The most important early contributions in the immunosorption field came from Pauling's laboratory by Campbell, Lenscher, and Lerman [15].

Lerman in 1953 [16] published the first paper in which an enzyme inhibitor was attached to a solid support (cellulose) and used for selective adsorption of the enzyme (Tyrosinase). In 1960 [17, 18] Arsenis and McCormick published a series of relevant papers in which flavin compounds were covalently bound to cellulose and used to purify flavin-dependent enzymes. Among other historically important contributions I would like especially to mention those of Fritz, Werle, and their associates in Germany [19, 20], who extensively purified proteases and protease inhibitors.

In Uppsala, Rhodén, Flodin, and I started in 1959 attempts to use cross-linked dextran (Sephadex) as support for blood group substances with the hope of purifying isoagglutinins. These studies were discontinued but taken up many years later by Kristiansen, Sundberg and myself (see, e.g., Refs. 21 and 22). Miller-Anderson and I [23] synthesized avidin-Sephadex and biotin-Sephadex and studied the adsorption behavior of labeled biotin on the former and avidin on the latter. Whereas biotin could be easily recovered from avidin-Sephadex, the avidin yield from egg-white was at best 35%. The purification factor was extremely high and could not be assessed with certainty. It was estimated to be in the range 20,000 to 100,000. On use, the adsorbent lost an intolerable part of its capacity.

The breakthrough came with the introduction of cyanogen bromide [24] as a coupling method and agarose as support [25]. Immobilized enzymes of high catalytic power were readily obtained, and the biospecific adsorbents gave very satisfactory results both in batch and column procedures.

We were very surprised indeed when the paper of Cuatrecasas, Wilchek, and Anfinsen [1] appeared only one year after we had published the BrCN method-a lagtime so short in comparison with the years it took until Sephadex was generally recognized! No doubt this fortunate situation reflects the adoptability of prepared minds.

In order to enrich to a maximum extent compounds present in trace amounts, it is necessary to suppress nonspecific interactions. The latter may otherwise predominate and decrease the separation power, thus making the procedure less attractive. Instead of an approximately 10,000-fold purification, the result may be two or three orders of magnitude lower. This very fact explains why it took so long a time until bioaffinity chromatography became a routine procedure for protein isolation.

We have studied in detail the requirements for increasing the specificity up to a maximum. Particular emphasis has been made on the solid support or matrix and the method of anchoring the ligand to become an efficient adsorption center. The development of effective adsorbents has been described in original papers [25, 48-50, 56] and in review articles [2, 3, 26-42].

Agarose has been selected as the starting material that most closely approaches the ideal one for preparing adsorbents for biopolymers. Agar and agarose chemistry is well-known, especially from the works of Araki in Japan [45] and Rees, Duckworth, and collaborators in England [46, 47]. The chemical structure is unique, as is also the physicochemical structure of agar(ose) gels. The former renders agar(ose) resistant to bacterial attacks and enzyme breakdown, which is unusual among polysaccharides. It also determines the macroporous gel structure. Figure 1(a) shows how the agarose polysaccharide chain is built up of agarobiose units (from Laas [48]). Figure 1(b) indicates schematically the structure of agarose gel (right) with junction zones of polymer aggregates. The gel structure composed of cross-linked separate polymer chains is shown for comparison (left).

The polysaccharide chains form parallel double helices in the gel state. The helix is internally stabilized by hydrogen bonds and, due to imperfections in the structure, each agarose chain will join 8 to 10 other molecules to form bundles. The consequence of this lining up of the polysaccharide chains will be a gel containing macroporous cavities supported by multiple strands of polymers. Such a structure appears to be unique and distinctly different from the macroporous gels of cross-linked synthetic polymers. It also explains some interesting features of importance for production of agar(ose)-based adsorbents with desirable properties.

If agarose is cross-linked in solution, e.g., at 65 to 70°C, a gel is formed which has much lower permeability for proteins than that common spontaneously formed (unpublished observation). The former kind of gel has a structure consistent with the one indicated in Figure 1(b), and similar to Sephadex and common cross-linked polymers.

If aged, spontaneously formed agar(ose) gels are cross-linked whereby the gel structure is largely preserved and fixed, so that it can survive heating to 100°C and above.

There is, however, apparently a latitude in free movement within the polysaccharide chain bundles which explains the elasticity and compressibility of the gel. By cross-linking with

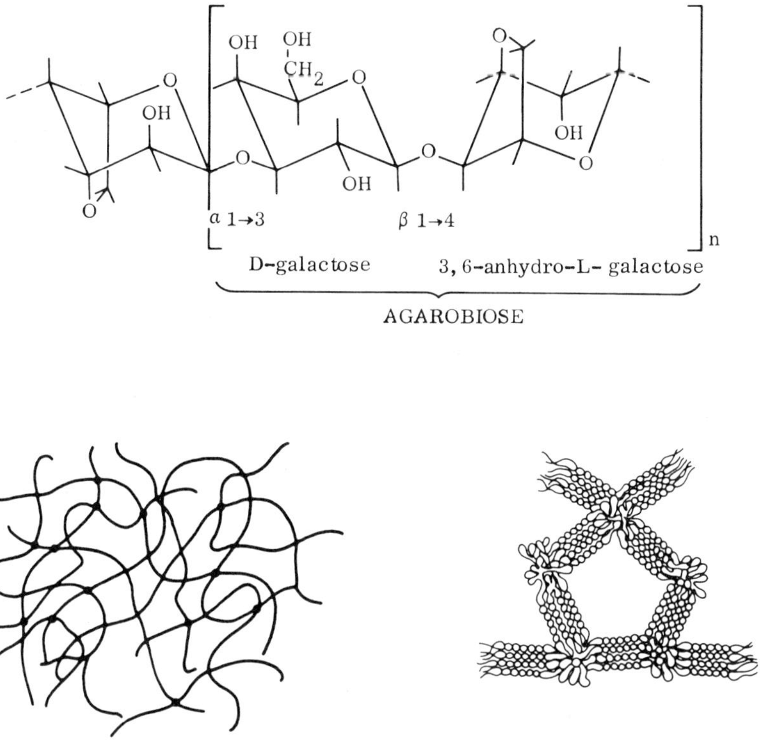

FIG. 1

(a) Main features of the structure of agarose. (b) Schematic structures of a synthetic gel (e.g., Sephadex) (left) and agaro(ose) (right) with polymer aggregates forming "junction zones" (after Låås).

bifunctional reagents of a certain chain length, the bundles will be locked and the strands will lose most of their motility. Consequently, the gel will be mechanically reinforced by such cross-links. Divinylsulfone is a cross-linking agent that will increase the gel strength to an extent that makes possible flow rates in beds of beaded gel 10 to 30 times higher than can be obtained with common agarose gels.

It is possible to cross-link an agar(ose) gel, consisting of 2% matrix, with divinylsulfone to increase the matrix density so it approaches twice the original value without significantly affecting the permeability. This can only be explained on the assumption that the large internal cavities are empty and cannot be bridged over either by single polysaccharide strands or by the cross-linker.

The unique structure is also reflected in the substitution conditions for small and large molecular size solutes. When reactive groups are introduced in the agar(ose) matrix, only a small part of them will be available for attachment of macromolecular substances. A major part will be available for substitution with small-sized solutes such as glycine, but a sizable portion of the groups may not be readily accessible even to glycine [50]. Presumably only reactive groups at the cavity linings will take part in protein coupling. The spacer effect, so much discussed, may be of importance for projecting the electrophilic groups of the activated matrix out into the bulk solution of the cavities and preventing coupling to groups located in regions where the ligands would be sterically hindered to interact with the counter ligands.

Every extra manipulation with the matrix should be regarded with suspicion. Occasionally the adsorbent may be improved, but frequently the opposite is true. The spacer may or may not relieve the obstruction imposed by the matrix. The concept of the spacer action was immediately accepted in most quarters. However, precautions are recommended in selection of spacer substance and interpretation of results in terms of biospecificity. It was an important discovery and a milestone when O'Carra and Barry [37] showed that the thiogalactoside-agarose lacks the presumed specificity toward β-galactosidase. The spacer itself was found to be responsible for the affinity. It is now clear that the spacer, if the introduction of such a structure is deemed necessary, should be hydrophilic, flexible, and neutral. Wilchek and Cuatrecasas have designed different kinds of spacer substances [51, 52]. A ligand spacer unit containing a spacer terminal nucleophilic group, NH_2 or SH, can be synthesized and subsequently be attached to the matrix. This technique is preferable to those where the spacer is built up from the matrix by solid phase synthesis. The latter will frequently result in the introduction of adsorption centers that cannot easily be removed. For example, the intermediates often contain amino or carboxyl groups. Condensation reactions such as those based on carbodiimide for water extraction will not go to completion and in addition they may introduce new substituents which in turn may cause nonspecific adsorption to occur. We have preferred to use a spacer forming reagent in the activation procedure. This is clearly demonstrated by the bisoxirane attachment [44, 50]. The weak point in the bisoxirane attachment is a sluggish reaction with many proteins in an acceptable pH-range. We are presently attempting to improve the conditions for coupling by inserting an intermediate stage with elevated reactivity compared to that of oxirane, e.g., over benzoquinone:

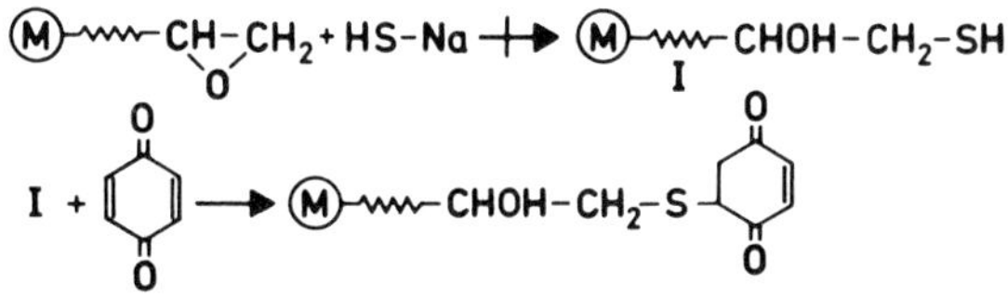

Coupling with proteins will then proceed over SH or NH_2 at lower pH:

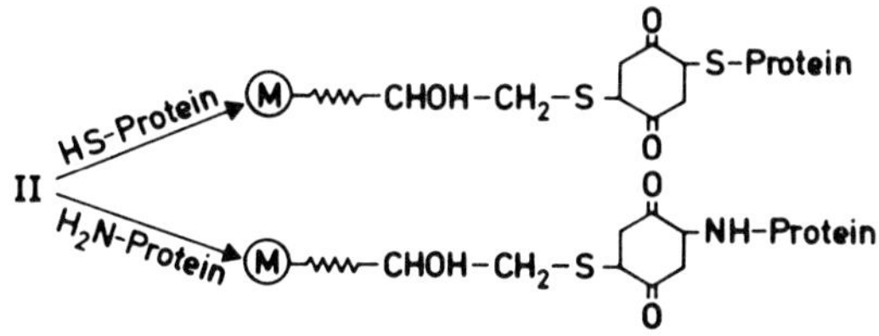

The method of attachment is frequently of crucial importance for the successful use of an immobilized ligand. Since the immobilization problem has been covered in detail in many reviews [3, 26-42], it may suffice in this context to point out that we are guided by the principle of avoiding introduction of substituents which will increase nonspecific adsorption. This implies that reagents which introduce ionic or ionogenic groups should be avoided. Also, highly hydrophobic or aromatic reagents are not permitted. A further restriction is mandatory: the linkages formed should be stable. When we consider all these requirements and the fact that an introduced spacer must have balanced hydrophilic (amphipatic) properties, we realize that very few reagents are possible. Divinylsulfone, dibromodiacetyl, n-butane-1, 4-diol-bisglycidyl ether, polymerized glutaraldehyde, and benzoquinone may serve as examples of such reagents. Cyanogen bromide, the most commonly used reagent for protein coupling, reacts in a complicated fashion to yield mixtures of imido carbonate-, isourea-, and carbamate linkages, where the two first mentioned are charge preserving. Bisoxiranes may be used for the coupling of proteins via thiol groups:

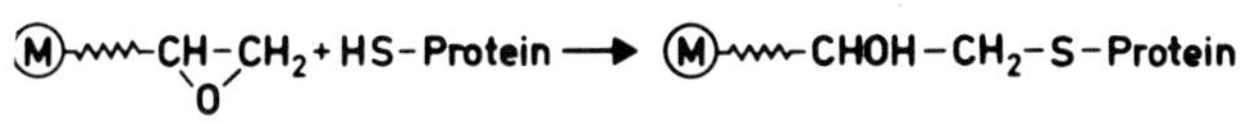

Another interesting coupling procedure is based on thiol-disulfide interchange. This method is introduced by Brocklehurst and collaborators in Professor Crook's laboratory in London [54]. A thiol containing gel is activated by 2,2'-dipyridyl disulfide:

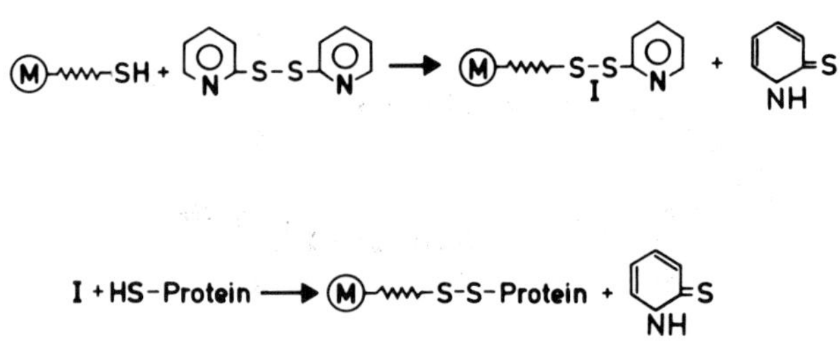

The method was further studied by Carlsson and Axén in my laboratory [55].

The leakage problem is of much concern, especially in connection with adsorbents for immunoassays. It arises as a consequence of unstable attachment, solubilization of matrix substance, or enzymic cleavage of ligand or parts of the ligand. We have cross-linked the agar to suppress leakage due to solubility. Cross-linked agarose is now commercially available (Pharmacia Fine Chemicals). Cyanogen-bromide-activated agarose yields a variety of connector groups, one or two of which appear not to be completely stable. Multipoint attachment can bring down the leakage of ligand substance, but this is likely to create new problems: steric hindrance and extinction of ability to form adsorption complex. Another way is to use an agent for activation that will secure formation of an extremely stable bond. Bisoxirane is such an agent.

Isolation of proteins and naturally occurring substances is just as much an art as a science. Bioaffinity chromatography evokes the hope to short-cut isolation procedures. Arguments in favor of such hopes are not lacking, but there are indeed still many problems to be solved.

Prefractionation may be necessary even if adsorbents are available for application of biospecific methods. Among possible alternatives the best fractionation route is seldom obvious. Also, employment of highly selective adsorption procedures may require additional fractionation methods, since the desorbed material may consist of a mixture of many substances, all with affinity for the ligand, but differing in molecular size, distribution of surface exposed ionic or hydrophobic groups, etc. Only experience makes possible the design of optimal fractionation conditions for each particular separation problem.

A problem of great importance for the fractionation strategy concerns the selection of ligand. Should the ligand be highly specific or selective, or should it exhibit group specificity? If the former is desirable, one has to synthesize or isolate the ligand substance from natural sources. This in itself may be just as difficult as to use nonspecific methods for isolation of the desired substance. If a reusable adsorbent can be obtained, it may be worthwhile to engage in lengthy procedures for the preparation of the efficient bioadsorbent. In other cases it may be more time and labor saving to prepare group specific adsorbent and instead increase the specificity in chromatography by employing specific elution techniques.

Mosbach [57, 58], Dean and Lowe [59], O'Carra [37], Tesser [60], and others have prepared nucleotide containing adsorbents for purification and isolation of enzymes with nucleotides as cofactors or modifiers. Such group-specific adsorbents are extremely versatile. From a practical point of view the gain in convenience far overshadows the negative consequences of a lower affinity. Cofactor adsorbents can be further developed in other areas, e.g., pyridoxial derivatives for transaminases [41] and specific elution techniques can be further advanced. The methods of attachments can also be much simplified.

We find numerous examples in the literature of group-specific adsorbents directed toward proteolytic enzymes with substrate analogs or inhibitors as ligands. Conversely, enzyme inhibitors may be isolated by chromatography on beaded immobilized enzymes (for references see the reviews 2, 3, 26-42).

Attachment of monosaccharides to the matrix will result in adsorbents which are group specific for the carbohydrate ligand. For example, by coupling galactose or lactose to agarose, a highly selective adsorbent is obtained for lectins which can react with terminal galactose. The saccharide coupling can be easily accomplished by divinylsulfone:

$$\text{Ⓜ-OH} + H_2C{=}CH{-}SO_2{-}CH{=}CH_2 \rightarrow \underbrace{\text{Ⓜ-O-}CH_2{-}CH_2{-}SO_2{-}CH{=}CH_2}_{\mathrm{I}}$$

$$\mathrm{I} + ROH \rightarrow \text{Ⓜ-O-}CH_2{-}CH_2{-}SO_2{-}CH_2{-}CH_2{-}O{-}R$$

The lectins in turn may serve as group-specific adsorbents for mono-, oligo-, and polysaccharides, glycoproteins and presumably also for glycolipids. The lectin-agarose gels may turn out to be extremely useful for isolation of membrane fragments and viruses [61, 62].

The reciprocal interrelationships between hormones and hormone receptors can be studied by bioaffinity chromatography and used to advantage for isolations of the latter [63-65].

Low molecular weight biologically active compounds such as vitamins and steroid hormones are frequently bound to carriers in blood and other fluids for nutrient and waste product distribution. In immobilized form these substances can be used for isolation or analysis of their corresponding complex partners, i.e., estradiol-agarose for isolation of estradiol carriers [66]. Since steroids are highly hydrophobic, we encounter in the corresponding adsorbents an inherent weakness or limitation of the bioaffinity methods. The ligand is amenable to interact in a nonspecific way with hydrophobic compounds including many proteins. If the matrix is highly substituted with steroid ligands, it will become a strongly hydrophobic adsorbent. Presumably the nonspecific interaction can be suppressed by mixing the solvent with moderately polar substances such as ethylene glycol and dimethylsulfoxide. The pitfall has not always been realized by those who have used the bioaffinity methods.

Immunosorption expands the affinity methods to almost unlimited fields of application. Antigen and antibody isolation by immunological recognition dependent procedures have been practiced since early years of this century. However, technical improvements in connection with enzyme-affinity chromatography have also greatly improved immunosorption methods. They can be used for purification of small molecular size compounds as well as for macromolecules. I may only mention the brilliant work of Wilchek, Gival, and co-workers on the development of affinity procedures for isolation of peptides from protein digests [67, 68].

.oints at the present stage in ...me of them may not be so serious ... soon be oversome. Other qualification ...so be met in industry, notably the economics. ...igid, inexpensive supports or matrices may not be ...produce. In attempts to find such materials it is likely ...t complications will arise due to extensive nonspecific adsorption. For example, ceramic matrix materials may show excellent mechanical properties but intolerable adsorption characteristics. Because of their unique properties, agar and agar derivatives are likely to continue to be the supports of choice, if other polysaccharides or even synthetic polymers such as Speron find their special fields of application.

REFERENCES

[1] P. Cuatrecasas, M. Wilchek, and C. B. Anfinsen, Proc. Natl. Acad. Sci. U. S. A., 61, 636 (1968).
[2] J. Porath, Biochimie, 55, 943 (1973).
[3] J. Porath, and T. Kristiansen, in The Proteins, Vol. 1, 3rd ed. (H. Neurath and R. L. Hill, eds.), Academic, London, 1975, pp. 95-178.
[4] S. G. Hedin, Biochem. J., 2, (1907).
[5] S. G. Hedin, Z. Physiol. Chem., 50, 497 (1907).
[6] E. Starkenstein, Biochem. Z., 24, 210 (1910).
[7] P. T. Bockestein, Acta Brevia Neerl., Physiol. Pharmacol. Microbiol., 2, 132 (1932).
[8] O. Holmberg, Biochem. Z., 258, 134 (1933).
[9] J. Blom, A. Bak, and B. Brae, Z. Physiol. Chem., 250, 104 (1937).
[10] P. S. Thayer, J. Bacteriol., 66, 656 (1953).
[11] S. Schwimmer and A. K. Balls, J. Biol. Chem., 179, 1063 (1949).
[12] S. Schwimmer and A. K. Balls, Ibid., 180, 883 (1949).
[13] L. F. Leloir and S. H. Goldenberg, Ibid., 235, 919 (1960).
[14] H. Svensson, Kem. Min. Geol., 17A (5), (1943).
[15] D. H. Campbell, E. Leuscher, and L. S. Lerman, Proc. Natl. Acad. Sci. U. S. A., 37, 575 (1951).
[16] L. S. Lerman, Ibid., 39, 232 (1953).
[17] C. Arsenis and D. B. McCormick, J. Biol. Chem., 239, 3093 (1964).
[18] C. Arsenis and D. B. McCormick, Ibid., 241, 330 (1966).
[19] H. Fritz, H. Schult, M. Hutzel, M. Wiedermann, and E. Werle, Hoppe-Seyler's Z. Physiol. Chem., 348, 308 (1966).

[20] H. Fritz, G. Wunderer, and B. Dittman, Ibid., 353, 893 (1972).
[21] T. Kristiansen, L. Sundberg and J. Porath, Biochim. Biophys. Acta, 184, 93 (1969).
[22] J. Porath, Nature (London), 218, 834 (1968).
[23] J. Porath, Gamma Globulins, Proceedings of the Nobel Symposium, 3rd., Halsted, New York, 1967, 1967, p. 287.
[24] R. Axen, J. Porath, and S. Ernback, Nature (London), 214, 1302 (1967).
[25] J. Porath, R. Axén, and S. Ernback, Ibid., 215, 1491, (1967).
[26] P. Cuatrecasas and C. B. Anfinsen, Methods Enzymol., 22, 345 (1971).
[27] J. Porath, and L. Sundberg in The Chemistry of Biosurfaces, Vol. 2 (M. L. Hair, ed.) Dekker, New York, (1972), p. 653.
[28] G. Feinstein, Biochim. Biophys. Acta, 236, 74 (1971).
[29] F. Friedberg, Chromatogr. Rev., 14, 121 (1971).
[30] R. H. Reiner and A. Walch, Chromatographia, 4, 578 (1971).
[31] J. Chudzik and A. Koj, Postepy Biochem., 18, 73 (1972).
[32] V. Kocemba-Sliwowska, Ibid., 18, 59 (1972).
[33] H. H. Weetall, in The Chemistry of Biosurfaces, Vol. 2 (M. L. Hair, ed.), Dekker, New York, (1972), p. 597.
[34] H. O. Orth and W. Brummer, Angew. Chem., Int. Ed. Engl., 11, 249 (1972).
[35] P. Cuatrecasas, Adv. Enzymol., 36, 29 (1972).
[36] O. Pihar, Chem. Listy, 65(7), 713 (1971).
[37] P. O'Carra, S. Barry, and T. Griffin, Biochem. Soc. Trans., 1, 289 (1973).
[38] H. Guilford, Chem. Soc. Rev., 1973, 249.
[39] H. H. Weetall, Sep. Purif. Methods, 2(2), 199 (1973).
[40] M. Wilchek, Adv. Exp. Med. Biol., 42, 15 (1974).
[41] P. O'Carra, in Industrial Aspects of Biochemistry (B. Spencer, ed.), Fed. of European Societies, North Holland, 1974, p. 174.
[42] J. Turkova, J. Chromatogr., 91, 267 (1974).
[43] M. J. Comer, D. B. Craven, A. Atkinson, and D. G. Dean, Eur. J. Biochem., 55, 201 (1975).
[44] W. B. Jakoby and M. Wilchek, eds., Methods Enzymol., 34B, (1974).
[45] C. Araki, Bull. Chem. Soc. Japan, 29, 543 (1956).
[46] D. A. Rees, Chem. Ind., 16, 630 (1972).
[47] M. Duckworth and W. Yaphe, Carbohydr. Res., 16, 189 (1971).
[48] T. Laas, "Agar Derivatives for Chromatography, Electrophoresis and Gel-bound Enzymes," Dissertation, Uppsala, 1975.
[49] J. Porath, T. Laas, and J.-C. Janson, J. Chromatogr., 103, 49 (1975).
[50] L. Sundberg and J. Porath, Ibid., 90, 87 (1974).
[51] P. Cuatrecasas, J. Biol. Chem., 245, 3059 (1970).
[52] M. Wilchek, FEBS Lett., 33, 70 (1973).
[53] J. Brandt, L.-O. Andersson, and J. Porath, Biochim. Biophys. Acta, 386, 196 (1975).
[54] K. Brocklehurst, J. Carlsson, M. P. J. Kierstan, and E. M. Crook, Biochem. J., 133, 573 (1973).

[55] J. Carlsson, R. Axen, K. Brocklehurst, and E. M. Crook, Eur. J. Biochem., 44, 189 (1974).
[56] J. Porath, J.-C. Janson, and T. Laas, J. Chromatogr., 60, 167 (1971).
[57] K. Mosbach, H. Guilford, P.-O. Larsson, R. Ohlsson and M. Scott, Biochem. J., 125, 20 (1971).
[58] K. Mosbach, H. Guilford, R. Ohlsson, and M. Scott, Ibid., 127, 625 (1972).
[59] C. R. Lowe and P. D. G. Dean, FEBS Lett., 14, 313 (1971).
[60] G. J. Tesser, H.-V. Fisch, and R. Schwyzer, Ibid., 23, 56 (1972).
[61] T. Kristiansen, in Protides of the Biological Fluids, XXIII Colloquium, Bruges 1975 (H. Peeters, ed.), Pergamon, Elmsford.
[62] M. J. Hayman, J. J. Skekel, and M. J. Crumpton, FEBS Lett., 29, 185 (1973).
[63] P. Cuatrecasas, Proc. Natl. Acad. Sci. U. S. A., 69, 1277 (1972).
[64] E. Karlsson, E. Heilbronn, and L. Widlund, FEBS Lett., 28, 107 (1972).
[65] J. Schmidt and M. A. Raftery, Biochem. Biophys. Res. Commun., 49, 572 (1972).
[66] J. C. Nicolas, M. Pons, B. Descomps, and A. Crastes de Paulet, FEBS Lett., 23, 175, (1972).
[67] D. Givol, Y. Weinstein, M. Gorecki, and M. Wilchek, Biochem. Biophys. Res. Commun., 38, 827 (1970).
[68] M. Wilchek, V. Bocchini, M. Becker, and D. Givol, Biochemistry, 10, 2828 (1971).

GENERAL LIGAND AFFINITY CHROMATOGRAPHY IN ENZYME PURIFICATION

Ligands, Affinity Chromatography, Enzyme Purification

Chi-Yu Lee and Nathan O. Kaplan
Department of Chemistry
University of California, San Diego
La Jolla, California 92093

INTRODUCTION

Recent developments in enzyme purification have revolutionized isolation techniques and at the same time provided a more profitable manner in which to obtain them. These developments are primarily concerned with the preparation and usability of immobilized ligands for the selective isolation and purification of biologically active substances[1-5], particularly enzymes. This technique is based on the unique principle that the biological macromolecule is able to be adsorbed to the immobilized ligand, specifically and reversibly, unlike the traditional methods by which proteins are separated by their differences in physicochemical properties. Numerous proteins have been purified since this technique was first applied by Campbell et al.[6] for the purification and isolation of antibodies. From these preliminary efforts it is apparent that further development of this method would greatly facilitate our understanding and utilization of enzymes.

Purification of proteins based on biospecific adsorption is generally termed "affinity chromatography"[1-3]. There are two basic procedures that one may use to purify enzymes via affinity column techniques. One is to prepare a ligand which is specific to the enzyme under study[1]. The ligands of high specificity are usually inhibitors, substrates, or perhaps antibodies which can bind a particular enzyme[1, 6, 7] to the column. The second method is the use of a general ligand which would have the capacity to adsorb a large number of enzymes or family of enzymes[8, 9]. Ligands exhibiting general specificity are able to bind a family of enzymes which have cofactors or inhibitors in common; compounds immobilized on Sepharose are usually derivatives of these.

We have applied the technique of affinity chromatography primarily to two classes of enzymes, namely, dehydrogenases and kinases. Dehydrogenases are pyridine nucleotide-dependent enzymes which require either DPN^+ (NAD^+) or TPN^+ ($NADP^+$) as cofactors. These enzymes which utilize nucleotide cofactors are usually involved in the enzymatic oxidation-reduction of alcohols, aldehydes, ketones, or acids. Kinases, on the other hand, catalyze phosphate transfers, where the cofactors ATP and ADP are phosphate donors and acceptors, respectively.

Since there are a large number of dehydrogenases and kinases, the question arises then as to whether a particular type of general ligand column can be employed to initially bind them and then specifically elute a given enzyme from the column. Specific elution would use either an inhibitor, substrate, or dead-end ternary complex which is specific only for this particular enzyme.

Adenine nucleotide derivatives, such as 5'-AMP, 5'-ADP, 5'-ATP, DPN^+, or TPN^+, are known either as inhibitors or cofactors of dehydrogenases as well as kinases[8-12]. Various adenine nucleotide derivatives immobilized on Sepharose are found to bind numerous dehydrogenases and kinases [10, 12-14]. Thus the success of using general ligand affinity columns for protein purification may very well depend on whether the bound enzyme can be eluted specifically from the column under the proper experimental condition. In this article we present a brief review on the purification of these enzymes using general ligand affinity column techniques.

VARIOUS ADENINE NUCLEOTIDE DERIVATIVES PREPARED FOR GENERAL LIGAND AFFINITY CHROMATOGRAPHY

We have mentioned that the ligands used as general ligands in affinity chromatography are able to retain a family of enzymes from crude extracts of various sources. Most of the general ligands prepared exhibit relatively high affinity either for dehydrogenases or kinases. However, some of these ligands have the capacity to adsorb more than one class of enzymes.

The Sepharose-bound ligands are usually derivatives of adenine mononucleotides as well as diphosphopyridine or triphosphopyridine coenzymes which are covalently attached to Sepharose through long carbon chain spacers. As shown in Table 1, at least twenty different types of derivatives have been prepared as general ligands for affinity columns. Based on the positions of the attachment of spacers to the ligand, five major groups of affinity adsorbants can be discerned. These general ligands are denoted by Roman numerals in Table 1 and in the following descriptions within the text of the paper.

Attachment Through C^8 Position of Adenine Mononucleotides or Dinucleotides

There are several affinity ligands in which the spacer 1,6-diaminohexane was attached to the C^8 position of adenine nucleotides[12, 13, 15, 16]. These adsorbants are listed as follows: 8-(6-aminohexyl)-amino-5'-AMP-Sepharose (I), 8-(6-aminohexyl)-amino-TPN^+-Sepharose (IV)[13], and corresponding 8-substituted ADP- and ATP-Sepharose (V and VI)[13, 15]. Ligands (I) and (II) exhibit good affinity to DPN^+-dependent dehydrogenases; ligands (III) and (IV) exhibit good affinity to TPN^+-dependent enzymes. 8-(6-Aminohexyl)-amino-5'-AMP-Sepharose exhibited high affinity not only to DPN^+-dependent dehydrogenases but also to other classes of enzymes such as nucleases, phosphorylase a, DPN^+-dependent isocitrate dehydrogenase from yeast, and other enzymes which are either activated or inhibited by 5'-AMP. Ligands (V) and (VI) exhibit good affinity to kinases.

TABLE 1. Nucleotide Derivatives as General Ligands for Affinity Columns

	Name of ligand	Position of space attachment	General specificity of enzymes
I.	8-(6-Aminohexyl)-amino-5'-AMP[12,15]	C^8 of adenine base	DPN^+-dependent dehydrogenases; nucleases, enzymes which are activated or inhibited by 5'-AMP
II.	8-(6-Aminohexyl)-amino-DPN^+ [12]	C^8 of adenine base	DPN^+-dependent dehydrogenases
III.	8-(6-Aminohexyl)-amino-2'-AMP[12]	C^8 of adenine base	TPN^+-dependent dehydrogenases from; transhydrogenase from Pseudomonas aeruginosa
IV.	8-(6-Aminohexyl)-amino-TPN^+ [12]	C^8 of adenine base	TPN^+-dependent dehydrogenases
V.	8-(6-Aminohexyl)-amino-ADP[15]	C^8 of adenine base	Dehydrogenases and kinases
VI.	8-(6-Aminohexyl)-amino-ATP[15]	C^8 of adenine base	Kinases
VIA.	P-8-Diazo-DPN^+[N-(6-aminohexyl)-benzamide][21, 22]	C^8 of adenine base	DPN^+-dependent dehydrogenases
VIB.	P-8-Diazo-ADP[N-(6-aminohexyl)-benzamide][24]	C^8 of adenine base	Kinases and a few dehydrogenases
VIC.	P-8-Diazo-ATP[N-(6-aminohexyl)-benzamide][24]	C^8 of adenine base	Kinases
VII.	N^6-(6-Aminohexyl)-5'-AMP[14]	N^6 of adenine bases	DPN^+-dependent dehydrogenases; enzymes activated or inhibited by 5'-AMP

VIII.	DPN^+-[N^6-N-(6-aminohexyl)-acetamide][22]	N^6 of adenine base	DPN^+-dependent dehydrogenases
IX.	6-Mercapto-purine riboside-5'-phosphate S^6-[N-(6-aminohexyl)-acetamide][22]	S^6 of 6-mercapto purine	DPN^+-dependent dehydrogenase; adenosine deaminase
X.	6-Mercapto-purine riboside-5'-triphosphate[22]	S^6 of 6-mercapto purine	Kinases
XI.	6-Mercapto-purine-DPN^+ [22]	S^6 of 6-mercapto purine	DPN^+-dependent dehydrogenases
XII.	5'-AMP (6-amino-caproiate)[9, 29]	Ribose of nucleo-tides	DPN^+-dependent dehydrogenases
XIII.	DPN^+-(6-amino-caproiate)[9, 29]	Ribose of dinucleo-tides	DPN^+-dependent dehydrogenases
XIV.	TPN^+-(6-amino-caproiate)[9, 29]	Ribose of dinucleo-tides	TPN^+-dependent dehydrogenases
XV.	DPN^+-(6-amino-adipic hydra-zide)[30, 31]	Ribose of nucleotides	DPN^+-dependent dehydrogenases
XVI.	TPN^+-(6-amino-adipic hydra-zide)[30, 31]	Ribose of nucleotides	TPN^+-dependent dehydrogenases
XVII.	ATP-(6-amino-adipic hydra-zide)[30, 31]	Ribose of nucleotides	Kinases; myosine
XVIII.	3-Amino-pyridine-DPN^+-N^3-[N-(6-amino-caproamide)]	N^3 of 3-amino-pyridine DPN^+	DPN^+-dependent dehydrogenases; kinases

TABLE 1 (Continued)

XIX.	3-N-(6-Aminohexyl)-nicotinamide-DPN^+	Nicotinamide of DPN^+	DPN^+-dependent dehydrogenases; kinases
XX.	3'-N-(6-Aminohexyl)-nicotinamide-deamino-DPN^+ [13]	Nicotinamide of deamino-DPN^+	DPN^+-dependent dehydrogenases; kinases
XXI.	3-[4-(2-Aminoethyl)-amino naphthyl]-azo-pyridine-deamino-DPN^+	C^3 of pyridine of pyridine-deamino-DPN^+	Dehydrogenases
XXII.	P^1-(6-Aminohexyl)-P^2-(5-adenosine)-pyrophate[15]	Terminal phosphate of 5'-ADP	DPN^+-dependent dehydrogenases; kinases
XXIII.	8-(6-Aminohexyl)-amino-desulfo-CoA[17, 20]	C^8 of adenine base	CoA-dependent enzymes
XXIV.	8-(6-Aminohexyl)-amino-2',5'-ADP[17]	C^8 of adenine base	TPN^+-dependent enzymes
XXV.	8-(6-Aminohexyl)-amino-3',5'-ADP[17]	C^8 of adenine base	CoA-dependent enzymes
XXVI.	TPN^+-N^6-[N-(6-aminohexyl)-acetamide][9]	N^6 of adenine base	TPN^+-dependent enzymes
XXVII.	N^6-(6-Aminohexyl)-2',5'-ADP[26]	N^6 of adenine base	TPN^+-dependent enzymes
XXVIII.	N^6-(6-Aminohexyl)-3',5'-ADP[26]	N^6 of adenine base	CoA-dependent enzymes
XXIX.	ATP-N^6-[N-(6-aminohexyl)-acetamide][28]	N^6 of adenine base	ATP-dependent enzymes; citrate synthase

Preparations of these ligands follow a relatively simple and straightforward procedure (Fig. 1). This usually involves a two-step synthesis, i.e., an initial direct bromination at the C^8 position of nucleotides and then substitution with 1,6-diaminohexane. However, some modifications are required for the preparations of 8-substituted DPN^+ or TPN^+ derivatives. Since the aqueous alkaline conditions are not favorable for oxidized coenzymes, 8-Br-DPN^+ or 8-Br-TPN^+ should be reduced enzymatically before the substitution reactions can proceed in an organic solvent such as DMSO. The yield for the preparations of these 8-substituted adenine nucleotide derivatives ranged from 50 to 95%[13].

Recently the direct synthesis of 8-(6-aminohexyl)-amino-ATP (compound III) has been developed in our laboratory[13]. This involves a direct bromination of ATP to 8-Br-ATP[12] and then a subsequent replacement reaction by 1,6-diaminohexane under alkaline conditions to form 8-(6-aminohexyl)-amino-ATP. Unlike the indirect synthesis reported by Trayer et al.[15], this improved procedure allows a relatively large-scale preparation of the general ligand for many ATP-dependent enzymes[17-19].

Based on a similar two-step procedure, 8-(6-aminohexyl)-amino-desulfo-CoA-Sepharose (XXIII) was also prepared in our laboratory[20]. This affinity gel has been shown to exhibit good affinity to many CoA-dependent enzymes including citrate synthase and succinate thiokinase from pig heart mitochondria[17] and phosphotransacetylase from Clostritium kluyveri[20]. 8-(6-Aminohexyl)-amino-2',5'-ADP- and 8-(6-aminohexyl)-amino-3',5'-ADP-Sepharose (XXIV and XXV) were also prepared from 5'-AMP by a four-step procedure starting with 5'-AMP. This involves first the bromination of 5'-AMP, phosphorylation of 8-Br-5'-AMP in the presence of $POCl_3$ and triethylphosphate, subsequent reaction with 1,6-diaminohexane to prepare mixtures of 8-(6-aminohexyl)-amino-2',5'-ADP and -3',5'-ADP and finally separation of these two derivatives by ion exchange resin. These two affinity gels have been shown to exhibit relatively good affinity to TPN^+-dependent dehydrogenases and CoA-dependent enzymes, respectively[17].

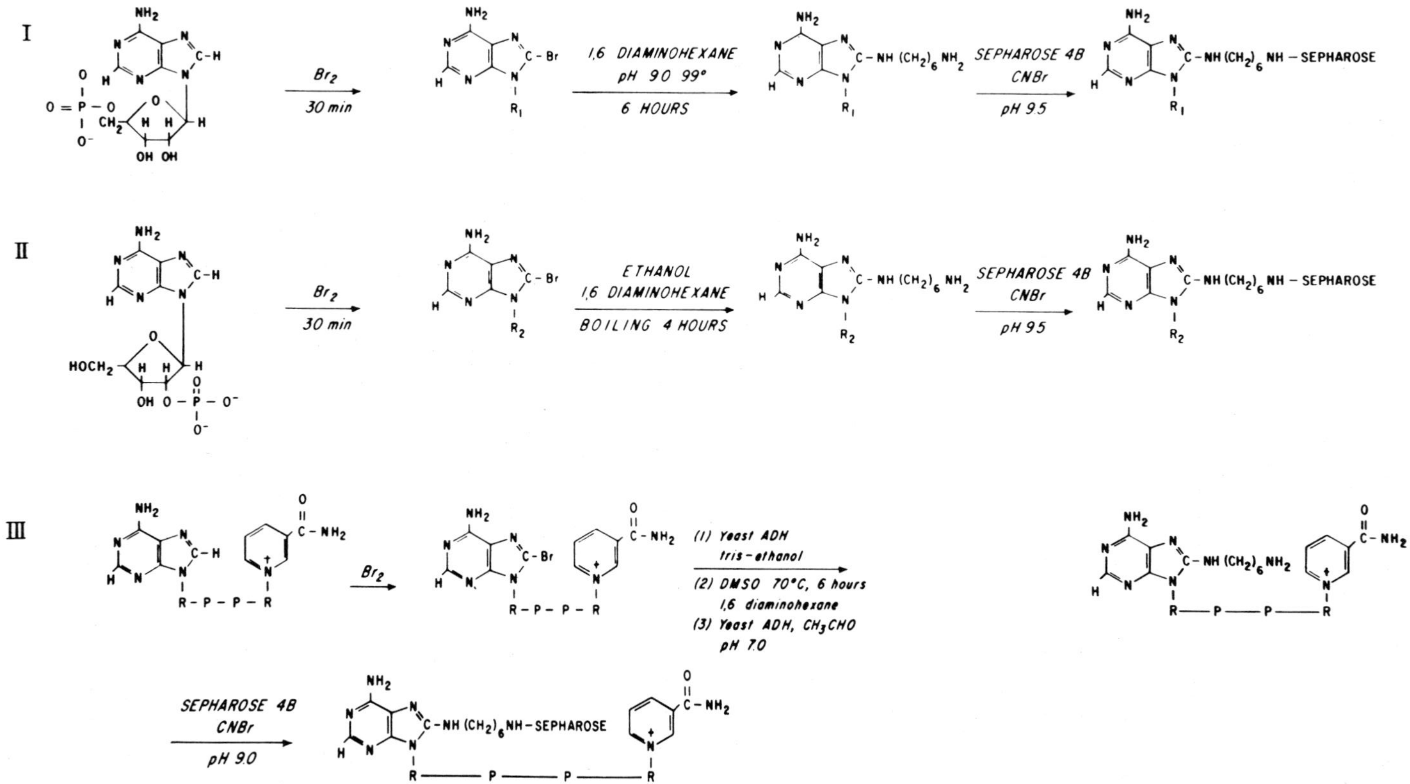

FIG. 1
A general scheme for the synthesis of (I) 8-(6-aminohexyl)-amino-5'-AMP-Sepharose, (II) 8-(6-aminohexyl)-amino-2'-AMP-Sepharose, and (III) 8-(6-aminohexyl)-amino-DPN$^+$-Sepharose[13].

The 8-substituted AMP-Sepharose is relatively stable over a wide pH range. However, DPN^+ and TPN^+ Sepharose are stable only under neutral and acidic conditions. These columns can be regenerated by washing the Sepharose with 8 M urea and high salt solutions. All these biospecific Sepharoses can be used repeatedly without any significant loss of affinity toward enzymes. An 8-substituted DPN^+ (VIA) coupled by means of an azo bond through the spacer to Sepharose was also reported[21, 22]. However, the detailed structure of the bound DPN^+ was not well characterized. This ligand showed good affinity to a number of dehydrogenases, such as lactate dehydrogenase and glyceraldehyde-3-phosphate dehydrogenase, but not malate dehydrogenase[22]. The limited application of this Sepharose-bound DPN^+ may be due to the fact that the bulky benzoyl substitution of the ligand may allow the binding of only a few enzymes. Preparation of 8-azo-ADP- and 8-azo-ATP-Sepharose columns was also reported by Lee and Kaplan[13] for the purification of kinases (Table 1, VIB and VIC) (Fig. 2).

Attachment Through N^6 Position of Adenine Mononucleotides and Dinucleotides

This type of ligand was first prepared by Guilford et al.[14] and Craven et al.[23] using N^6-(6-aminohexyl) group as a spacer for 5'-AMP (VII) and N^6-[N-(6-aminohexyl)-acetamide] group as a spacer for DPN^+ (VIII)[24]. Other types of nucleotide derivatives using 6-mercapto-purine riboside 5'-phosphate (IX) and 5'-triphosphate (X) and 6-mercapto-purine DPN (XI) were also reported[25]. N^6-(6-amino-hexyl)-5'-AMP was prepared either from (1) 6-chloropurine riboside via a two-step synthesis involving phosphorylation and substitution with 1,6-diaminohexane, or (2) direct substitution of 6-mercapto-5'-AMP with 1,6-diaminohexane[23] (Fig. 3). The starting material is usually expensive when obtained from commercial sources. A large-scale preparation of 6-substituted AMP usually starts from adenosine via several steps of synthesis. This usually involves diazotization of adenosine to inosine and protection of the ribose hydroxyl group followed by chlorination with $POCl_3$ to prepare 6-chloropurine riboside.

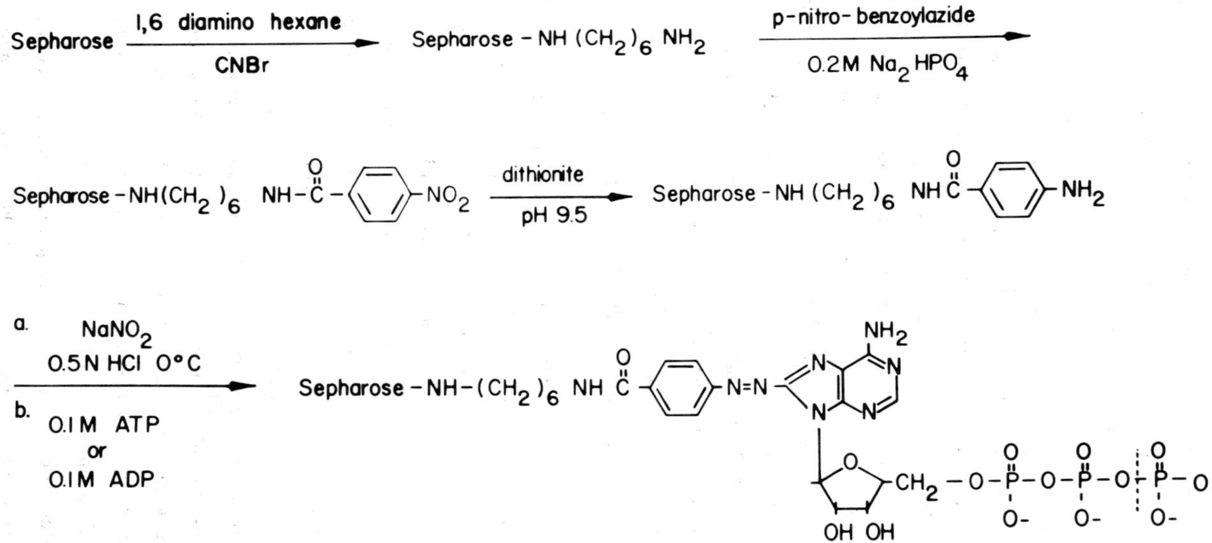

FIG. 2

A general scheme for the synthesis of 8-substituted ADP or 8-substituted ATP-Sepharose[13].

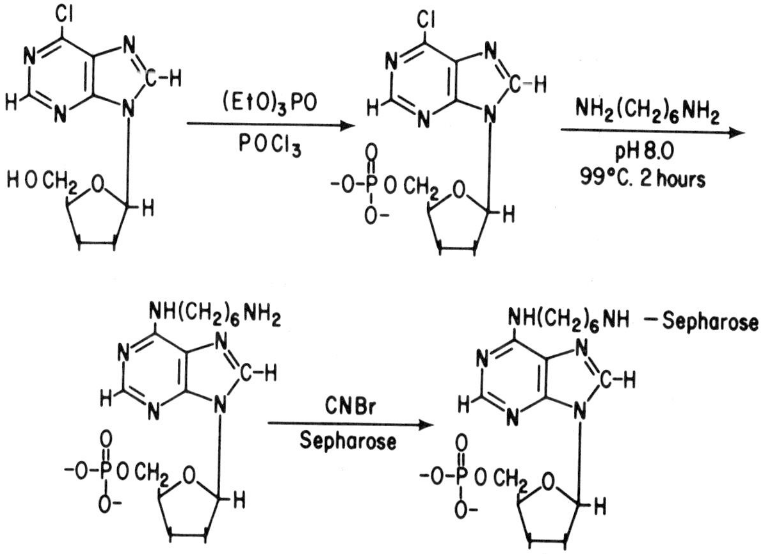

FIG. 3
A general scheme for the synthesis of N^6-(6-aminohexyl)-5'-AMP-Sepharose[14].

DPN^+-N^6-[N-(6-aminohexyl)-acetamide] and TPN^+-N^6[N-(6-amino-hexyl)-acetamide] (XXVI) were prepared by a three-step procedure. This involves the coupling and the rearrangement of iodoacetate to the N^6 position of DPN^+ and TPN^+ and then the covalent attachment of 1,6-diaminohexane by DCC coupling. The yield of this compound was reported to be low. These two N^6-substituted ligands exhibited relatively good affinity for various DPN^+-dependent dehydrogenases. The stability of these biospecific gels is similar to that reported for the 8-substituted nucleotide Sepharose[12, 13]. Attachment of 6-mercapto-purine nomonucleotide or dinucleotide derivatives was achieved by the condensation reaction between the ligand and Sepharose bound bromoacetyl group at alkaline pH[25]. However, no detailed studies regarding the nature of these ligands were reported. Preparation of N^6-(6-amionhexyl)-2',5'-ADP- and N^6-(6-aminohexyl)-3',5'-ADP-Sepharose (XXVII and XXVIII) were independently reported by Brodelius et al.[26] and Morelli and Benatti[27]. This involves a three-step procedure starting with 6-chloropurine riboside. N^6-(6-aminohexyl)-2',5'-ADP-Sepharose was shown to exhibit relatively good affinity to many TPN^+-dependent enzymes from yeast extracts. ATP-N^6-[N-(6-aminohexyl)-acetamide]-Sepharose (XXIX) was prepared by Lindberg and Mosbach[28]. This 6-substituted ATP-Sepharose has been shown to exhibit relatively good affinity to citrate synthase and could be used for a single step purification of this enzyme[28].

Attachment Through Ribosyl Position of Nucleotides

Ligands bound through the ribosyl moiety were first prepared by Larsson and Mosbach[29]. They are prepared by coupling the appropriate nucleotides such as 5'-AMP (XII), DPN^+ (XIII), or TPN^+ (XIV) to ε-aminocaproyl-Sepharose by carbodiimide reaction in 80% pyridine. The formation of the ester linkage between the ribose hydroxyl group and Sepharose-bound spacers was suggested[29]. Since the attachment was relatively simple, the method has been widely adopted for enzyme purification. However, the ester linkage between the cofactor and the spacer was reported to be unstable[27].

Another type of ligand involves the periodate oxidation of ribose hydroxyl group by periodate oxidation followed by the covalent attachment with Sepharose-bound adipic hydrazide derivatives of AMP, ADP, ATP, DPN^+, and TPN^+ as well as their derivatives (XV, XVI, XVII). These were first prepared by Lamed et al.[30, 31]. Since the ribosyl groups of the ligands are modified by oxidation, only a limited number of dehydrogenases or kinases show good affinity to these Sepharose-bound ligands.

Attachment Through Pyridine Moiety of Coenzymes

Because the adenine moiety of pyridine coenzymes is normally essential for the binding of coenzymes to dehydrogenases, modification by various substitutions on the adenine ring may effect the binding of some dehydrogenases to the Sepharose-bound ligands[32]. Therefore the attachment of coenzyme derivatives to Sepharose through pyridine moiety showed good affinity to the binding of some dehydrogenases such as transhydrogenase from chicken heart mitochondria and aldehyde dehydrogenases from Clostritium kluyveri. The preparation of this type of ligand is relatively simple (Fig.

4). Normally 3-substituted pyridine analogs of DPN^+ or deamino-DPN^+ are used for coupling to Sepharose. Examples of these ligands are 3-aminopyridine-DPN^+ (XVIII) (Fig. 4), 3-nicotinic acid-DPN^+ (XIX), 3-nicotinic acid-deamino-DPN^+ (XX), and 3-azopyridine deamino-DPN^+ (XXI).

These ligands not only exhibit good affinity for some dehydrogenases but also for some kinases such as adenylate kinase as well as creatine kinase, since the free ADP moiety may be essential for the binding of kinases.

Attachment Through the Phosphate Group of the Adenine Nucleotides

Harvey et al.[33] and Trayer et al.[15] reported the preparation

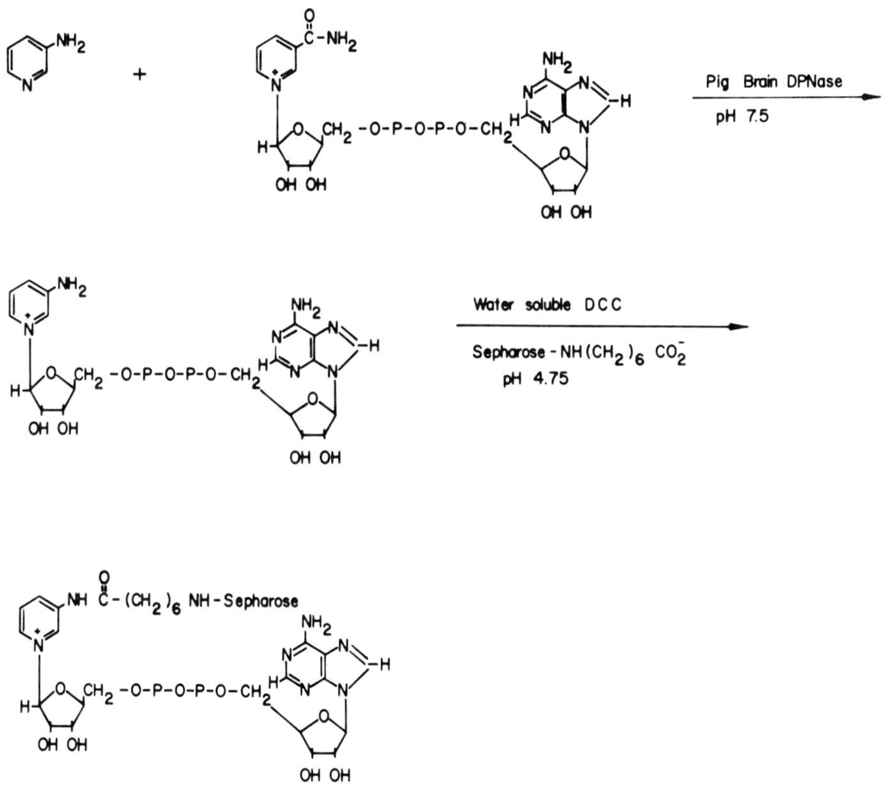

FIG. 4
Preparation of 3-amino pyridine-DPN^+-N^3-[N-(6-amino-caproamide)-Sepharose (see Table 1).

of P^1-(6-aminohexyl)-P^2-(5'-adenosine)-pyrophosphate Sepharose (XXII). The Sepharose-bound ligand was made through the covalent attachment to the pyrophosphate groups, and showed good affinity to both dehydrogenases and a few kinases[15, 33]. The preparation of this ligand is similar to that of UDP derivatives reported earlier by Barker et al.[34].

PHYSICOCHEMICAL AND BIOLOGICAL PROPERTIES OF FREE AND SEPHAROSE-BOUND LIGANDS

Any Sepharose-bound ligand which exhibits high affinity for enzymes must be either an inhibitor, a substrate, or a coenzyme of the respective enzyme. This can be shown by steady-state kinetic studies using these modified ligands in various enzymatic systems[13]. In the case of 8-(6-aminohexyl)-amino-DPN^+ and -TPN^+, these derivatives usually exhibited a higher apparent K_m than the natural coenzymes with most of the enzymes (Table 2). However, the V_{max}'s are less than those using DPN^+ or TPN^+. This indicates that the modified coenzymes may be bound to the enzymes similar to the natural coenzymes. However, in contrast there is one exceptional case: N^6-(6-aminohexyl)-5'-AMP exhibits a K_i (1×10^{-4} M) for lactate dehydrogenase which can be one order of magnitude lower than 5'-AMP itself. The activity of Sepharose-bound DPN^+-N^6-[N-(6-aminohexyl)-acetamide] in an enzymatic process has been demonstrated by Lindberg et al.[24] using a recycling multiple enzyme complex (malate dehydrogenase-citrate synthase-lactate dehydrogenase).

Effect of Chemical Modifications on the Structure of Adenine Nucleotide Derivatives

Significant changes in the interproton coupling constants of ribosyl protons were observed in the C^8-alkyl substitutions of adenine nucleotides[13]. In the case of 8-(6-aminohexyl)-amino-5'-AMP, the coupling constant between the 1'- and 2'-ribosyl protons, $J_{1'-2'}$, was observed to be 7.9 Hz in contrast to 5.5 Hz observed for 5'-AMP. This indicates that a conformational change in the ribose and/or in the glycosydic bond between the base and sugar ring occurs in the modified nucleotide derivatives. In addition to the steric hindrance caused by alkyl substitution, this may be the reason that some enzymes exhibit relatively low affinity to these modified ligands.

Physicochemical Studies of General Ligands

Physicochemical characteristics of the Sepharose-bound ligand, especially N^6-(6-aminohexyl)-5'-AMP, were studied in detail by Harvey et al.[33, 35-37]. The concentration of the Sepharose-bound ligand has been shown to be a critical factor in determining the nature and strength of the enzyme-ligand interaction, and the capacity of the adsorbants was related to column geometry, total amount of ligand, time required for the equilibration, as well as pH of the loading buffer[35, 36]. Biospecific absorption of enzymes on Sepharose-bound ligands is also strongly temperature dependent[37]; i.e., the binding decreases drastically with increasing temperatures. The dissociation constants of the complex between Sepharose-bound ligand and several enzymes were also determined by Lowe et al.[38]. In the case of the binding of lactate dehydrogenase and glycerokinase to Sepharose-bound N^6-(6-aminohexyl)-5'-AMP, the dissociation constants were measured to be as low as 1.1 μM under batchwise conditions. This is significantly lower than that reported for the complex between N^6-(6-aminohexyl)-5'-AMP and lactate dehydrogenase (K_d = 100 μM)[14]. The apparent high affinity of the enzymes to Sepharose-bound ligands may reflect a hydrophobic effect superimposed on the intrinsic affinity of the enzyme for this ligand[38].

In short, these physicochemical studies demonstrate the fact that Sepharose-bound ligands could exhibit very different properties from those of the free ligand in terms of binding to enzymes. The structural characteristics of Sepharose could also allow the immobilized ligands to bind enzymes more effectively during its application for affinity chromatography.

Since the Sepharose-bound adenine nucleotide derivatives usually carry a negatively charged phosphate or pyrophosphate group in addition to the long hydrocarbon spacer, they could also nonspecifically adsorb a number of proteins due to electrostatic

TABLE 2. Biochemical Studies of Various Substituted Adenine Nucleotide Derivatives

Enzyme	Natural or modified ligand	Rate relative to the natural compound	K_m or K_i (M)	Relative V_{max}
Adenylate kinase (rabbit muscle)	5'-AMP[13]	1.00		
	8-(6-Aminohexyl)-amino-5'-AMP	$\leq 1 \times 10^{-3}$		
Transhydrogenase (Pseudomonas aeruginosa)	2'-AMP[13]	1.00		
	8-(6-Aminohexyl)-amino-2'-AMP[13]	~1.00		
Lactate dehydrogenase (chicken muscle)	DPNH[13]		1×10^{-5}	1.00
	8-(6-Aminohexyl)-amino-DPNH[13]		8×10^{-6}	0.60
Glucose-6-phosphate dehydrogenase yeast)	TPN^+ [13]		3.6×10^{-5}	1.00
	8-(6-Aminohexyl)-amino-TPN^+		4.4×10^{-5}	0.42
Transhydrogenase (Pseudomonas aeruginosa)	TPNH[13]	1.00		
	8-(6-Aminohexyl)-amino-TPNH[13]	~1.00		

Alcohol dehydrogenase (liver)	DPN^+ [24]	1.00	
Lactate dehydrogenase (pig heart)	DPN^+-N^6-[N-(6-aminohexyl)-acetamide]	∿1.00 (ADH)	
Malate dehydrogenase (pig heart)		0.50 (LDH)	
		0.75 (MDH)	
Lactate dehydrogenase	5'-AMP		1×10^{-3}
	N^6-(6-Aminohexyl)-5'-AMP [a]		1×10^{-4}

[a] K. Mosbach, personal communication.

and/or hydrophobic interactions. The elimination of these nonspecific proteins from an affinity column is an important factor in selective protein purification. Usually the nonspecific charge interaction between proteins and immobilized ligands are related to the surface charge distributions on the protein which are strongly pH and ionic strength dependent. However, these nonspecific charge interactions can be minimized by the proper adjustments of the pH as well as the ionic strength of the solution during the specific elution of proteins from the affinity columns. The length of spacer arms between Sepharose and the ligand was also found to be an important factor in determining the relative affinity of the Sepharose-bound ligand to enzymes[39]. A comparative study was performed with the binding of chicken H_4LDH to Sepharose-bound 8-substituted 5'-AMP using various lengths of carbon chain spacers[40]. It was found that the affinity of the Sepharose-bound ligand toward the enzyme increases with increasing spacer length over the range from a two- to a five-carbon chain. However, high affinity was first obtained when the spacer was longer than five carbons. Although 1,6-diaminohexane was usually employed as the optimum spacer arm for the affinity ligands in most of the binding studies carried out, the proper length of spacer arm between the Sepharose and the ligand could vary with the enzymes studied. This result indicates that Sepharose and the surface of the active site of the enzyme should be at least 10 to 12 Å away from each other in order to achieve maximum affinity. Spacer arms longer than a six-carbon chain usually do not increase the affinity between the enzyme and the Sepharose-bound ligand and, in some cases, the affinity is decreased.

APPLICATIONS OF GENERAL LIGAND AFFINITY CHROMATOGRAPHY FOR SELECTIVE PROTEIN PURIFICATION

The essential approach to the development of general ligand affinity chromatography was based upon the premise that a single, immobilized derivative might be useful in the selective purification of a number of enzymes only if the proper conditions for specific elution could be achieved[14, 24]. Based upon such principles and criteria, about 60 enzymes have indeed been highly purified from crude extracts of various sources. The results of this purification are presented in Table 3. Brief descriptions of the purification of each enzyme are given in this section.

Since the Sepharose-bound general ligand is able to retain a family of enzymes based on their general binding characteristics, selective purifications of enzymes can be achieved only when the elution utilizes an inhibitor, substrate or dead-end ternary complex which is specific for only one of the enzymes. This possibility has been demonstrated by a model study by Kaplan and his coworkers[11]. Three different pure dehydrogenases, lactate dehydrogenase (LDH), malate dehydrogenase (MDH), and alcohol dehydrogenase (ADH), were mixed and diluted in 0.1 M phosphate buffer at pH 7.5. The mixture was then passed through an N^6-(6-aminohexyl)-5'-AMP column. After all three enzymes were adsorbed on the column, MDH, LDH, and ADH were specifically eluted from the column as separate enzymes with reduced oxalacetate-DPN adduct, pyruvate-DPN adduct, and a mixture of DPN^+ and hydroxylamine, respectively. Excellent separations of these three dehydrogenases were obtained since each element is known to be a specific inhibitor of the respective enzyme. The recovery of all three enzymes, after the appropriate dialysis to remove inhibitors, was almost quantitative (90%). Hence a mixture of at least three purified enzymes could be effectively resolved using the techniques and principles of specific elution from an AMP-Sepharose column (Fig. 5).

Selective purification of two enzymes from one crude extract was demonstrated in our laboratory[41]. An extract of chicken heart was passed through an 8-(6-aminohexyl)-amino-DPN^+-Sepharose column in 0.01 M maleate buffer, pH 6.0. After the enzymes were adsorbed on the column, it was extensively washed with buffer to eliminate all nonspecifically bound proteins. Specific elution of MDH was carried out with 1.2×10^{-4} M reduced oxalacetate-DPN adduct and LDH was then eluted with 1×10^{-4} M reduced pyruvate-DPN adduct.

TABLE 3. Enzyme Purifications Using Derivatives of Adenine Nucleotides as General Ligands in Affinity Chromatography

Enzyme	Source of crude extracts	Ligand in affinity chromatography	Fold of purification	Purity
Lactate dehydrogenase[12]	Dogfish muscle	8-(6-Aminohexyl)-amino-5'-AMP	37	∿90%
Lactate dehydrogenase[12]	Dogfish muscle	8-(6-Aminohexyl)-amino-DPN$^+$	44	∿95%
Lactate dehydrogenase[11]	Dogfish muscle	N^6-(6-Aminohexyl)-5'-AMP	34	∿90%
Lactate dehydrogenase[a]	Rabbit muscle	8-(6-Aminohexyl)-amino-5'-AMP	271	$\geq$90%
Lactate dehydrogenase[a]	Beef heart	8-(6-Aminohexyl)-amino-5'-AMP	147	$\gtrsim$80%
Lactate dehydrogenase[a]	Beef liver	8-(6-Aminohexyl)-amino-5'-AMP	167	$\gtrsim$90%
Lactate dehydrogenase[a]	Lobster tail	8-(6-Aminohexyl)-amino-5'-AMP	370	$\gtrsim$90%
Lactate dehydrogenase[12]	Losbter tail	8-(6-Aminohexyl)-amino-DPN$^+$	86[i]	$\gtrsim$80%
D-Lactate dehydrogenase[b]	Abalone	8-(6-Aminohexyl)-amino-5'-AMP	17[i]	∿90%
D-Lactate dehydrogenase	Limulus polyphemus	8-(6-Aminohexyl)-amino-DPN$^+$	10[i]	∿15%
Lactate dehydrogenase[12]	Chicken heart	8-(6-Aminohexyl)-amino-DPN$^+$	33	∿80%
Lactate dehydrogenase[41]	Human heart	8-(6-Aminohexyl)-amino-5'-AMP	90	∿80%
Lactate dehydrogenase[41]	Human liver	8-(6-Aminohexyl)-amino-5'-AMP	145	∿90%

Lactate dehydrogenase[41]	Human erythrocytes	8-(6-Aminohexyl)-amino-5'-AMP	19,000	~99%
Lactate dehydrogenase[41]	Human heart	8-(6-Aminohexyl)-amino-DPN^+	11	~10%
Lactate dehydrogenase[41]	Human liver	8-(6-Aminohexyl)-amino-DPN^+	65	~50%
Lactate dehydrogenase[41]	Human erythrocytes	8-(6-Aminohexyl)-amino-DPN^+	2,430	~15%
Lactate dehydrogenase[11]	Human serum	8-(6-Aminohexyl)-amino-DPN^+	17,000	5%
Transhydrogenase[12]	Pseudomonas aeruginosa	8-(6-Aminohexyl)-amino-2-AMP	110[j]	25%
Transhydrogenase[11]	Pseudomonas aeruginosa	N^6-(6-Aminohexyl)-amino-5'-AMP	314	25%
Transhydrogenase[c]	Chicken heart mitochondria	8-(6-Aminohexyl)-amino-5'-AMP	15	
Transhydrogenase[c]	Chicken heart mitochondria	3-Azo-pyridine-deamino-DPN^+	20	
Glutamate dehydrogenase[d]	Chicken liver	8-(6-Aminohexyl)-amino-5'-AMP	700	90%
Glucose-6-phosphate dehydrogenase[13]	Yeast	8-(6-Aminohexyl)-amino-TPN^+	4,000	98%
Glucose-6-phosphate dehydrogenase[13]	Human erythrocytes	8-(6-Aminohexyl)-amino-TPN^+	3,900	10%

TABLE 3 (Continued)

Glucose-6-phosphate dehydrogenase[11]	Human erythrocytes	TPN^+-(6-amino-caproiate)	4,600	10%
Glucose-6-phosphate dehydrogenase	Rat liver	8-(6-Aminohexyl)-amino-TPN^+	200	5%
Malate dehydrogenase[e]	Chicken heart	8-(6-Aminohexyl)-amino-5'-AMP	110[i]	99%
Alcohol dehydrogenase[f]	Yeast	8-(6-Aminohexyl)-amino-DPN^+	21	95%
Creatine kinase[18]	Chicken heart	3-Amino pyridine DPN^+ (compound XVIII)	80	80%
D-Mannitol-1-phosphate dehydrogenase[g]	Escherichia coli	8-(6-Aminohexyl)-amino-fructose-6-phosphate-DPN^+ adduct	100	50%
Aldehyde dehydrogenase[h]	Clostridium kluyveri	8-(6-Aminohexyl)-amino-DPN^+	40	30%
Adenylate kinase[13, 18]	Rabbit muscle	8-Azo-ATP (compound VIC)	100	∿40%
Adenylate kinase[13, 18]	Beef liver	8-Azo-ATP (compound VIC)	20	∿10%
Pyruvate kinase[18]	Rabbit muscle	8-Azo-ADP (compound VIB)	10	∿30%
Hexokinase[18]	Yeast	3-Amino pyridine-DPN^+ (compound XVIII)	30	∿20%
Creatine kinase[18]	Human brain	8-Hexyl-ATP (compound VI)	200	∿60%

Creatine kinase[18]	Chicken muscle	8-Azo-ADP (compound VIB)	20	30%
Creatine kinase[18]	Beef heart mitochondria	8-Azo-ATP (compound VIC)	80	90%

[a]This work was carried out in collaboration with Dr. S. S. Taylor.

[b]This work was carried out in collaboration with Dr. G. Long.

[c]This work was carried out in collaboration with Drs. L. Lazarus and Y. Levin.

[d]This work was carried out in collaboration with Mr. F. Esch.

[e]This work was carried out in collaboration with Mr. B. Bachman and Mr. F. E. Stolzenbach.

[f]This work was carried out in collaboration with Dr. C. Wills.

[g]C.-L. T. Lee and N. O. Kaplan, manuscript in preparation.

[h]This work was carried out in collaboration with Mrs. L. T. Smith.

[i]After ammonium sulfate fractionation.

[j]After DEAE cellulose column.

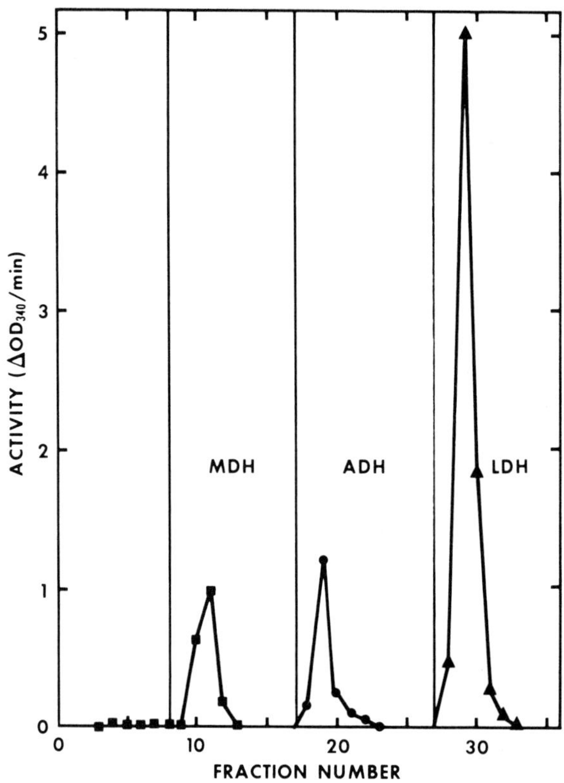

FIG. 5.
Elution pattern of MDH, ADH, and LDH from the N^6-(6-aminohexyl)-5'-AMP-Sepharose column. One-half a milliliter of a solution containing the three enzymes in 0.1 M phosphate buffer, pH 7.5, was applied to the column and the column was subsequently washed with buffer. The MDH was eluted using 0.12 mM reduced DPN-oxaloacetate adduct in 0.1 M phosphate buffer, pH 7.5; the ADH was eluted with a mixture of 0.5 mM DPN^+ and 3 mM hydroxylamine in buffer. The fractions at which the elution buffers were changed are indicated by vertical lines. Recoveries of each enzyme were greater than 90%. The activities obtained were after removal of the adduct of hydroxylamine by dialysis. Column size: 6.0 x 2.5 cm[11].

Little overlap of these two enzyme fractions was observed in this preparation. The purity of LDH and MDH obtained from this affinity column was greater than 50%. However, similar experiments using 6-substituted of 8-substituted AMP-Sepharose columns did not yield satisfactory results for the crude extracts from chicken heart. Both MDH and LDH fractions overlap even in the presence of eluents specific for one of the enzymes[42].

Purification of Dehydrogenases from Crude Extracts

Using either an N^6-(6-aminohexyl)-5'-AMP or 8-(6-aminohexyl)-amino-5'-AMP-Sepharose column, M_4 lactate dehydrogenase was first purified from the muscle extract of dogfish (Fig. 6). The enzyme was readily eluted from the column with 1×10^{-4} M reduced pyruvate DPN^+ adduct[11, 12]. The specific activity of the purified enzyme appears to be equivalent to that obtained from the lengthy multi-step procedures. Subsequently, LDH's from other sources, such as chicken heart, beef liver, rabbit muscle, and lobster tail, were also purified using either 8-(6-aminohexyl)-amino-5'-AMP or $-DPN^+$-Sepharose columns. The purity of the enzymes eluted from these

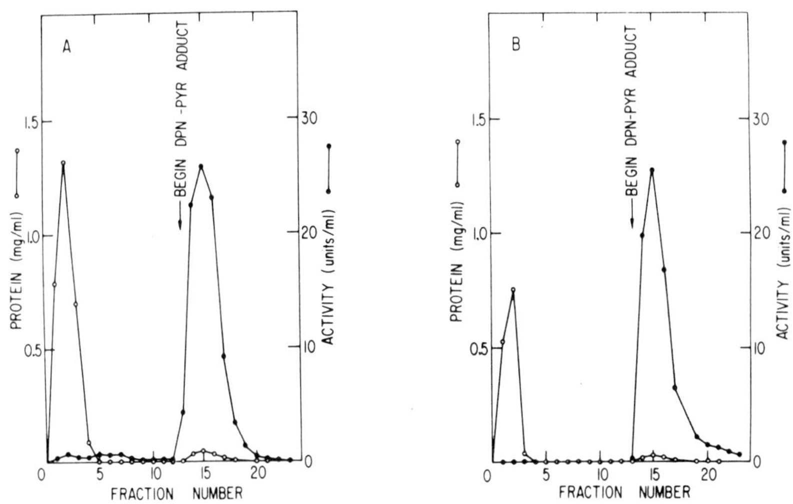

FIG. 6
Illustration of chromatographic separation of lactate dehydrogenase from an extract of dogfish muscle. Specific activities and gel electrophoresis of peak tubes revealed the enzymes to greater than 90% pure. (A) 8-(6-Aminohexyl)-amino-DPN^+-Sepharose column. (B) 8-(6-Aminohexyl)-amino-5'-AMP-Sepharose column. Elution of enzyme was made with DPN pyruvate adduct at a concentration of 2×10^{-4} M[13].

columns was greater than 90%. It is worthwhile to mention that 8-(6-aminohexyl)-amino-5'-AMP-Sepharose exhibits much better affinity for LDH from lobster tail or D-LDH from abalone than N^6-(6-aminohexyl)-5'-AMP-Sepharose. Selective purification of these two enzymes from crude extract could be demonstrated by use of the 8-substituted 5'-AMP or -DPN^+-Sepharose columns[12]. Purification of LDH from various human tissues was first carried out by Bachman and Lee in our laboratory (Table 3)[41]. Both 8-(6-aminohexyl)-amino-5'-AMP and -DPN^+-Sepharose columns were used for comparison. Under the identical conditions for elution, the 8-(6-aminohexyl)-amino-5'-AMP-Sepharose column yielded a much higher fold of purification for LDH from heart tissue and erythrocytes than 8-DPN^+-Sepharose column. However, similar purification of human liver LDH was obtained for both affinity columns. In the case of LDH from human heart and erythrocytes, 8-(6-aminohexyl)-amino-DPN^+-Sepharose exhibited a greater affinity for the enzyme than 8-AMP-Sepharose. Therefore the enzyme could not be specifically eluted from the 8-DPN^+-Sepharose column using relatively low concentrations of reduced pyruvate-DPN^+ adduct. As a result, the enzyme was eluted from the 8-AMP column in a relatively sharp peak, whereas a broad peak was obtained from the elution of the 8-DPN^+ column. This observation indicates that LDH isoenzymes from human heart (mostly H-type isozyme) and liver (mostly M-type isozymes) show different affinities for 8-DPN^+-Sepharose. Thus, using 8-(6-aminohexyl)-amino-5'-AMP-Sepharose, LDH can be purified in one step from three different human tissues. This study demonstrated that affinity chromatography techniques can be applied to clinical and medical research in the near future. Subsequently, lactate dehydrogenases from at least ten other different sources were also purified by this one-step affinity column procedure (Table 3).

Malate dehydrogenase from chicken heart was also successfully purified with an AMP-Sepharose column (Table 3). Because of the difficulties in the specific elution of MDH and LDH from this general ligand column, these two enzymes were first separated by ammonium sulfate precipitation before being applied to the column. Malate dehydrogenase was eluted readily from the affinity column with 0.2 mM of reduced oxalacetate-DPN^+ adduct at pH 6.0, 1 mM phosphate buffer. The specific activity of the eluted enzyme appeared to be equivalent to that obtained from the conventional ten-step procedures.

The purification of glutamate dehydrogenase from chicken liver using an 8-(6-aminohexyl)-amino-5'-AMP-Sepharose column has been studied in our laboratory (Table 3). It was found that the purity of the eluted enzyme strongly depended upon the conditions of elution. This requires the proper choice of pH and ionic strength. The use of reduced-α-ketoglutarate-DPN adduct alone could not elute the enzyme even at high pH (∿7.5). The column was first washed with 0.06 M phosphate at pH ∿6.0 to remove contaminating proteins, and the enzyme was then eluted with 1 mM adduct in 0.03 M potassium phosphate buffer, pH 7.5. The specific activity of the purified enzyme was as high as 42 units/mg (∿90% purity) and the overall recovery was greater than 90%.

Glucose-6-phosphate dehydrogenase (G-6-PDH) from crude extracts of various sources was also purified using the approach of general ligand affinity chromatography. This enzyme was purified from human erythrocytes using either 8-(6-aminohexyl)-amino-TPN^+-Sepharose[13] or TPN^+-[5-aminocaproiate]-Sepharose[11]. Approximately 4000-fold purification was achieved via one-step elution by a 0 to 1 mM TPN^+ gradient. However, since this elution scheme is rather nonspecific, the purity of the enzyme was only 10%. Subsequently, the enzyme was further purified over a DEAE-Sephadex column and, as a result, a homogeneous enzyme was obtained in the second step[11]. Glucose-6-phosphate dehydrogenase from Neurospora crassa was also purified using TPN^+-(6-aminocaproiate)-Sepharose (Table 3). After removing the nonspecifically bound protein with high salt (0.1 M NaCl) at high pH (7.5), the enzyme was readily eluted with 0.1 M potassium phosphate at pH 7.5. The purity of the eluted enzyme ranged from 50 to 100%. Elution with phosphate buffer seems to be rather specific since the phosphate ion was shown to be an inhibitor of this enzyme[43].

Glucose-6-phosphate dehydrogenase from yeast extract was purified by using an 8-(6-aminohexyl)-amino-TPN^+-Sepharose column[13] (Table 3). Purity of the eluted enzyme was very sensitive to the conditions during the washing of the column. If the crude extract was loaded on the column and washed with 0.1 M phosphate buffer at pH 6.0 in the presence of 10% glycerol, the purity of the enzyme obtained by elution with a 0 to 1 mM TPN^+ gradient was only approximately 10%. However, if the column was thoroughly washed with 0.1 M phosphate buffer at pH 6.0 in the presence of 20% glycerol, most of the weakly bound TPN^+-dependent enzymes were eliminated with about 70% retention of glucose-6-phosphate dehydrogenase. It could then be readily eluted with a 0 to 1 mM TPN^+ gradient from the column with a purity of approximately 95%. Multi-enzyme purification of TPN^+-dependent enzymes was demonstrated by Brodelius et al.[26] and by Lee and Kaplan[13]. Upon the adsorption of yeast extract onto either an N^6-(6-aminohexyl)-amino-2',5'-ADP-[26] or an 8-(6-aminohexyl)-amino-TPN^+-Sepharose column[13], four to five different TPN^+-dependent enzymes were subsequently eluted from the affinity column by a proper TPN^+ gradient[13, 26].

An attempt was made to purify glucose-6-phosphate dehydrogenase and 6-phospho-gluconic acid dehydrogenase from rat liver extracts. A 200-fold purification was obtained when the enzymes were eluted with 1 mM TPN^+ from the column[13], although the enzymes were not homogeneous after this procedure.

The pyridine nucleotide transhydrogenase from Pseudomonas aeruginosa was purified by either 8-(6-aminohexyl)-amino-2'-AMP-Sepharose[12] or N^6-(aminohexyl)-5'-AMP-Sepharose columns[11]. In the case of the 6-substituted-5'-AMP-Sepharose column, the enzyme was readily eluted with 1 mM DPNH at neutral pH with an observed purity of only 10% due to the nonspecific elution of this enzyme. However, a homogeneous enzyme could be obtained by passing this preparation through a G-200 Sephadex column[11]. This enzyme is known to form an aggregated species and can be eluted from the column in the void volume, while the contaminating proteins were excluded by the Sephadex.

Transhydrogenase from chicken heart mitochondria was also purified by about 20-fold using affinity chromatographic techniques. This enzyme can only be properly purified by using various 3-substituted pyridine-DPN$^+$-Sepharose. The extreme lability of this enzyme makes affinity chromatographic techniques useful methods for which this enzyme can be purified.

Yeast alcohol dehydrogenase was purified in cooperation with Dr. C. Wills at UCSD, using 8-(6-aminohexyl)-amino-DPN$^+$-Sepharose. Neither an 8-substituted nor 6-substituted 5'-AMP-Sepharose column exhibits good binding for this enzyme from crude extracts and therefore does not yield satisfactory results in terms of purifications. After the crude extract was loaded on an 8-substituted-DPN$^+$-Sepharose column, extensive washing with salt at low pH (6.0) eliminated the nonspecifically bound protein. The enzyme was readily eluted with 0.1 m*M* reduced acetaldehyde-DPN adduct. Purity was greater than 90% with a recovery of almost 100%. The application of affinity chromatography techniques for purification is likely to make possible the rapid isolation of isoenzymes of alcohol dehydrogenase from various mutants of yeast. Multi-enzyme purification from a single crude extract was first demonstrated by using an 8-(6-aminohexyl)-amino-5'-AMP-Sepharose column[19]. After passing the rabbit muscle extracts through the affinity column, at least four different dehydrogenases (MDH, α-GPD, LDH and TPD) were biospecifically eluted from the column by a 0 to 1 m*M* DPNH gradient. The purity of the isolated enzymes ranges from 50 to 90%. It is worthwhile mentioning that only dehydrogenases were adsorbed on the 8-substituted-AMP-Sepharose column whereas most kinases were recovered in the eluent during the loading of extracts. They could be purified subsequently by an ATP-Sepharose column[19].

There is one interesting observation which deserves further enumeration here. Saturation of the affinity column with respect to a given enzyme often results in the exclusion of other weakly bound enzymes. Readsorption of the unbound enzymes on a second affinity column usually gives a higher degree of purification. An example of this procedure is the purification of three different dehydrogenases from chicken muscle extract. LDH, α-GPD and MDH were purified separately by two 8-substituted-AMP and one NAD-Sepharose columns, respectively, based on this saturation and re-adsorption technique. This concept was also successfully applied to the purification of several kinases[18, 19].

Isoenzyme purification of horse liver alcohol dehydrogenase was first reported by Andersson et al.[44]. EE, SS and ES isoenzymes were eluted specifically and separately from an N^6-(6-aminohexyl)-5'-AMP-Sepharose column. The homogeneous isoenzymes could be obtained by this affinity column procedure.

<u>Purification of Kinases from Crude Extracts</u>

Recently it was found in our laboratory that the Sepharose-bound ligands which possess a free 5'-ADP moiety exhibit relatively high affinity for various kinases. In the case of a 3-amino-pyridine-DPN^+-N^3-[N-(6-amino-caproamide)]-Sepharose column, it exhibits low affinity for most dehydrogenases, but good affinity for several kinases, such as hexokinase from yeast, creatine kinase from chicken heart[18], and protein kinase from pig muscle[40]. The adsorbed enzymes can be readily eluted from the column with proper concentrations of ATP. For example, using a linear ATP gradient for elution, creatine kinase from chicken heart extracts can be purified from this column with a purity greater than 80% (Table 3).

Hexokinase was also purified by using a 3-amino-pyridine-DPN^+-Sepharose column (Table 1, Compounds X and VIII). The adsorbed enzyme was readily eluted either with 0.10 <u>M</u> phosphate buffer at pH 7.0 or with a 0 to 10 m<u>M</u> ATP gradient at pH 6.0. The purity of the enzyme obtained in this preparation is about 20%, which corresponds to a 30-fold purification of the protein (Table 3).

3-Amino-pyridine-DPN$^+$-Sepharose column does not exhibit good affinity to adenylate kinase, pyruvate kinase from rabbit muscle, and creatine kinase from chicken muscle. Purification of these enzymes can be achieved only with other types of general ligand affinity columns.

With the use of 8-azo-ATP-Sepharose column (Table 1, Compounds VIB and VIC), adenylate kinase from rabbit muscle can be purified 100-fold by an elution of 0 to 10 m*M* ATP linear gradient. The specific activity of the enzyme obtained from this affinity column can be as high as 700 units/mg which corresponds to 40% in purity. With the use of an 8-azo-ADP-Sepharose column, pyruvate kinase from rabbit muscle can be purified by using an elution with an ATP gradient. As a result, the purity of the enzyme can be as high as 30% after this affinity column. The summary of purification of several kinases is presented in Table 3.

Multi-enzyme purification of kinases from a single extract has been demonstrated by using either an 8-azo-ATP- or 8-(6-aminohexyl)-amino-ATP-Sepharose column[18, 19]. At least three different kinases plus aldolase were eluted biospecifically by 0 to 10 m*M* ATP gradient. Adenylate kinase and creatine kinase were coeluted at an ATP concentration of 0.5 m*M*, whereas pyruvate kinase and aldolase were eluted together at 2 to 3 m*M* ATP. After the single-step elution, a sole mixture of two enzymes was obtained in the peak fraction of each enzyme[18]. These four enzymes can be further purified to homogeneity by one more step of protein purification with conventional procedures[18].

A similar approach has been successfully applied for the purification of creatine kinase and adenylate kinase from many different sources. These include human heart, brain, liver, beef heart mitochondria and fish muscle (Table 3). Generally speaking, a unified two-step procedure, affinity chromatography and isoelectric focusing were required to obtain homogeneous enzymes from all these tissue extracts.

Recently Dr. S. Taylor has purified the holoenzyme from of 3', 5'-cyclic AMP-dependent protein kinase by employing 8-(6-aminohexyl)-amino-ATP-Sepharose[40]. The enzyme activity can be eluted either with a salt gradient or with ATP and recoveries range from 60 to 90%. The eluted enzyme is fully dependent on cyclic AMP for enzymatic activity and has a molecular weight indistinguishable from that of the holoenzyme[40].

Special Applications of General Ligand Affinity Chromatography

It is known that most dehydrogenases are strongly and specifically inhibited by the adduct prepared from the alkaline condensations of an oxidized coenzyme (DPN^+) and their respective substrates[45]. A Sepharose column prepared from the adduct of 8-(6-aminohexyl)-amino-DPN^+ and the oxidized substrate of a given dehydrogenase becomes a "specific ligand" affinity column for this particular enzyme. This method is extremely useful when the enzyme does not bind well to the "general ligand" column. This is demonstrated by the purification of D-mannitol-1-phosphate dehydrogenase using a DPN^+-fructose-6-phosphate adduct Sepharose column. This column exhibits an extremely high capacity with respect to the binding of this enzyme from crude extracts of E. coli (Table 3). The enzyme, which was purified 100-fold after this affinity column, could be further purified to a homogeneous state after passage through Sephadex G-100. This work was carried out in our laboratory by Mrs. C.-L. Lee.

Purification of Other Classes of Enzymes by General Ligand Affinity Chromatography

Recently attempts have been made to purify CoA-dependent enzymes and amino acyl-t-RNA synthetases by the approach of general ligand affinity chromatograhy. With the use of either N^6-substituted ATP-Sepharose[28] or 8-(6-aminohexyl)-amino-ATP-Sepharose column[17], citrate synthase was purified by a single-step procedure. In the case of 8-substituted-ATP-Sepharose column, a crude extract of pig heart mitochondria was passed through affinity column with a capacity of 3 mg of enzyme per ml of affinity gel. Biospecific elution was made with a 0.5 m*M* CoASH. Since very few CoA-dependent enzymes were inhibited by ATP and were adsorbed on the ATP-Sepharose, citrate synthase was obtained with greater than 95% in purity in the peak fractions. On the other hand, elution with ATP did not result in a satisfactory purification of this enzyme, since the obtained citrate synthase was contaminated with many other ATP-dependent enzymes[17]. Purification of citrate synthase with an 8-(6-aminohexyl)-amino-desulfo-CoA-Sepharose column was also reported[17]. However, other CoA-dependent enzymes such as succinate thiokinase were co-eluted with citrate synthase by an elution with 0.5 m*M* CoASH. As a result, the obtained citrate synthase was not homogeneous by this affinity column procedure.

Recently general ligand affinity chromatography was applied to the purification of amino-acyl-t-RNA synthetases[46]. With the use of 8-(6-aminohexyl)-amino-ATP-Sepharose, the valine enzyme from E. coli (partially purified by DEAE column) was adsorbed on the affinity column. Valyl-t-RNA-synthetase was eluted specifically with a 0 to 5 m*M* ATP gradient in the presence of 5 m*M* Mg^{++} and 0.1 m*M* valine. As a result, the obtained enzyme is greater than 85% pure[46].

COMMENTS ON ENZYME PURIFICATION USING GENERAL LIGAND AFFINITY CHROMATOGRAPHY TECHNIQUES

Although the basic principles of using "general ligand" for protein purification are known, successful purification of a given enzyme from the affinity column strongly relies upon the following factors: (1) proper choice of a general ligand and carrier[39], (2) conditions of washing, and (3) specific elution. The proper combinations of these conditions will permit an efficient purification from crude homogenates.

It is apparent that the proper choice of an affinity ligand must be made for the successful application of affinity column techniques. The binding affinity between the ligand and the enzyme of interest must be great enough to prevent its elution with nonspecific eluents[39]. However, if the binding is too strong, the tightly bound enzyme cannot be readily eluted under mild conditions. Therefore successful application of affinity chromatographic techniques cannot be achieved in either case.

The strength of binding of a given immobilized ligand to the enzyme is also influenced by several factors. These include: (1) the distance of the ligand from the surface of the solid support as discussed previously[39], (2) proper choice of the solid-support carrier, and (3) the coupling methods.

Sepharose is generally accepted as one of the best carriers for affinity columns[1, 2, 47-49]. There are several advantages for using this carrier in the synthesis of affinity columns: (1) weak nonspecific interactions of its surface with proteins, (2) good hydrophilic and flow properties, (3) a high degree of porosity and high concentration of functional groups which can easily be activated usually by means of cyanogen bromide for the attachment with the desired ligand[50-54], and (4) a high degree of physical and chemical stability.

In addition to the selection of proper ligands and solid-support matrix, there are several important steps which have to be taken into account before proceeding with purification of enzymes from crude extracts.

(1) Before passing crude extracts through the Sepharose columns, they should be dialyzed against a low ionic strength and low pH buffer (0.01 $\underline{M}$ salt and pH $\leq$ 7.0) in order to prevent interference from the contaminating nucleotides in the original extracts. Some tissue extracts should also be diluted in order to decrease the ionic strength effects arising from high protein concentrations.

Experimentally, it is generally observed that the binding of enzymes to the nucleotide Sepharose columns is better at a low pH and low ionic strength buffer. Since the surface charge distribution and the conformation of protein strongly depend on these two parameters, solutions of low pH and low ionic strength in general decrease the nonspecific repulsion force between the bound ligand and proteins; binding is therefore more readily facilitated.

(2) After the crude extract is loaded on the column, it is important that nonspecifically bound proteins be extensively removed before the specific elution of an enzyme commences. This is usually accomplished by extensive washing of the column with buffer. A buffer of high ionic strength and high pH is preferred provided that the enzyme to be purified can be preferentially retained on the column.

(3) Since an entire family of enzymes are usually retarded by the general ligand on the Sepharose columns, purification of the enzyme of interest can be achieved only when a specific inhibitor or substrate is applied to elute the enzyme from the column. In the case of dehydrogenases, it was reported by Everse et al.[45] that some dehydrogenases are specifically inhibited by an addition product of DPN^+ with their respective substrate, such as reduced DPN-pyruvate adduct for lactate dehydrogenases and reduced DPN-α-ketoglutarate adduct for glutamate dehydrogenase. Under certain circumstances the enzyme can be eluted specifically with DPN^+ and its oxidized substrate or inhibitor. Since the formation of a ternary complex often prevails, resulting in a strong inhibition, the enzyme can then be eluted. In case a specific eluent is not available for a given enzyme, a proper coenzyme or substrate gradient can be used for selective elution[55, 56]. The enzyme can frequently be purified with a high degree of purity in this manner. This is exemplified by the purification of glucose-6-phosphate dehydrogenase from yeast extracts as we have previously described[13].

Although not all the enzymes we have studied can be purified in one step by these affinity columns, homogeneous enzyme preparations can usually be obtained by an additional step of conventional procedures, such as ammonium sulfate fractionation, DEAE cellulose, or Sephadex chromatography. General ligand affinity chromatographic techniques are now extensively applied to the large-scale preparation of lactate dehydrogenase from various sources. With a half liter column of 8-(6-aminohexyl)-amino-5'-AMP-Sepharose, one can process and crystallize 1 gram of purified

LDH from crude extracts in a day or so, unlike the traditional procedures for which a week or a month are required to achieve the same goal. Large-scale purification of other dehydrogenases and kinases using the developed affinity column technique is expected in the near future. The idea that a single general ligand affinity column can be used to purify a number of enzymes has now become a reality. There is no doubt that this method appears to be simple and inexpensive; in many cases the purification involves a single adsorption and elution.

ACKNOWLEDGMENTS

The authors wish to thank the following persons in our laboratory for their sincere collaboration as well as kindly providing unpublished experimental data for this manuscript: Drs. L. H. Lazarus, B. Wermuth, Y. Levin, J. Everse, S. S. Taylor and C. Wills, Mr. F. E. Stolzenbach, Mrs. C.-L. T. Lee, Mr. D. Lappi, Mr. B. Bachman, Mrs. L. T. Smith, and Miss K. Berger. Collaboration in various aspects of this work **is** highly appreciated. Parts of this work were supported in part by grants from the American Cancer Society, (BC-60) and the National Science Foundation (NSF RANN GI 36249).

REFERENCES

1. P. Cuatrecasas and C. B. Anfinsen, Ann. Rev. Biochem., 40, 259 (1971).
2. P. Cuatrecasas, J. Agric. Food Chem., 19, 600 (1971).
3. P. Cuatrecasas, in Biochemical Aspects of Reactions on Solid Supports (G. R. Stark, ed.), Academic, New York, 1971, p. 79.
4. I. H. Silman and E. Katchalski, Ann. Rev. Biochem., 35, 873 (1972).
5. H. H. Weetall in Chemistry of Biosurfaces, Vol. 2 (M. L. Hair, ed.), Marcel Dekker, New York, 1972, p. 597.
6. D. H. Cambell, E. Leuscher, and L. S. Lerman, Proc. Nat. Acad. Sci. U. S. A., 37, 575 (1951).
7. P. Cuatrecasas and C. B. Anfinsen, Methods Enzymol., 22, 191 (1971).
8. K. Mosbach, H. Guilford, R. Ohlsson, and M. Scott, Biochem. J., 127, 625 (1972).
9. C. R. Lowe and K. Mosbach, Eur. J. Biochem., 49, 511 (1974).
10. C. R. Lowe and P. D. G. Dean, FEBS Lett., 14, 313 (1971).

11. N. O. Kaplan, J. Everse, J. Dixon, F. Stolzenbach, C.-Y. Lee, C.-L. Lee, S. Taylor, and K. Mosbach, Proc. Nat. Acad. Sci. U. S. A., 71, 3450 (1974).
12. C.-Y. Lee, D. A. Lappi, B. Wermuth, J. Everse, and N. O. Kaplan, Arch. Biochem. Biophys., 161, 561 (1974).
13. C.-Y. Lee and N. O. Kaplan, Ibid., 168, 665 (1975).
14. H. Guilford, P. O. Larsson, and K. Mosbach, Chem. Scr., 2, 165 (1972).
15. I. P. Trayer, H. R. Trayer, D. A. P. Small, and R. C. Bottomley, Biochem. J., 139, 609 (1974).
16. B. Jergil, H. Guilford, and K. Mosbach, Ibid., 139, 441 (1974).
17. C.-J. Johanson and C.-Y. Lee, FEBS Lett., in press (1976).
18. C.-Y. Lee, L. H. Lazarus, and N. O. Kaplan, Arch. Biochem. Biophys., submitted (1976).
19. C.-Y. Lee, L. H. Lazarus, and N. O. Kaplan in Enzyme Engineering, Vol. 3, Plenum Press, New York, in press, 1976.
20. C.-Y. Lee, L. T. Smith and N. O. Kaplan, unpublished results.
21. M. K. Weibel, H. H. Weetall, and H. J. Bright, Biochem. Biophys. Res. Commun., 44, 347 (1971).
22. S. Barry and P. O'Carra, Biochem. J., 135, 595 (1973).
23. D. B. Craven, M. J. Harvey, C. R. Lowe, and P. D. G. Dean, Eur. J. Biochem., 41, 329 (1974).
24. M. Lindberg, P.-O. Larsson, and K. Mosbach, Ibid., 40, 187 (1973).
25. S. Barry and P. O'Carra, FEBS Lett. 37, 134 (1973).
26. P. Brodelius, P.-O. Larsson, and K. Mosbach, Eur. J. Biochem., 47, 81 (1974).
27. A. Morelli and U. Benatti, Ital. J. Biochem., 23, 279 (1974).
28. M. Lindberg and K. Mosbach, Eur. J. Biochem., 53, 481 (1975).
29. P.-O. Larsson and K. Mosbach, Biotech. Bioeng., 13, 393 (1971).
30. R. Lamed, Y. Levin, and M. Wilchek, Biochim. Biophys. Acta, 304, 231 (1973).
31. R. Lamed, Y. Levin, and A. O. Platka, Ibid., 305, 163 (1973).
32. J. Everse and N. O. Kaplan, Adv. Enzymol., 37, 61 (1973).
33. M. J. Harvey, C. R. Lowe, D. B. Craven, and P. D. G. Dean, Eur. J. Biochem., 41, 335 (1974).
34. R. Barker, K. W. Olsen, J. H. Shaper, and R. L. Hill, J. Biol. Chem., 247, 7135 (1972).
35. C. R. Lowe, M. J. Harvey, and P. D. G. Dean, Eur. J. Biochem., 41, 341 (1974).
36. C. R. Lowe, M. J. Harvey, and P. D. G. Dean, Ibid., 41, 347 (1974).
37. M. J. Harvey, C. R. Lowe, and P. D. G. Dean, Ibid., 41, 353 (1974).
38. C. R. Lowe, M. J. Harvey, and P. D. G. Dean, Ibid., 42, 1 (1974).
39. H. H. Weetall, Purif. Methods, 2, 199 (1973).
40. S. S. Taylor, C.-Y. Lee, L. Swain, and P. H. Stafford, Biochem. Biophys. Res. Commun., in press (1976).
41. B. Bachman and C.-Y. Lee, Anal. Biochem., in press (1976).

42. This overlap, however, does not occur with the crude extracts from dogfish muscle as well as other sources, so that most of the LDH's can be homogeneously obtained in one step.
43. M. Marguet, C. R. Acad. Sci., 256, 4997 (1963).
44. L. Andersson, H. Jörnvall, Å. Åkeson, and K. Mosbach, Biochim. Biophys. Acta, 364, 1 (1974).
45. J. Everse, R. E. Barnett, C. J. R. Thorne, and N. O. Kaplan, Bioorg. Chem., 1, 207 (1971).
46. C.- Y. Lee, M. Lindberg, and K. Mosbach, Int. Cong. Biochem., Abstract, in press (1976).
47. P. Cuatrecasas, M. Wilchek, and C. B. Anfinsen, Proc. Nat. Acad. Sci. U. S. A., 61, 636 (1968).
48. P. Cuatrecasas, J. Biol. Chem., 245, 3059 (1970).
49. I. Kato and C. B. Anfinsen, Ibid., 244, 1004 (1969).
50. J. Porath, R. Axen, and S. Ernbach, Nature (London), 215, 1491 (1959).
51. D. R. Axen and J. Porath, Ibid., 210, 367 (1966).
52. R. Axen, J. Porath, and S. Ernbach, Ibid., 214, 1302 (1967).
53. R. Axen, E. Hielbron, and A. Winter, Biochim. Biophys. Acta, 191, 478 (1969).
54. R. Axen, P. A. Myris, and J. C. Janson, Biopolymers, 9, 401 (1970).
55. A gradient of DPNH can also be utilized to separate different dehydrogenases as well as different isoenzymes of lactate dehydrogenase[56].
56. P. Brodelius and K. Mosbach, FEBS Lett., 35, 223 (1973).

APPLICATION OF IMMUNOADSORBENTS FOR ISOLATION OF PLACENTAL ALKALINE PHOSPHATASE, CARBOXYPEPTIDASE G-1, AND SERUM HEPATITIS ANTIGEN

Stanley E. Charm and Bing L. Wong

New England Enzyme Center
Tufts University School of Medicine
Boston, Massachusetts 02111

INTRODUCTION

The extraordinary specificity of antigen-antibody reactions forms the basis of a simplified isolation method for antigens. The use of antigens to isolate antibodies has been employed by immunologists for many years [1, 2]. The early use of immunoadsorbents to isolate antibodies was not always successful because the physical adsorption used to fix the antigen permitted "leaking". However, with the various methods available today for covalently bonding to solid supports, this is no longer a serious problem.

The use of immobilized antibody to isolate antigens is rapidly becoming an accepted laboratory technique [3, 4]. However, attempts in the past, to isolate biologically active molecules such as enzymes have not been successful due to low yields, usually less than 25% [5].

On the other hand, the isolation of nonbiologically active substances seems to be quite adaptable to this process.

It was our purpose to determine whether immunoadsorption could be employed as a simplified enzyme isolation method.

ISOLATION OF HUMAN PLACENTAL ALKALINE PHOSPHATASE

Human placental alkaline phosphatase was isolated from a crude suspension in a two-step immunoadsorbent procedure. The placenta had first been extracted with butanol and subjected to an ammonium sulfate fractionation to remove some of the extraneous protein. The immunoadsorbent in this case was agarose (Sepharose 2B) with rabbit serum immunoglobulin covalently bonded through the CNBr method [6].

A batch reaction gently mixing the immunoadsorbent with the crude enzyme solution was employed to effect the antibody-enzyme complex. It was felt that this is easier to control than a column and more acceptable to a large-scale system. After a sufficient reaction time the immunoadsorbent with the complexed enzyme was separated from the suspension and washed to remove nonspecifically adsorbed protein. Placental alkaline phosphatase was dissociated from the immunoadsorbent using 0.2 M Na_2CO_3, pH 11.4 [7].

In this case the usual agents for dissociated antibody-antigen complexes (e.g., KI, extremes of pH, KCNS) do not result in an active enzyme.

Yield and purification depend to a large extent on the affinity of the antibody for the antigen. A low affinity antibody permits greater dissociation. The greater the concentration of crude antigen exposed to the immunoadsorbent, the greater the degree of purification and yield (see Table 1). Immunoadsorbent 7 is made from serum drawn 7 weeks after initial inoculation, and immunoadsorbent 64 was drawn after 64 weeks. The pH at which crude antigen is associated with the immunoadsorbent also influences to a lesser extent how strongly the antigen is complexed. It was possible by using an anti-alkaline phosphatase immunoadsorbent to effect a purification of up to 100-fold [8].

Antibody to the alkaline phosphatase impurity, which in this case is human plasma, was also produced. By using a second immunoadsorbent against the impurity, it is possible to effect another 2-fold purification. Thus in a two-step procedure, a 200-fold purification with about a 90% yield is obtained.

CARBOXYPEPTIDASE G-1

Carboxypeptidase G-1 also has been successfully isolated by immunoadsorption [9]. The immunoadsorbent and antigen are complexed at pH 6.5 and the complex dissociated by 4 M KCNS or by 6 M guanidine HCl (see Tables 2 and 3). The yield from this procedure was about 85% with a 35-fold purification. It is possible to account for about 95% of the activity entering the immunoadsorption process.

Antibody against the impurities was also produced and made into an immunoadsorbent for removing impurity from the partially purified carboxypeptidase. It was possible to effect another 1.5- to 2-fold purification doing this (see Table 4). The effectiveness of using an anti-impurity immunoadsorbent is limited because of the nonspecific adsorption of the enzyme which is in relatively high concentration as compared with the complexed impurities.

The antigen used to produce the carboxypeptidase antibody in this case was not pure. This probably accounts for reduced purification by immunoadsorbent.

With conventional purification procedures (about 10 steps with a final substrate affinity step which is delicate), it is possible to obtain an enzyme with a specific activity of 700. Using the two-step immunoadsorbent procedure, a specific activity of 225 was obtained. It is expected that with a purer antibody, a higher purification of enzyme would result from the first immunoadsorbent step.

REMOVAL OF SERUM HEPATITIS ANTIGEN FROM BLOOD PLASMA BY IMMUNOADSORPTION

It has been possible by immunoadsorption to remove serum hepatitis antigen from blood and blood plasma. Our initial work employed goat antibody to serum hepatitis antigen. The

TABLE 1

Influence of pH, Antigen Load, and Antibody Date on Yield and Purification of Placental Alkaline Phosphatase

Immuno-adsorbent	pH of association reaction	Tube no.	Ag Available S.A. 25 (units per mg Ig)	Ag associated (units per mg Ig)	Ag dissociated (units per mg Ig)	Percent Ag dissociated of total associated (%)	Specific activity of dissociated antigen	Degree of purification (fold)
7	7.4	1	2.7	2.6	1.6	61.7	61.3	24.4
		2	5.4	4.9	4.3	87.9	114.5	45.8
		3	10.7	8.9	6.6	74.9	134.6	53.8
		4	26.8	15.2	8.9	58.5	157.0	62.8
		5	48.2	21.1	11.0	52.2	169.8	68.0
		6	91.1	35.4	12.9	36.4	152.3	61.0
	8.6	1	2.7	2.3	1.1	46.7	18.9	7.5
		2	5.6	4.7	4.6	96.2	90.2	36.0
		3	11.3	7.9	7.4	94.5	119.5	48.0
		4	28.2	12.7	12.2	96.0	122.7	49.0
		5	50.7	15.1	15.0	99.3	112.2	45.0
		6	95.7	19.1	17.2	91.1	110.8	44.3
64	7.4	1	5.5	5.0	2.3	45.0	124.1	49.6
		2	11.1	8.8	3.3	37.6	162.3	64.9
		3	44.3	15.2	5.9	38.6	225.9	90.3
		4	60.9	18.9	6.7	35.8	259.4	103.7
		5	72.5	20.4	6.7	32.7	155.8	62.3

8.6	1	0.7	0.6	0.2	33.3	10.8	4.3
	2	2.1	1.8	0.7	37.3	40.6	16.2
	3	8.6	5.0	2.0	40.3	102.5	41.0
	4	13.6	4.5	2.0	44.1	73.0	29.2
	5	21.4	7.2	3.0	41.7	133.8	53.5
	6	51.3	13.8	4.2	30.7	180.3	72.1
	7	74.5	17.4	2.6	14.8	97.6	39.0

TABLE 2

Carboxypeptidase G-1 Activity and Protein Balances for Typical Batch Immunoadsorption and Elution Experiments using Guanidine HCl[a]

	INTO SYSTEM			OUT OF SYSTEM		
Run	Protein (mg)	Enzyme units	Specific activity		Enzyme activity (units)	Protein (mg)
#1	209.3	966	4.6	Primary filtrate and washes	568	201.4
				Amount of activity complexed with immunoadsorbent	378	——
				Dissociated	288	4.37
				% Recovery of complexed enzyme	72.4	
				Specific activity	167.4 μ/mg	
				Degree of purity	36.1	
#2	209.3	966	4.6	Primary filtrate and washes	620.7	197.7

				Amount of activity complexed with immunoadsorbent	345.3	——
				Dissociated	321.7	4.7
				% Recovery of complexed enzyme	93.2	
				Specific activity	156.0 μ/mg	
				Degree of purity	33.8	
#3	209.3	966	4.6	Primary filtrate and washes	599.3	189.9
				Amount of activity complexed with immunoadsorbent	366.7	——
				Dissociated	292.9	3.4
				% Recovery of complexed enzyme	79.9	
				Specific Activity	166.7 μ/mg	
				Degree of purity	36.1	

[a]Antibody #37. 9/26/73. Titer 1/16. Serum attached to Sepharose: 19 mg/cc. 5 cc samples about 0.95 g Sepharose/cc. Incubation time: 2 hrs. 3.6 cc of 6 M guanidine/g Sepharose for dissociations.

TABLE 3

Carboxypeptidase G-1 Activity and Protein Balances for Typical Batch Immunoadsorption and Elution Experiments Using 4M KSCN Followed by 6M Guanidine[a]

	INTO SYSTEM			OUT OF SYSTEM		
Run	Protein (mg)	Enzyme units	Specific activity		Enzyme activity (units)	Protein (mg)
#1	156	657.6	4.6	Primary filtrate and washes	149.3	139.1
				Amount of activity complexed with immunoadsorbent	508.3	——
				Dissociated enzyme		
				4 M KSCN	345.6	2.0
				6 M Guanidine	118.5	1.2
				Total recovery	464.1	
				% Recovery of enzyme associated	91.2	
				Specific activity		
				4 M KSCN	169.4 μ/mg	

				6 M Guanidine	98.8 μ/mg	
				Degree of purity		
				4 M KSCN	36.7	
				6 M Guanidine	21.4	
#2	156	657.6	4.6	Primary filtrate and washes	152.1	139.1
				Amount of activity complexed with immunoadsorbent	505.5	—
				Dissociated enzyme 4 M KSCN	310.5	1.7
				6 M Guanidine	147.6	1.3
				Total recovery	458.1	
				% Recovery of enzyme associated	69.6	
				Specific activity		
				4 M KSCN	184 μ/mg	
				6 M Guanidine	115 μ/mg	
				Degree of purity		
				4 M KSCN	39.1	

TABLE 3 (Continued)

				6 M Guanidine	24.9	
#3	156	657.6	4.6	Primary filtrate and washes	172	147.1
				Amount of activity complexed with immunoadsorbent	485.6	——
				Dissociated enzyme 4 M KSCN	284.2	1.7
				6 M Guanidine	135.3	1.3
				Total recovery	419.5	
				% Recovery of enzyme associated	86.42	
				Specific activity		
				4 M KSCN	165.0 μ/mg	
				6 M Guanidine	105.0 μ/mg	
				Degree of purity		
				4 M KSCN	35.7	
				6 M Guanidine	22.7	

#4	156	657.6	4.6	Primary filtrate and washes	148.6	137.5
				Amount of activity complexed with immunoadsorbent	509	——
				Dissociated enzyme		
				4 M KSCN	280.8	1.7
				6 M Guanidine	132.0	1.3
				Total recovery	412.8	
				% Recovery of enzyme associated	80.1	
				Specific activity		
				4 M KSCN	166.4 μ/mg	
				6 M Guanidine	105.0 μ/mg	
				Degree of purity		
				4 M KSCN	36.1	
				6 M Guanidine	22.7	

[a]Antibody #38. 7/13/73. Titer 1/32. IgG fraction attached to Sepharose, 19.89 mg/cc. 2cc samples. Incubation time: 2 hrs. Ratio of 10 ml elution solution/g Sepharose.

TABLE 4

Enzyme Activity and Protein Balances for Typical Batch Immunoadsorption of Contaminating Proteins

	INTO SYSTEM			OUT OF SYSTEM		
Run	Protein (mg)	Enzyme units	Specific activity		Enzyme activity (units)	Protein (mg)
#1	1.2	172.5	143.6 μ/mg	Primary filtrate and washes	172.8	0.780
				% Recovery of mass	65	
				% Recovery of activity	100	
				Specific activity	222.0	
				Increase in purity	1.55	
#2	1.2	172.5	143.6 μ/mg	Primary filtrate and washes	159.8	0.719
				% Recovery of mass	65	
				% Recovery of activity	93	
				Specific activity	222	
				Increase in purity	1.55	

immunoadsorbent may be prepared by attaching anti-immunoglobulin or antisera to Sepharose 2B by the CNBr method. It may also be made using glass beads or any other suitable support. The antigen is complexed by immunoadsorbent in a gently mixed batch system (see Fig. 1). Antigen is reduced by 90% each 2 hr. at 4°C. Thus in the first 2 hr., 90% is complexed; in 4 hr., 99%; in 6 hr., 99.9%; etc. At higher temperature the complexing rate is more rapid, e.g., at 25°C, 90% is complexed in 1.5 hr., and at 40°C in about 1 hr. (see Fig. 2). The immunoadsorbent-antigen complex may be dissociated with 0.23 M NH_4OH or pH 2.8 glycine-HCl and the immunoadsorbent reused (see Fig. 3).

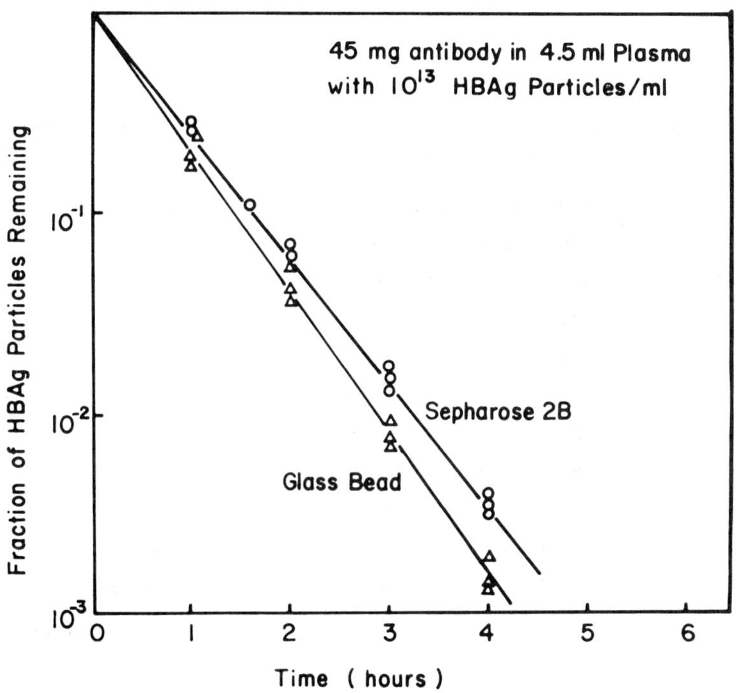

FIG. 1

Comparison of solid supports: Sepharose 2B gel and zirconia-clad glass beads.

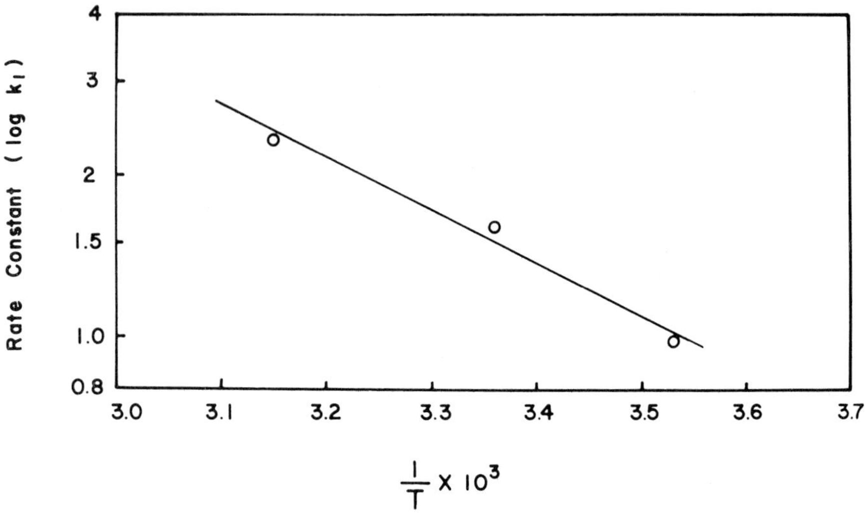

FIG. 2

The Arrhenius plot: activation energy of the HBAg immunoadsorption. Temperature ranges from 8 to 45°C.

With current detection methods, e.g. radioimmunoassay, it is possible to detect hepatitis antigen to levels of about 5×10^8 particles/ml. However, it is expected that 10^3 particles could result in disease [10].

In a general processing practice where only negative test material was employed, it may be assumed the initial concentration is not more than 5×10^8 particles/ml since this is the lowest concentration presently detectable by test.

In a 100-liter plasma pool there would be $5 \times 10^8 \times 10^5$ particles initially that should be reduced to a level where the probability is such that less than 1 antigen particle will remain in the pool. This would theoretically require about 27 hr. of mixing with the immunoadsorbent.

It must also be considered that equilibrium between the complexed and free antigen may occur at some point. Equilibrium was not reached over a 4 log cycle reduction (from 10^{13} to 10^9 particles/ml, the limit of detection). Therefore, regeneration of immunoadsorbent should be carried out after every

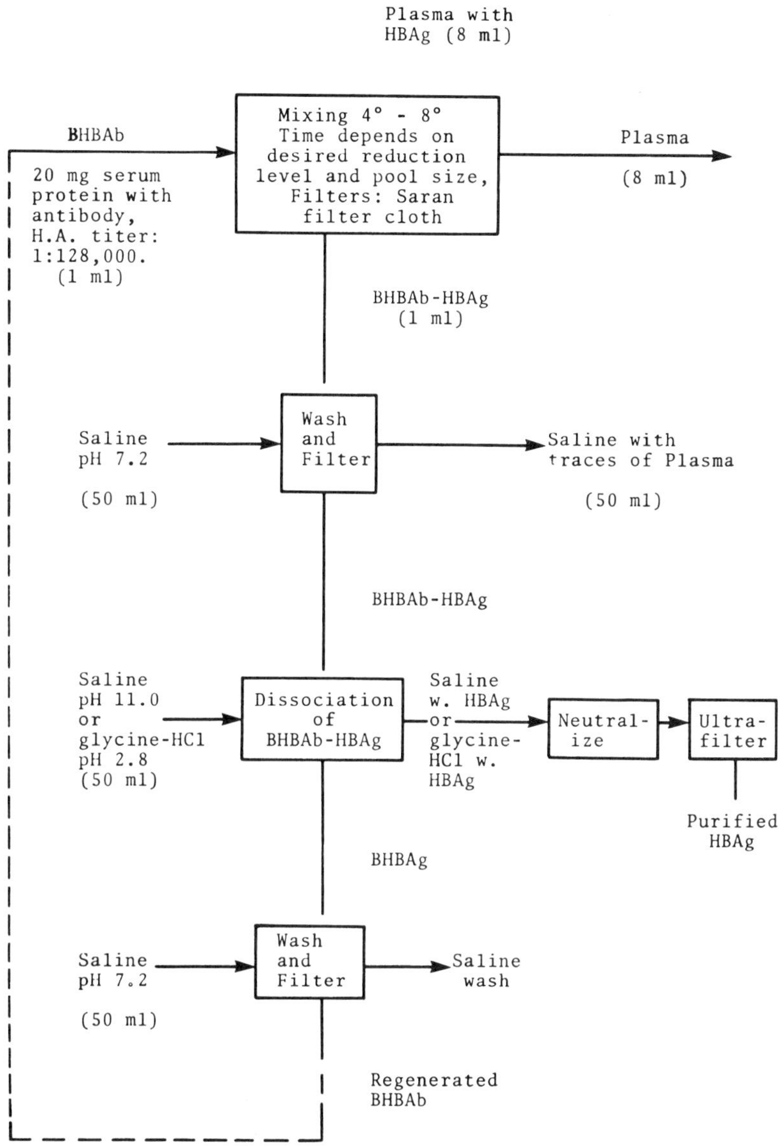

FIG. 3
Flow diagram of batch process for removing HBAg from plasma and recovering purified HBAg.

4 log cycle reduction to prevent equilibrium.

Increasing the concentration of antibody in the immunoadsorbent increases the rate of complexing up to a point. Then

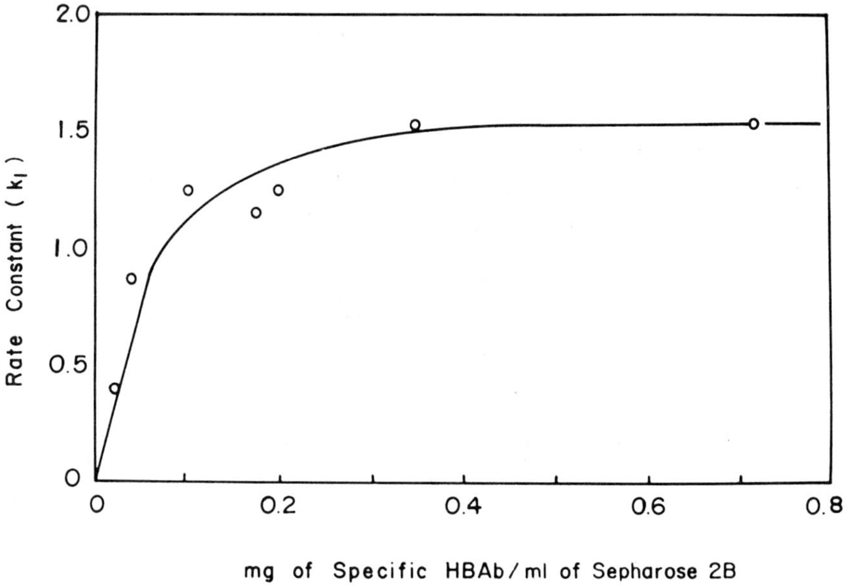

FIG. 4

Antibody concentration effect on the complexing rate.

other factors limit the complexing rate (see Fig. 4). Possibly the rate of diffusion of antigen to the immunoadsorbent becomes the limiting factor.

It has been possible to remove hepatitis infection from Factor IX (a plasma product with a high incidence of hepatitis associated) by this immunoadsorption process [10]. In this case a gibbon was used as the test animal. The gibbon was first inoculated with Factor IX that had been subjected to the immunoadsorption process. After 8 months, when it did not show any signs of infection, the gibbon was inoculated with unprocessed Factor IX. Within 14 weeks signs of hepatitis were observed. (This study was carried out by Dr. Hoofnagle, Department of Biologics, National Institutes of Health.)

Use of an immunoadsorbent made with animal antibody presents certain problems. Care must be taken that none of the animal protein leaks into the plasma. This could result in an allergenic reaction. For this reason, human antibody obtained from hemophiliacs was tried.

The kinetics are similar but the affinity constant is much higher than that of animal antibody:

$$\text{affinity constant} = \frac{\text{free antigen at equilibium}}{\text{complexed antigen}}$$

The affinity constant from animal antibody is not known, but for human antibody it is about 10^{-2}.

Thus after using human antibody, the rate of complexing hepatitis antigen levels off after a reduction of about 2 log

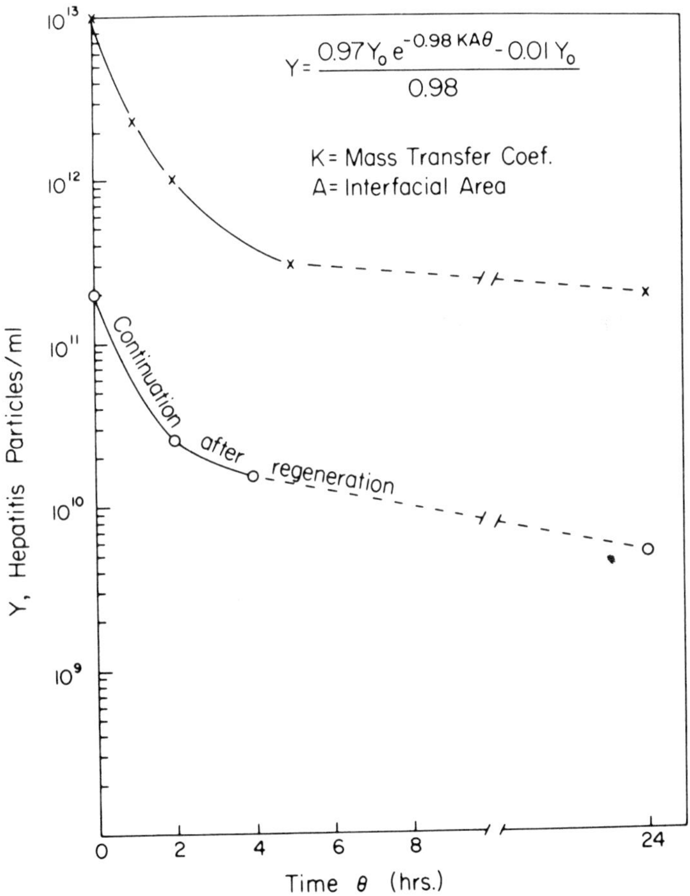

FIG. 5

Complexing of hepatitis antigen with human antibody immunoadsorbent.

cycles (see Fig. 5). For further reduction, the immunoadsorbent must be regenerated. Using human immunoadsorbent, it seems that plasma passage through an immunoadsorbent column would be more convenient that a batch adsorption of antigen that would require more regenerations of the immunoadsorbent to achieve the desired reduction of antigen than when using a higher affinity animal antibody. Serum hepatitis antigen recovered from infected plasma by this process is used to produce antibody in animals. The dissociated antigen is subjected to an antiplasma immunoadsorbent to remove traces of plasma. The final antigen is purified about 350X.

CONTINUOUS IMMUNOADSORPTION

A continuous immunoadsorption process has been developed using an endless belt of immunoadsorbent. The immunoadsorbent belt is driven through various compartments a) crude antigen → b) wash → c) dissociating solutions → d) wash → a) etc. (see Fig. 6).

FIG. 6

Endless belt of immunoadsorbent for continuous antigen isolation.

The belt is made from high fiber bond paper protected by wide mesh fiberglass screening. The major problems associated with this system is the carry over from one compartment to another. This is overcome to some extent by employing "squeezers" between compartments.

This type of system could theoretically be employed for the continuous removal of serum hepatitis antigen from plasma using human antibody, avoid the equilibrium problem between complexed and free antigen associated with the batch processing.

ACKNOWLEDGMENTS

Our studies were supported by National Science Foundation Grant Numbers GI 36527 and GB 40554.

REFERENCES

[1] D. Pressman, D. H. Campbell, and L. Pauling, J. Immunol., 44, 100 (1942).
[2] D. H. Campbell, E. Leuscher, and S. C. Lerman, Proc. Natl. Acad. Sci. U. S. A., 37, 575 (1951).
[3] S. Avrameas and T. Ternynck, Immunochemistry, 6, 53 (1953).
[4] P. Cuatrecasas and C. B. Anfinsen, Ann. Rev. Biochem., 40, 279 (1971).
[5] D. M. Livingston, E. M. Scolnick, W. P. Parks, and G. J. Todaro, Proc. Natl. Acad. Sci. U. S. A., 69(2), 373 (1972).
[6] P. Cuatrecasas, M. Wilchek, and C. B. Anfinsen, Ibid., 61, 636 (1968).
[7] G. C. Pitarra, S. E. Charm, and S. Green, Biotech. Bioengin., 17, 607 (1975).
[8] S. Hoag, S. E. Charm, and S. Ramm, Immunochemistry, 12, 833 (1975).
[9] R. Cornell and S. E. Charm, Submitted to Biotechnology and Bioengineering, 1976.
[10] S. E. Charm and B. L. Wong, Biotech. Bioengin., 16, 593 (1974).

Affinity Chromatography and Characterization of the Acetylcholine Receptor from Torpedo californica*

R. L. VANDLEN, J. SCHMIDT,** and M. A. RAFTERY

Church Laboratory of Chemical Biology
California Institute of Technology
Pasadena, California 91125

INTRODUCTION

Affinity chromatography has become one of the most important tools available to the biochemist for the selective isolation and purification of interesting biological macromolecules. The pioneering work of Porath [1,2] and subsequent developmental work by Porath [3] and Cuatrecasas [4] have served as practical guides for the successful application of affinity chromatography in many

**Current address: Department of Biochemistry, State University of New York at Stony Brook, Stony Brook, New York 11790

*Contribution No. 4968

areas of biochemical research. Specific ligands can be covalently attached to an insoluble, stationary support such as agarose or polyacrylamide beads and used to "fish out" those components which show an appreciable affinity for the ligands.

The successful application of affinity chromatography to a particular problem is governed by many variables, some of which are not fully understood at present, although progress is being made along these lines. (See other papers in this symposium.) Some of the design variables which must be optimized for efficient use of affinity chromatography are (1) the type of ligand, (2) the length of the arm or "leash" by which the ligand is coupled to the insoluble support, (3) the chemical nature of the leash, either polar or nonpolar or some combination thereof, and (4) the concentration of the ligand on the support. Normally, ligands are chosen for use in affinity chromatography because of their known specific interaction with the active center of the macromolecule of interest. Occasionally, however, ligands which exploit other chemical characteristics of the molecule can be utilized; for example, simple long-chain alkyl groups can be used to separate molecules based on their hydrophobic characteristics [5].

The purification of membrane proteins presents a special problem. Many membrane-bound protein molecules which can be solubilized by mild detergents have similar molecular weights, sedimentation coefficients, isoelectric points, and other chemical properties. However, by exploiting differences in the chemical specificity of these proteins, the techniques of affinity chromatography have been especially useful for separating and purifying specific membrane proteins. This has been especially so for neurochemistry. Substantial interest has been focused on the molecular events which occur during the neuron-specific functions of synaptic transmission and axonal conduction. In order to understand the mechanisms which involve the release of neurotransmitter and the interaction of the transmitter with a receptor to produce ionic conductance changes in the postsynaptic membrane, it is necessary to separate, purify, and characterize the various components of the synapse.

The neuromuscular junction utilizes acetylcholine (AcCh)* as its transmitter substance. Acetylcholine is synthesized in the nerve terminal by the enzyme choline acetyltransferase and stored in vesicles. When an impulse arrives at the nerve terminal, AcCh is released, diffuses across the synapse, and interacts with an integral membrane protein called the acetylcholine receptor (AcChR). Thereupon, the postsynaptic membrane displays an increase in cation permeability with concomitant depolarization of the membrane. Also located in the synaptic cleft is the powerful enzyme acetylcholinesterase (AcChE) which rapidly hydrolyzes acetylcholine and effectively controls the duration of the postsynaptic response.

Because of its central role in neuromuscular transmission, the AcChR has been the subject of numerous studies and reviews [6-8]. This protein can be removed from the membrane and purified by affinity chromatographic techniques. The function of this article is to describe in detail some of our experiences with affinity chromatography encountered during the purification of the acetylcholine receptor and to describe some of the structural characteristics of the purified protein molecule as they relate to the functioning of the molecule in the postsynaptic membrane.

---

*Abbreviations: Acetylcholine receptor, AcChR; acetylcholine, AcCh; acetylcholinesterase, AcChE; diisopropylfluorophosphate, DFP; diethylaminoethyl, DEAE; (diethyl-2-hydroxypropyl)aminoethyl, QAE; ethylenediaminetetraacetate, EDTA; phenylmethanesulfonylfluoride, PMSF; sodium dodecyl sulfate, SDS; bis-(3-aminopyridine-1,10) decane diiodide, DAP; 5,5'-dithio-(2-nitrobenzoic acid), DTNB; α-bungarotoxin, α-BuTx; 4-(N-maleimido)-benzyltrimethylammonium iodide, MBTA: trimethyloxonium tetrafluoroborate, TMO.

CHEMICAL SPECIFICITY OF THE ACETYLCHOLINE RECEPTOR

The chemical specificity of the AcChR is best described in terms of the molecules which interact strongly with the receptor. Some of the cholinergic ligands which bind to the AcChR are depicted in Fig. 1. These can be grouped into three general classes [9]: (1) agonists - those molecules which cause an immediate depolarization of the postsynaptic membrane; (2) depolarizing antagonists - molecules which cause an immediate but prolonged depolarization and render the membrane insensitive to subsequent application of AcCh; and (3) antagonists and blocking agents - those ligands which act simply as competitive inhibitors. Included in this last class are small proteins known as α-toxins found in the venom of many poisonous snakes. These neurotoxins, purified from several species of snakes, have molecular weights around 7,000 to 8,000 daltons, have similar amino acid sequences, and bind specifically and with high affinity to the AcChR [10]. The agonists of the receptor all contain a single quaternary trimethylammonium moiety while the antagonists are generally bifunctional or bisonium ligands. The reactive group or groups on the neurotoxins have not yet been definitively identified.

The binding of AcCh or one of the other reversible ligands is the natural method to assay for the receptor. Such binding can easily be demonstrated by equilibrium dialysis employing radioactive ligands. However, all of the small ligands in Fig. 1 also bind to the acetylcholinesterase molecule with substantial affinities. The esterase has a very rapid turnover, and even trace amounts of the enzyme are sufficient to make reversible binding studies with AcCh very difficult. Equilibrium dialysis can be employed, however, if the esterase activity is inhibited by diisopropylfluorophosphate (DFP) [11] or other organophosphate molecules [12].

The high specificity of the α-toxins for AcChR and not for the AcChE is surprising, especially since the active sites of the two distinct molecules must be similar for each to bind the many chol-

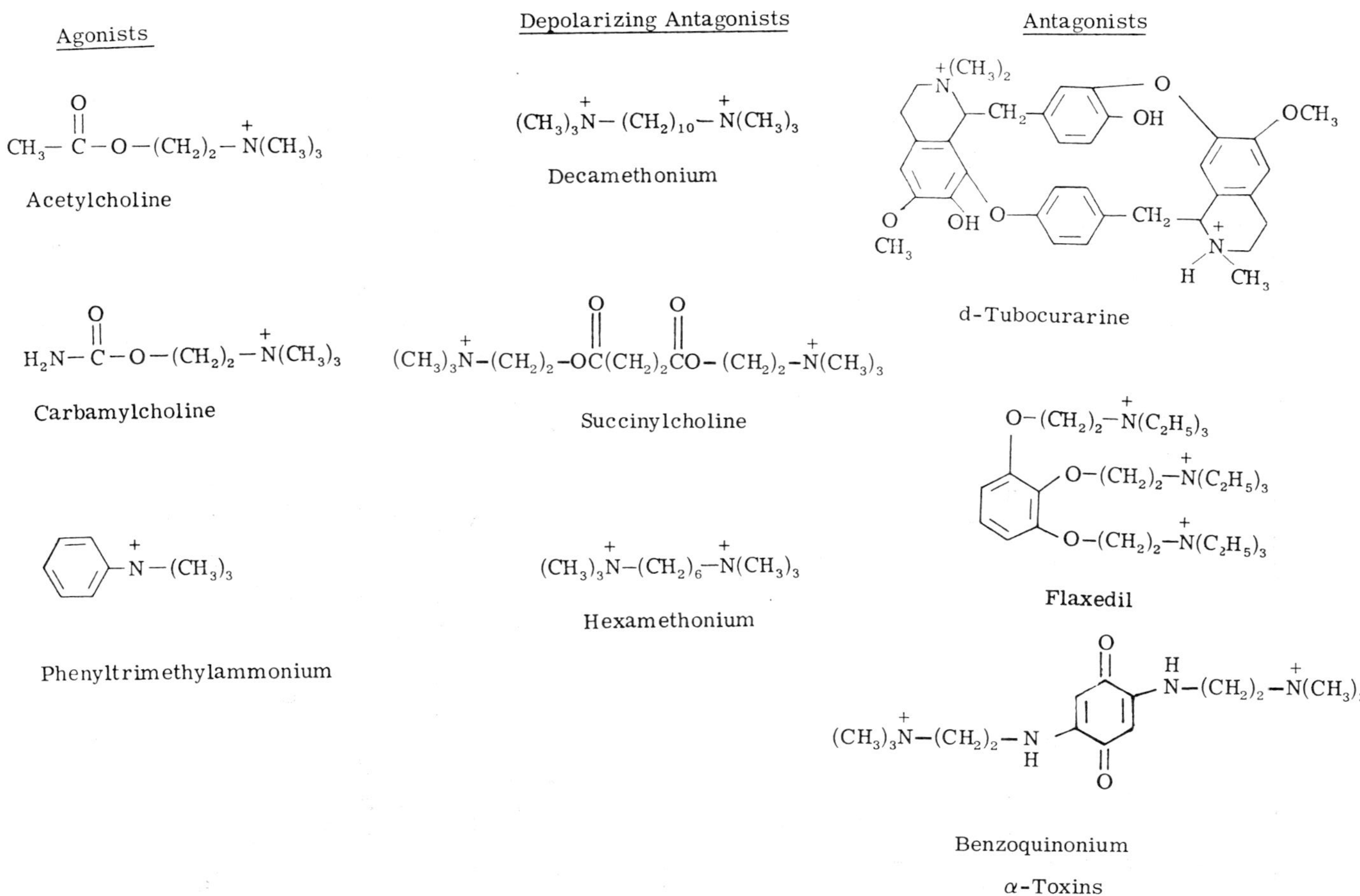

FIG. 1. Cholinergic ligands which bind to the acetylcholine receptor and to the acetylcholinesterase.

inergic ligands. Because these toxins bind with such specificity and high affinity, radioactively labeled derivatives of the toxins can be used as specific probes of the AcChR. One such toxin which has found wide use is α-bungarotoxin (α-BuTx), the major protein component of the Formosan krait Bungarus multicinctus [13]. The toxin can be purified from the crude venom [14], labeled with radioactive iodine [^{125}I], and used for a simple and extremely sensitive assay for the AcChR [15], since it binds essentially irreversibly. The receptor-toxin complex is acidic while the toxin is basic [10]. The complex can be adsorbed onto DEAE paper disks and unbound toxin washed away. The radioactivity trapped on the disks can be counted in a scintillation counter with the bound radioactivity proportional to the receptor concentration. The amount of toxin bound per unit mass of AcChR protein (n mole/mg) is defined as the specific activity of the receptor and is a measure of its purity. The specific activity can also be expressed as the amount of protein to which one molecule of toxin binds, or daltons AcChR/toxin molecule.

The quantity of AcChR in mammalian systems is extremely small due to the relatively few numbers of synapses. However, the electric organs of certain fish are composed of modified muscle cells containing large numbers of synapses. These synapses are typically cholinergic in nature and contain a large number of AcChR molecules [16]. Two common sources of electric organs are the electric eel from the Amazon River, Electrophorus electricus, and the marine electric ray of the Torpedo family. Torpedo organs comprise 20 to 25% of the total body weight (up to 200 lbs. for some species) and contain 10 to 20 times the concentration of acetylcholine receptor as does the eel electroplax.

RECEPTOR ISOLATION AND PURIFICATION

Membrane fragments enriched in AcChR from Torpedo californica are obtained by the procedure shown in Fig. 2 [17]. All electric organs are obtained fresh and used immediately or stored at -90°C. A marked deterioration of the organs is observed if the organs are

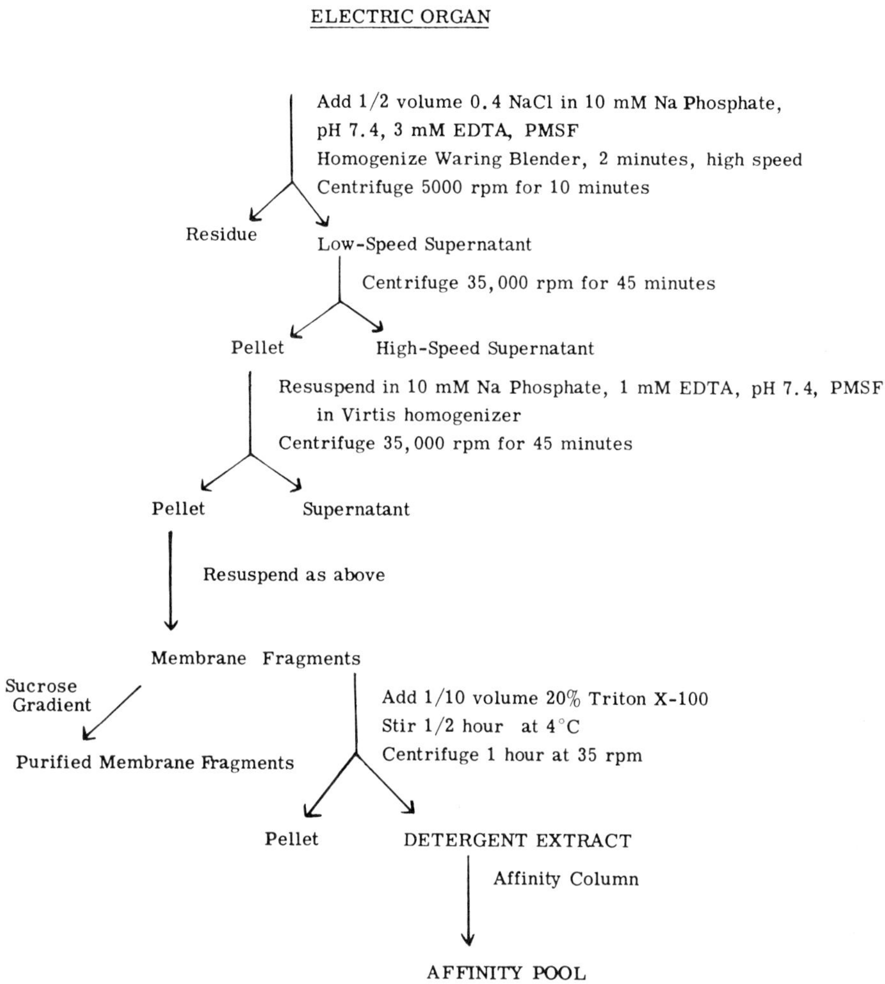

FIG. 2.

Extraction and purification procedure for the AcChR from Torpedo californica.

stored at higher temperatures for any length of time. The first low-speed centrifugation removes connective tissue and other large particles. The supernatant after the first high-speed spin contains all the water-soluble proteins and about 50% of the total

AcChE. A second wash with low ionic strength buffer further removes soluble proteins and reduces the ionic strength of the preparation. The membrane fragments can then be further enriched in AcChR by centrifugation on sucrose gradients or the AcChR can be solubilized from the membrane by use of the nonionic detergent Triton X-100. The receptor is quantitatively extracted from these fragments with 2% Triton in less than one-half hour at 4°C. After extraction, the solution is centrifuged once more to pellet the unextracted proteins. The solubilization and centrifugation step alone represents a 2-fold purification step.

At this point, conventional biochemical techniques were initially used to further purify the receptor in detergent solution. Figure 3 shows the protein and activity profile for gel filtration chromatography on Sepharose 6B in 0.1% Triton X-100 with 10 m$\underline{M}$ Na phosphate buffer at pH 7.4. The peak of toxin binding activity is

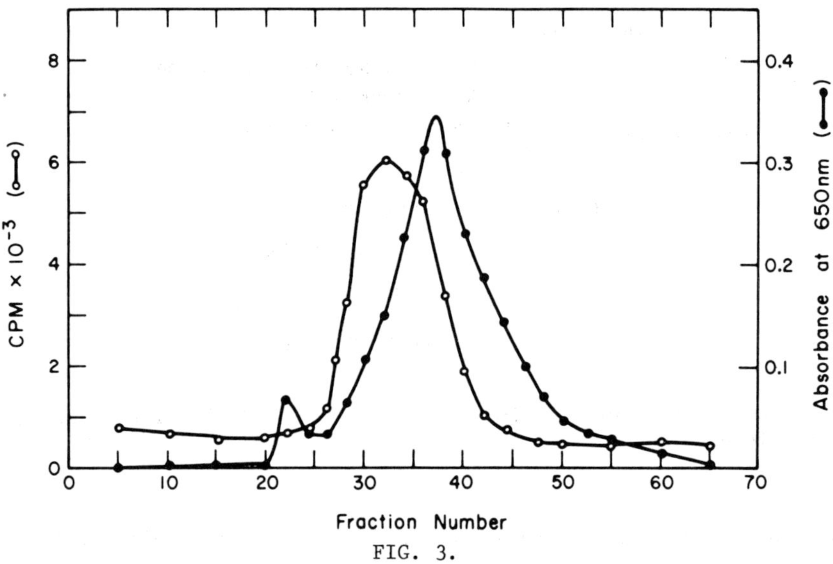

FIG. 3.

Sepharose 6B chromatography (5 X 80 cm column) of Triton X-100 extract. Eluting solvent was 10 m$\underline{M}$ Na phosphate pH 7.4, 0.1% Triton. Fractions of 20 ml were collected and analyzed for toxin-binding activity (o—o) and protein (●—●).

partially separated from the bulk of the detergent-solubilized proteins. The separation is not sufficient, however, to use gel filtration in a large-scale isolation procedure. Isoelectric focusing of detergent extracts was also performed with the results presented in Fig. 4 [18]. It was observed that the toxin-receptor complex has a pI of about 5 and that there was some separation of the toxin-binding material from the bulk of the other detergent solubilized protein. However, at its isoelectric point the AcChR tends to aggregate and denature, which limits the usefulness of this technique. Similarly, sucrose density gradient centrifugation does not provide a clean separation [18].

In addition to the structural information derived from these studies, the experiments with conventional biochemical techniques showed clearly that a different approach must be used if significant progress was to be made in the purification of the AcChR. Because many of the membrane bound proteins in the Torpedo electro-

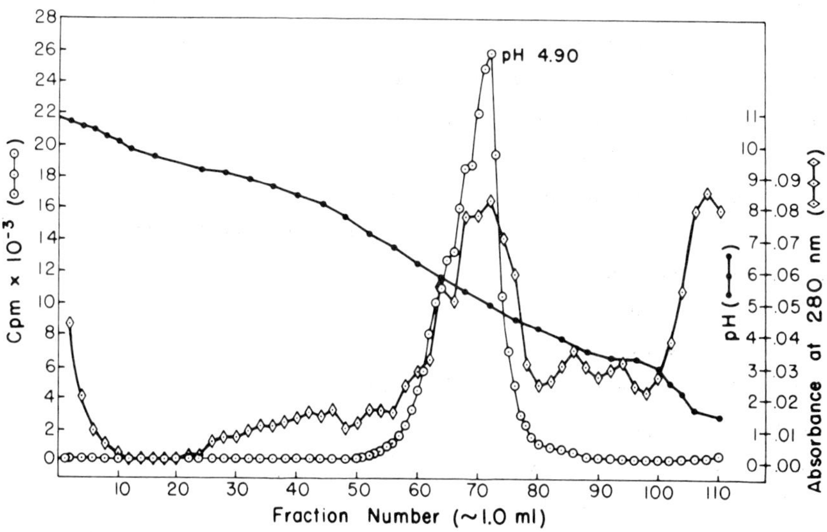

FIG. 4.

Isoelectric focusing of [^{125}I]-α-BuTx-receptor complex after Sepharose 6B chromatography (Fig. 3).

plax solubilized by Triton have similar molecular weights, isoelectric points, and hydrodynamic properties, one must use a technique which depends on some specific properties of the receptor molecule rather than its gross properties. Hence affinity chromatography was considered and because there were many different ligands which show affinity for the receptor, and which could be coupled to an insoluble support, there was a reasonable chance for success.

AFFINITY CHROMATOGRAPHY OF THE ACETYLCHOLINE RECEPTOR

Since all cholinergic ligands have quaternary nitrogen groups, it was thought that DEAE-cellulose or QAE-Sephadex might be useful as affinity chromatography resins [19]. It became readily apparent that little significant prufication could be achieved due mainly to the ion-exchange properties possessed by these resins, the concentration of functional groups on these resins being about 10^{-1} M. At much lower ligand concentrations, around 10^{-2} M, other resins which contained as an affinity ligand the phenyltrimethylammonium moiety, similar to that used by Berman and Young [20] and Dudai et al. [21] for the purification of the AcChE, were tested. These showed considerably more promise but sill retained a substantial amount of nonspecific protein which was eluted with the receptor by a sodium chloride gradient.

The affinity resin which has been most useful in the purification of the AcChR is resin A (Table 1). The ligand was synthesized [19] and coupled to Sepharose according to Cuatrecasas' procedure for the coupling of amines to agarose [4]. A series of resins with varying concentrations of ligand were made and tested for their ability to selectively adsorb the AcChR from the detergent extract and, upon elution, yield purified material. At ligand concentrations above 4×10^{-4} M, the resins displayed typical ion-exchange properties; in addition, only a small amount of the receptor could be eluted with low concentrations of cholinergic ligands. At concentrations below 4×10^{-4} M the capacity of the resin was too low to allow use as a feasible purification step. The resin

Table 1. Affinity Resins

A. Sepharose 2B - $NH(CH_2)_5CONH(CH_2)_3\overset{+}{N}(CH_3)_3$

B. 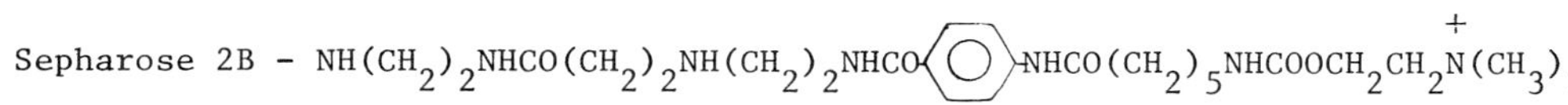

C. 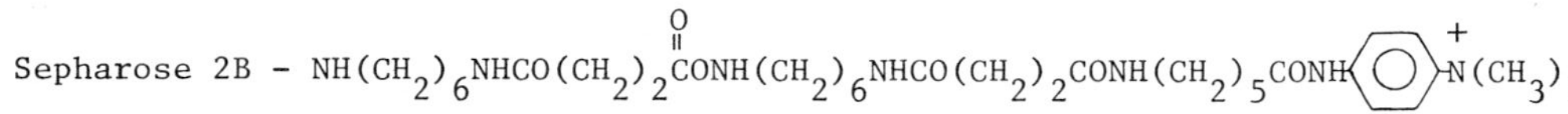

D. 

E. Sepharose 2B-Cobrotoxin

with 4 X 10^{-4} M ligand concentration was chosen as the best compromise between the observed ion-exchange effects at high concentration and the low capacity at low concentrations. The experimental conditions were optimized so as to yield a preparation with reasonable specific activities. Elution of the receptor was achieved by a sodium chloride gradient from 0 to 0.1 M NaCl in a buffer of 10 mM Na phosphate, 0.1% Triton, pH 7.4. This affinity chromatography step represented 10- to 20-fold purification over the detergent extracts. A large column, 5 X 80 cm, containing 2 liters of resin was prepared and has been routinely used for 2 years with no noticeable deterioration or change of its chromatographic properties. This column has been incorporated into a semiautomatic system for large-scale purification of the AcChR [17] employing an LKB Ultragrad gradient mixer (model 1300). The gradient mixer automatically controls the application of the detergent extract to the affinity column, washes the column with starting buffer (10 mM Na phosphate, 0.1% Triton, pH 7.4) until the nonspecific proteins are eluted, elutes the AcChR with a salt gradient, elutes the AcChE and other tight-binding proteins with a 1 M NaCl buffer wash, and reequilibrates the column with starting buffer. One complete cycle requires 24 hr. In addition, because of the highly reproducible performance of the affinity column when controlled by the gradient maker system, only the column effluent which contains the AcChR, and sometimes the AcChE, is routinely collected, the rest being shunted into a waste container. The total elapsed time from the homogenization of the organ to the elution of up to 100mg of purified receptor from the column is less than 30 hr. A typical profile of a chromatographic run with the automatic system is shown in Fig. 5.

A clean separation of receptor activity, as judged by toxin binding, from the bulk of the nonspecific protein, as well as from the AcChE is achieved. The purification procedure is summarized in Table 2. Several modifications to the originally published procedure [19], namely the inclusion of EDTA and the proteolytic enzyme inhibitor PMSF, and shorter extraction times, resulted in a substantial in-

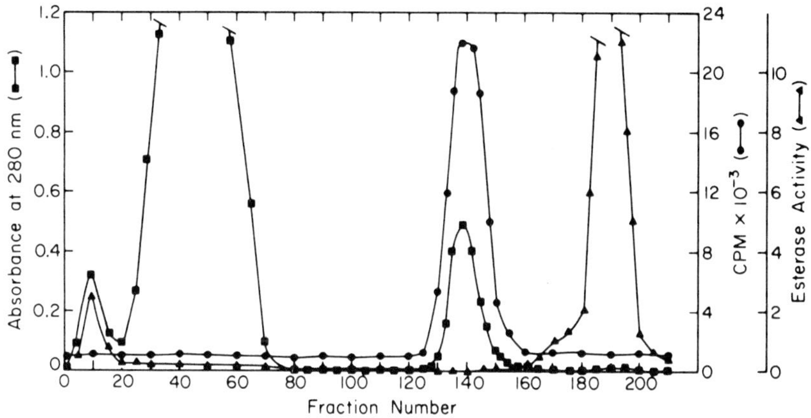

FIG. 5.

Large-scale purification of AcChR. The Triton X-100 extract was applied to the column and elution controlled by LKB Ultragrad gradient mixer. Fractions of 25 ml were collected and analyzed for toxin-binding activity (o—o), protein (●—●), and AcChE activity (Δ—Δ).

crease in the specific activity at all stages of the purification procedure. The yields in Table 2 would be even greater after affinity chromatography if the whole of the activity peak were pooled.

In practice, elution of AcChR from an affinity column should be possible with low concentrations of cholinergic ligands. It was somewhat surprising, then, that low concentrations of NaCl (about 50 mM) would elute the receptor in high purity. This observation led to the eventual discovery that ligand binding to AcChR was sensitive to the concentration and charge of cations [22], the inhibitory effect being greater than that due to ionic strength or electrostatic interactions between oppositely charged species. The inhibition constant for monovalent ions has been estimated to be about 6 to 8 mM and for divalent ions about 0.4 to 0.6 mM.

Ligand elution experiments with carbamylcholine, decamethonium, and hexamethonium with this affinity resin have been attempted. It was found that AcChR elution by cholinergic ligand could only be achieved at high ligand concentrations, those concentrations at which the ionic strength was roughly equivalent to 30 mM NaCl.

Table 2. Purification Table - AcChR

	Protein (mg)	Activity (nmole)	Specific activity (nmole/mg)	Purification fold	Recovery (%)
Homogenate	18,600	1,300	0.07	-	100
Membrane suspension	3,250	1,300	0.40	5.7	100
Triton extract	1,350	1,300	0.96	13.7	100
Affinity pool resin A	95	820	8.63	125	63
Affinity pool resin B	73	730	10.00	143	56

Thus it was suspected that instead of an affinity column in the strict sense of the word, this column was acting as a highly efficient and selective ion-exchange column. To test this hypothesis several additional experiments were conducted on a small scale and their results are shown in Fig. 6. A detergent extract was heated for a period of time known to destroy all toxin and small ligand-binding to the AcChR. The extract was applied to the column and the receptor protein eluted (Fig. 6b) in the same way as the control (Fig. 6a). The protein profile was nearly identical for the two runs, but the heat-treated sample had no binding activity. An additional experiment was conducted with an extract which had been saturated with radioactive toxin. Some of the activity was not retained by the column but some was eluted at the usual salt concentration as above. This experiment was not as definitive as the former, since the toxin (pI of 9.5) would tend

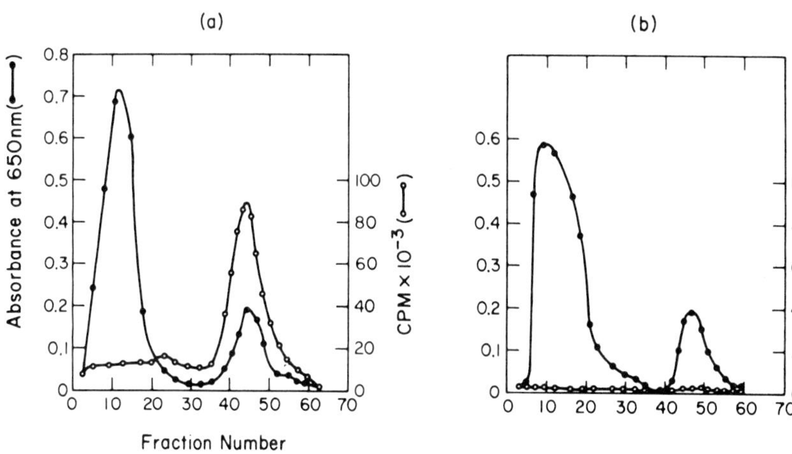

FIG. 6.

(a) Affinity chromatography of detergent extracts on Resin A. After application of the extract to the column equilibrated in 10 m$\underline{M}$ Na phosphate, 0.1% Triton, pH 7.4, the column was washed with the same buffer before a linear gradient of 0 to 0.1 $\underline{M}$ NaCl in buffer was begun. Fractions of 8 ml were collected and analyzed for toxin-binding activity (o—o) and protein (●—●). (b) Affinity chromatography of detergent extracts after heat treatment.

to change the ionic characteristics of the receptor (pI of 4.9). However, these experiments showed that the interaction of the affinity ligand at the active site of the receptor is not the only variable which is important in this affinity chromatographic procedure.

One possible rationale for the above behavior concerns the length of the "leash" or spacer arm which separates the affinity ligand from the agarose support. Cuatrecasas et al. [23] have demonstrated the relative ineffectiveness of Sepharose-bound D-tryptophan methyl ester as compared to the ε-aminocaproyl-D-tryptophan ligand for the purification of α-chymotrypsin. The importance of the arm extension is even more dramatic for systems which involve ligand-protein interactions of low affinity [4]. Hopff et al. [24] found that their spacer had to be at least the equivalent length of 30 carbon atoms for adequate affinity chromatography of the AcChE on a trimethylammonium ligand. When the spacer was shorter than this, the enzyme could be eluted with buffers of sufficient ionic strength. The importance of an adequate spacer length cannot be overemphasized. It would logically follow that for resin A, the ε-aminocaproic spacer is insufficient for real affinity chromatography of AcChR to occur. However, in other experiments to be described presently, the AcChR can still be eluted with NaCl even with very long spacer arms. Hence the observed effect of cations on the affinity of cholinergic ligands [22, 25] is still an important consideration in this system.

After affinity chromatography on resin A, further purification can sometimes be achieved. The AcChR protein obtained from resin A usually has the highest specific activity that has been routinely obtained (9 to 10 nmole α-BuTx/mg AcChR) and further purification steps are unnecessary. Occasionally, however, the specific activity of the pool is lower, about 8 nmoles α-BuTx/mg. Chromatography on Sepharose 6B, DEAE-cellulose, or hydroxylapatite will yield AcChR protein of higher specific activity. The difference in specific activities is apparently due to inactivated receptor molecules rather than to other nonspecific proteins. This conclusion was reached after numerous studies of the SDS gel electrophoresis patterns of preparations of high and low specific activities. (A more complete description of the SDS electrophoresis studies follows.) This is consistent with the observation that resin A retains inactive receptor produced artificially by heat treatment. This occasional variability in quality is apparently related to the condition or state of the organ when used, since the homogenation and purification procedures are constant from preparation to preparation.

It was desired to have an affinity resin which would separate active from inactive receptor molecules; resins B and C were synthesized and found to do just that. Resin B is a carbamylcholine analog linked to Sepharose by a long spacer arm at a concentration of 0.3 μ mole ligand/ml of packed resin. The pooled material from resin A is filtered through resin B followed by a short wash with buffer. The AcChR can be eluted with a buffer containing 1 m<u>M</u> succinylcholine. After dialysis to remove the succinylcholine, the specific activity of the receptor protein is consistently the highest we have ever obtained, about 10 nmole α-BuTx/mg protein. This corresponds to about one toxin molecule per 100,000 daltons of receptor protein.

Resin B has been used successfully to purify receptor from detergent extracts also. During initial experiments with the resin it was observed that the capacity of the resin was decreasing as a function of use. The cause of the diminishing capacity was, apparently, due to the hydrolysis of the carbamylcholine analog off the spacer arm by the acetylcholine esterase, which is known to hydrolyze carbamylcholine slowly [26] and is present in significant amounts in the detergent extracts. This problem could be effectively circumvented by inhibiting the esterase with DFP or other organophosphate inhibitors.

Resin C is composed of a phenyltrimethylammonium ligand used by Dudai et al. [21] for AcChE purification attached to a long arm at a concentration of 0.1 μmole/ml resin. This resin has a substantially lower capacity than resin B, but is not subject to the difficulties discussed above for resin B. Elution of the receptor is accomplished by a shallow gradient from 0 to 5 mM of succinylcholine with an equivalent purification as with resin B. The separation of esterase from receptor under these conditions is not as complete as observed with resin A; this resin was therefore not routinely used for receptor purification from detergent extracts but only for rechromatography after the esterase had been removed by resin A.

Even though resins B and C display desired affinity chromatography characteristics, such as elution of the receptor by cholinergic ligands at low concentrations and essentially quantitative recovery of toxin activity, sufficient concentrations of NaCl will also elute the receptor with the same yield and purity as succinylcholine, especially when the resins are used in rechromatography of the affinity pool from resin A. In contrast to the observed behavior of the esterase on affinity resins with different spacer arm lengths [24], the use of long spacer arms did not prevent the elution of the receptor with salt.

During the course of these studies, other affinity resins were prepared and tested. d-Tubocurarine was attached to a spacer arm by a diazo linkage [27] at a concentration of 0.2 μmole/ml resin. This resin was highly specific and showed great affinity for the receptor. The capacity of the resin was limited; only about 10% of the total amount of d-tubocurarine coupled to the resin was accessible to bind receptor. Only small amounts of active receptor (10 to 20% of the total amount bound) could be eluted with molar concentrations of cholinergic ligands, NaCl, or with millimolar concentrations of α-BuTx. Additional receptor in higher yields was obtained by reductive cleavage of the diazo bond with 0.1 M sodium dithionite solutions in a 0.2 M borate buffer (pH 9.0). Extensive dialysis was required to remove the d-tubocurarine before further studies were conducted on this material. An obvious drawback to the use of this ligand is the requirement to synthesize a new affinity resin before each use and the additional steps involved.

Because of the great specificity of the neurotoxins for the AcChR, one of the first attempts at affinity chromatography of the receptor utilized a resin (Resin E, Table 1) with the α-toxin from the cobra snake Naja naja atra coupled directly to Sepharose [28]. Initial experiments using α-BuTx to desorb the receptor resulted in the isolation and characterization of α-BuTx-receptor complexes. A high degree of purification with modest yields of 20 to 40% of the total adsorbed receptor activity can be obtained with this resin. Elution of the receptor could also be achieved by high concentrations of cholinergic ligands but not by sodium chloride. As observed with resin D, only about 10% of the total cobrotoxin coupled to the Sepharose is available to bind receptor.

Affinity resins containing other species of toxins have been used to isolate AcChR from the eel Electrophorus electricus as well [29-34]. In each case, high concentrations of ligands and/or elaborate procedures were used to desorb the receptor, with yields ranging from 10 to 50% of the adsorbed activity. The use of high concentrations of ligands to elute the toxin resins is accompanied by several possible disadvantages:

1. Extensive dialysis or other methods are required to remove the ligands from the receptor solutions.

2. High concentrations of acetylcholine and other ligands cause a phenomenon termed desensitization or receptor inactivation [35], a conversion of AcChR to an inactive form. The structural modifications of the receptor which accompany desensitization are unknown. This might be the origin of the observation by Eldefrawi and Eldefrawi [33] that AcChR desorbed by benzoquinonium will bind toxins but not small ligands.

3. Lindstrom observed that high concentrations of decamethonium (greater than 10^{-2} M) irreversibly denatured the receptor due to the high ionic strength [31]. Also, toxin-receptor complexes were eluted from the affinity column using benzoquinonium as elutant [36]. Apparently the toxin or a toxin-agarose piece is removed from the affinity resin during elution with high concentration of

ligands. Such a resin would have to be regenerated after usage and does not yield a satisfactory product.

Although affinity chromatography utilizing toxin resins would theoretically yield the purest receptor material because of the high specificity of toxin for receptor, in practice such chromatographic procedures are relatively less satisfying than affinity resins utilizing readily reversible ligands. Not only are the yields generally greater and the conditions milder with reversible ligands, but the ease with which large quantities of receptor can be obtained for subsequent biochemical studies is an important design variable to be optimized for affinity chromatography of the AcChR.

CHARACTERIZATION OF THE ACETYLCHOLINE RECEPTOR

Structural Characteristics

The conventional biochemical techniques gel filtration, isoelectric focusing, and sucrose density gradient centrifugation, while not being useful as purification steps, did nevertheless yield important information about the structural properties of the receptor. On gel filtration, the receptor-toxin complex migrates with an apparent Stokes radius somewhat less than that of β-galactosidase (MW = 540,000) [18]. On sucrose density gradients, the S value obtained for the complex suggests a molecular weight somewhat less than catalase (MW = 250,000). This dichotomy between molecular weights determined by two different methods has also been observed for the AcChR solubilized by detergents from the electric eel (Refs. 37 and 38, and for AcChE, Ref. 39). Such differences can be due to a variety of causes: differential amounts of detergent bound to the receptor and to the standard proteins (all of which are water-soluble), a shape radically different from a globular shape assumed for the standard proteins, unusual frictional coefficients, or other artifacts inherent in these experimental methods. There is preliminary evidence from electron microscopy and x-ray diffraction studies that the AcChR molecule is somewhat rod-shaped, but the deviation from globular shape would not be sufficient to account for the substantial differences between the apparent molecular weights determined by gel filtration and sucrose density gradients [40]. Another possible explanation is that there is a large amount of detergent bound to the receptor as compared to the standard proteins. It is known that about 100 molecules of Triton X-100 are bound for every 40,000 daltons of AcChR (D. Michaelson and M. A. Raftery, unpublished observations), while bovine serum albumin, a typical hydrophilic protein frequently used as a molecular weight standard, has four high affinity sites for Triton per molecule of 68,000 daltons [41]. The large amount of detergent bound to the receptor could drastically alter the apparent density and possibly the overall shape of the receptor molecule and give rise to anomalous molecular weights.

The question of purity of the AcChR eluted from affinity columns is still under investigation. Detergent-solubilized receptor of the highest specific activity exhibits a single sharp protein band on acrylamide gels run under nondenaturing conditions. A single band is also observed on gel filtration columns or on sucrose density gradients. There is, however, considerable discussion as to the subunit composition of the AcChR. When acrylamide gels are run in the presence of SDS [42] the most highly purified material exhibits a complex gel pattern (Fig. 7). There is a major component with an apparent molecular weight of about 40,000 daltons and other components with approximate molecular weights of 50,000, 60,000 and 65,000 daltons which cannot be eliminated by further purification steps. These subunits are in an approximate ratio of 4:1:1:1, respectively, which suggests a molecular weight for the AcChR complex of around 350,000 daltons. The relation of the minor subunits to the major subunit is unknown at present. The AcChR has two functions which might require a complex subunit structure, first, to bind acetylcholine and then, second, to open a gate or ionophore to allow ions to traverse the postsynaptic membrane and

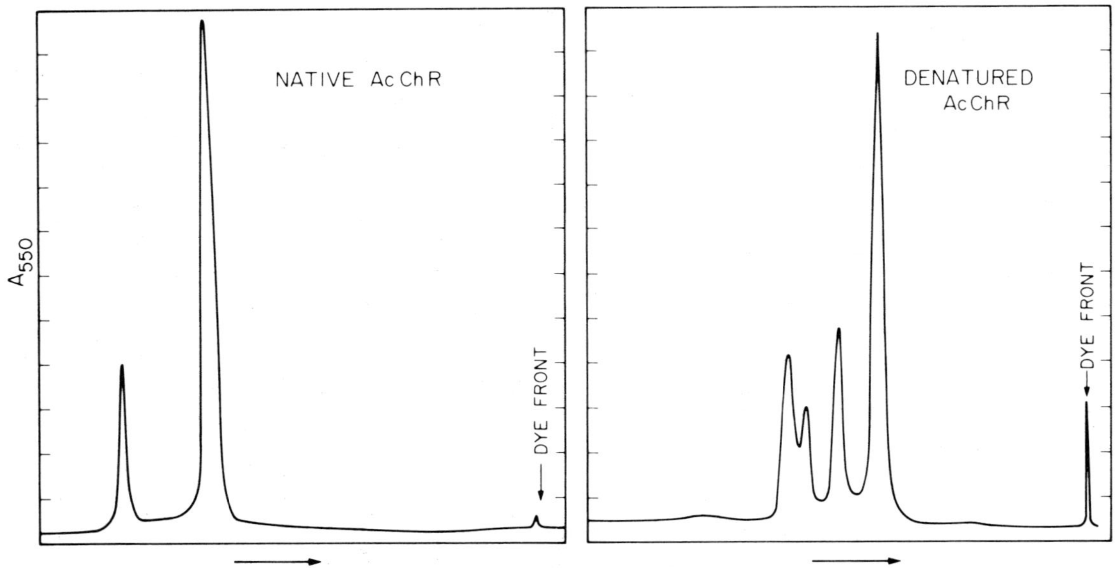

FIG. 7.

Gel scan analysis of purified AcChR. Left: Protein (10 μg) was electrophoresed in 4% acrylamide gels with 0.1% Na cholate and stained for protein [42]. The smaller peak to the left of the major peak corresponds to a dimer of the AcChR. Right: Same protein material (50 μg) was electrophoresed in 1% SDS acrylamide gels using the method of Fairbanks et al. [42]. The molecular weights of the major protein bands are indicated.

cause depolarization. Different subunits might be required for the binding of acetylcholine and for the opening of the ionophore. Toxin binding is a convenient assay for the binding component but there is no assay for the ionophore, if it even exists as a separate entity. It is known that the 40,000 subunit carries the ligand binding site [43]; the other subunits might be necessary for the organization of receptor molecules in the postsynaptic membrane.

Alternatively, the complex structure may be an artifact of detergent extraction. It is possible that during solubilization discrete membrane proteins may associate together to shield hydrophobic areas on the proteins from the aqueous solvent. Hence such aggregates would migrate as a single species under nondissociating conditions but would show a complex SDS gel electrophoresis pattern. While there is no direct evidence that would support this possibility, it should not be dismissed until more is known about the structure and function of membrane proteins.

One other factor which must be reconciled when a description of the subunit composition of the AcChR is made is the observation that even in the purest preparations to date, the ratio of the toxin sites to receptor is about one toxin per 100,000 daltons of receptor protein. Whether this means that more than one subunit is required to bind toxin or that the receptor is still only partially purified is open to debate and is under extensive study at the current time.

In spite of the open questions concerning receptor homogeneity and the subunit composition of the receptor, meaningful physical and chemical studies have been performed. The AcChR from Torpedo californica has been shown to carry carbohydrate moieties. When both SDS gels and nondenaturing gels of the purified receptor are stained for carbohydrate by the periodic acid-Schiff base procedure [42], all of the detectable polypeptides are found to contain sugar moieties [44]. Proteins which are known not to have carbohydrate do not stain by this procedure. The neutral sugar content of the receptor, as measured by the phenolsulfuric method

[45], with appropriate controls for detergent interference, is about 3%. The major neutral sugars are mannose, galactose and glucose in the approximate ratio of 7:2:1, respectively. In addition, 3 to 4 moles of N-acetylglucosamine per 40,000 daltons of protein have been detected in amino acid hydrolysates of the purified receptor. Purified AcChR can be completely adsorbed to affinity resins of Sepharose concavalin A or can be precipitated by concavalin A as observed by Meunier et al. [46]. The distribution of neutral sugars and the positive staining for carbohydrate on SDS gels rules out the possibility that the carbohydrate arises from a degradation of the Sepharose columns used during affinity chromatography [33, 36].

Functional Characteristics of AcChR

The functionality of the purified receptor material with regards to its binding of cholinergic ligands has been probed by direct binding methods employing radioactive ligands and equilibrium dialysis [47], by the inhibition of the rate of formation of toxin receptor complexes, and by the fluorescent probe DAP [25]. These studies have shown that the toxin-binding material purified by affinity chromatography binds cholinergic ligands with high affinity in a noncooperative fashion. Table 3 lists the ligand binding parameters obtained by these different techniques. Equilibrium dialysis and the fluorescence studies allow quantitation of the number of ligand binding sites; it was observed that there are nearly twice as many sites for toxin as for small ligands, or one small ligand per 200,000 daltons of AcChR. Similar findings have been reported for the receptor from eel [48] and for purified membrane fragments from Torpedo californica [49]. Hence the "half-site" reactivity is not an artifact due to solubilization of the receptor with detergents. Recent experiments have shown that there are two classes of ligand binding sites on the receptor molecule. One half of the toxin sites bind ligands with higher affinity, while the other half bind ligands with lower affinity, usually

TABLE 3. Ligand Binding Parameters of AcChR

Ligand	Equilibrium dialysis, a K_d ($\underline{M}$)	Fluorescence, b K_d ($\underline{M}$)
Acetylcholine	2.3×10^{-6}	5×10^{-6}
d-Tubocurarine	6.4×10^{-6}	5.2×10^{-7}
Decamethonium	55×10^{-6}	3.5×10^{-6}
Carbamylcholine		10.3×10^{-5}
DAP		1.3×10^{-7}
Na^+		11×10^{-3}
Ca^{2+}		2×10^{-4}

a Measured by direct binding of radioactive ligands in 200 m$\underline{M}$ NaCl, 10 mM Na phosphate, pH 7.4, 0.1% Triton X-100 [47].

b Measured by competitive inhibition of DAP binding in 15 m$\underline{M}$ NaCl, 10 m$\underline{M}$ Na phosphate, pH 7.4, 0.1% Triton X-100 [25].

by a factor of 5 to 25 fold. Toxin-inhibition studies have shown that cholinergic ligands slow the rate of complex formation; that concentration of ligand which slows the initial rate of complex formation by 50% is termed its protection constant, Kp. These protection constants are not true dissociation constants but do agree reasonably well with corresponding dissociation constants (Table 4).

TABLE 4. Protection Constants Determined by Toxin Inhibition

Ligand	K_p
Acetylcholine	2.5×10^{-6}
Carbamylcholine	7.7×10^{-5}
Nicotine	8.0×10^{-5}
Decamethonium	2.2×10^{-6}
Hexamethonium	2×10^{-6}
d-Tubocurarine	1.2×10^{-7}
Flaxedil	2×10^{-7}
Tetramethylammonium	9×10^{-4}
Tetraethylammonium	6.5×10^{-4}
Phenyltrimethylammonium	8×10^{-5}
Acetylthiocholine	1.6×10^{-4}
Choline	1.7×10^{-3}
α-Bungarotoxin	$<10^{-9}$

A comparison of the data from equilibrium dialysis obtained at a NaCl concentration of 200 m*M* and that from the fluorescence studies conducted at 15 m*M* NaCl shows a large difference for the apparent dissociation constants for decamethonium and d-tubocurarine. These discrepancies are due to the inhibitory effect of salt on the affinity of some ligands. The affinities of other ligands, especially acetylcholine and carbamylcholine, are not appreciably affected by salt. In general, the salt inhibition is greatest for those cholinergic ligands with two quaternary nitrogen functions and least for the monofunctional ligands. The apparent inhibition constants for sodium and calcium ions are given in Table 3 and are similar for both the inhibition of DAP and for the inhibition of the rate of toxin binding. These effects are also observed with the purified membrane fragments [49].

The inhibition of the rate of formation of the toxin-receptor complex provides a simple, specific method for estimating apparent inhibition constants of ligands for the AcChR. The method employs the disk assay [50] and measures the amount of radioactive toxin-receptor complex formed as a function of the ligand concentration. Table 4 lists the measured protection constants for a variety of cholinergic and noncholinergic ligands.

CHEMICAL MODIFICATION STUDIES

In an attempt to determine some of the functional characteristics of the purified AcChR, various chemical modification studies have been conducted. The effect of a change in pH on the binding of DAP by fluorescence measurements [25] revealed that the dissociation constant for DAP decreased from pH 5.9 to pH 7.0 and remained constant at values above pH 7.0. Conversely, equilibrium dialysis measurements tend to indicate that the number of sites for small ligands changes with pH [51]. At pH 5 or lower there is essentially no ligand binding with a gradual increase in the number of sites relative to toxin. At pH 7.4, there are an equal number of high and low affinity sites for small ligands. As the pH was increased, there was a conversion of low affinity sites into high affinity sites so that at pH 9 there were an equal number of high affinity sites as toxin sites. From these studies it appears that there may be a group with a pK about 6 to 7 which affects small ligand binding. The rate and extent of binding of radioactive toxin to AcChR was also investigated as a function of pH [22] (see Fig. 8). The results indicate that protonation of an ionizable group with a pK of about 6 inhibits binding. Inhibition is also seen at higher pH values and may be due to denaturation of the receptor or to effects on the toxin.

Amino acid composition studies on purified material indicated a small number of sulfur-containing amino acids. Quantitation of sulfhydryl content with DTNB revealed the presence of one or two

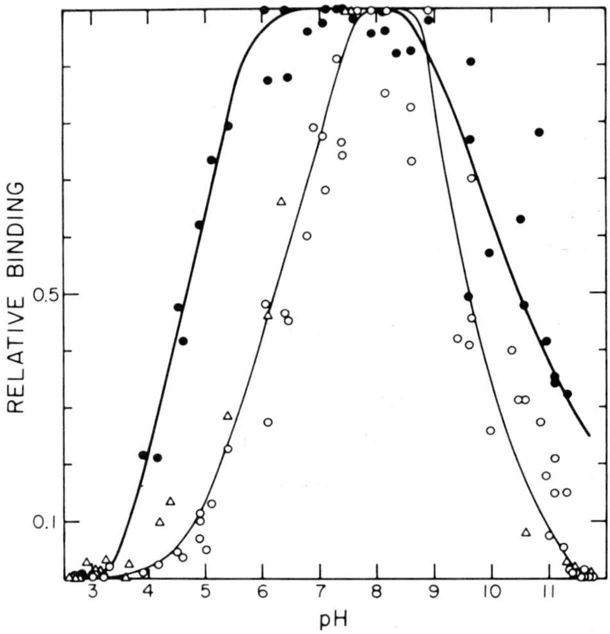

FIG. 8.

Effect of the pH on the rate and extent of binding of α-BuTx to purified AcChR. Receptor was incubated with a mixed citrate-acetate-phosphate buffer of constant sodium concentration and complex formation was started by the addition of toxin. Aliquots of 0.1 ml were withdrawn after 0.2 min (o) and 30 min (•) as described elsewhere [22].

sulfhydryl groups per 40,000 daltons of AcChR protein [52]. Treatment of the protein with β-mercaptoethanol or dithiothreitol to reduce any disulfide bonds, with subsequent removal of these reagents under nonoxidizing conditions, revealed two additional sulfhydryl groups. No additional sulfhydryl groups are found when the reactions are performed under protein denaturing conditions. Reaction of the free sulfhydryl group with ρ-mercuribenzoate or iodoacetamide has no effect on toxin or small ligand binding to the receptor.

Likewise, reduction of the disulfide bond does not affect toxin binding appreciably. Subsequent modification of the resulting sulfhydryl groups with the specific affinity label, 4-(N-male-imido-benzyltrimethylammonium iodide, first used by Karlin et al. [53], partially reduces toxin binding. Additionally, the specific incorporation of ^{3}H -MBTA at the reduced disulfide can be prevented by preincubation with toxin. These results generally agree with those published by Karlin and Cowburn for the AcChR from the eel [54]. Physiologically this disulfide bond seems to play an important structural role; reduction of the disulfide causes substantial changes in the electrophysiological response of the receptor to cholinergic ligands [55]. No effect is seen on the SDS gel patterns whether this disulfide bridge is reduced or not.

Other chemical modification studies have centered around the use of the highly reactive methylating agent, trimethyloxonium ion ($(CH_3)_3O^+$). This reagent, or the triethyl derivative, has been used to selectively label specific carboxyl groups in lysozyme [56] and has been shown to inactivate AcChE [57] (Fig. 9, top). Structurally similar to the trimethylammonium group common to all cholinergic ligands, trimethyloxonium ion (TMO) has been used to probe the active site of the AcChR [58].

Incubation of AcChR solutions with millimolar concentrations of trimethyloxonium tetrafluoroborate produces little or no change in the binding of toxin to the receptor. However, the effect on the binding of decamethonium is substantial. Reaction of the receptor with increasing amounts of TMO, as depicted in Fig. 9 (bottom), results in a progressive decrease in the amount of activity as determined by decamethonium or DAP binding. Complete inhibition of the esterase occurs at substantially lower concentrations of TMO than for the receptor. No appreciable effect on the binding of acetylcholine is seen at concentrations of TMO at which decamethonium binding is substantially decreased. Furthermore, the inhibition of decamethonium binding by TMO is antagonized by high concentrations of cations or by decamethonium. If it is assumed that there are two negative subsites in the active site, one each to bind the positive quaternary nitrogen groups on bisonium ligands,

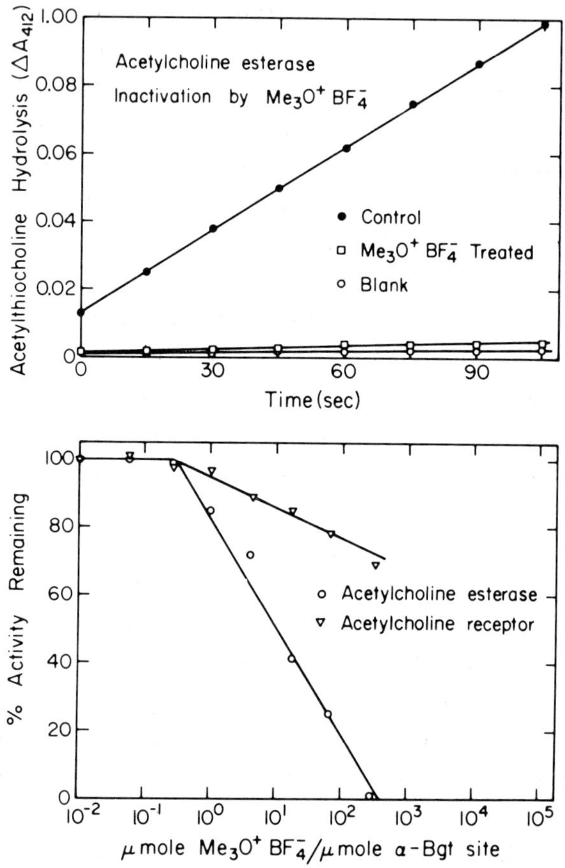

FIG. 9.

Inhibition of AcChR and AcChE by trimethyloxonium ion. Top: Inactivation of AcChE by TMO [58]. Esterase activity measured by method of Ellman et al. [62]. Bottom: Inactivation of decamethonium binding to AcChR and of acetylthiocholine hydrolysis by AcChE as a function of TMO concentration.

it appears that TMO reacts with a group (probably a carboxyl group) near the second negative subsite in a manner similar to that observed for cations and not at the subsite which binds the acetylcholine quaternary nitrogen atom. Those ligands whose affinities are affected by cations (except toxin) are especially affected by TMO.

Other studies are underway which should further characterize the active site of the AcChR.

OTHER USEFUL AFFINITY LIGANDS FOR AcChR PURIFICATION

An article on the affinity chromatography of the acetylcholine receptor would be incomplete without a description of some of the other affinity ligands which have been used in the purification of AcChR from various sources. A summary of the various ligands used, eluting conditions, the specific activity of eluted material, and subunit molecular weights as determined by SDS gel electrophoresis is given in Table 5. (See also Refs. 8 and 43.) Immobilization of cholinergic ligand analogs containing quaternary ammonium functions have been used for routine AcChR purification by Raftery *et al.* [17], Changeux *et al.* [61], Karlin and Cowburn [54], and Biesecker [34], while neurotoxin affinity ligands have been exploited by Changeux *et al.* [47], Reich *et al.* [32], Eldefrawi and Eldefrawi [33], Lindstrom and Patrick [31], and Heilbronn *et al.* [30, 59]. The best preparations have specific activities corresponding to one toxin per 100,000 to 150,000 daltons of AcChR protein. Regardless of the apparent purity of the preparation, two or more bands are usually observed with SDS gel electrophoresis studies except as noted in Table 5. While affinity chromatography techniques have resulted in substantial purification of the receptor from detergent extracts and give rise to similar molecular species, the rather substantial differences in specific activities among the various groups indicate that there are other parameters which must be considered. The complex SDS patterns have prevented a rigorous definition of purity or even a description of the subunit composition although it is apparent that one main component has a molecular weight somewhere around 40,000 to 45,000 daltons. Even though many molecular properties of the AcChR are still to be conclusively established, important and substantial information has been gained concerning the function of the acetylcholine receptor in synaptic transmission.

Table 5. Affinity Chromatography of AcChR

Source	Affinity Ligand	Eluant	Activity[a]	Subunits[b]	Ref.
T. californica	|-spacer-$NH(CH_2)_3\overset{+}{N}(CH_3)_3$	0 to 0.1 M NaCl gradient	100 c	40,50 60,65	
T. marmorata	|-acetylated naja naja siamensis toxin	0 to 1 M carbamylcholine gradient	430	2 bands 45 to 50	59
	|-naja naja siamensis toxin	1 M carbamylcholine	83 d	-	33
	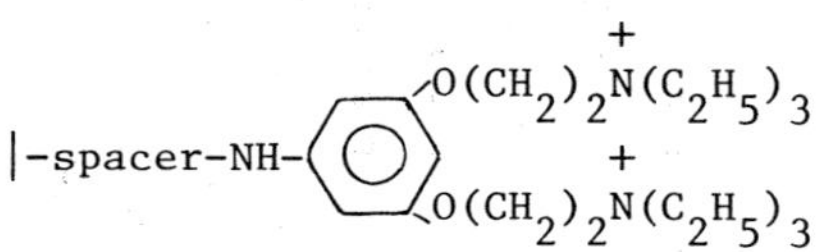	0 to 0.1 M flaxedil gradient	165	41.5	60
E. electricus	"		145	45,54	46,61

|-naja naja siamensis toxin	50 m$\underline{M}$ hexamethonium	90	50[e]	32
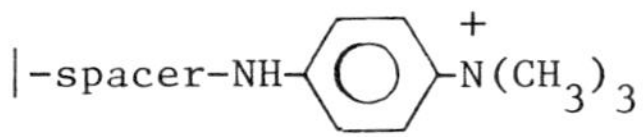	1 m$\underline{M}$ decamethonium	220	44,50	34
|-naja naja toxin	1 m$\underline{M}$ benzoquinonium	133	42,54	31
	50 m$\underline{M}$ carbamylcholine	250[f]	40,47 53	54

[a]Units are 10^3 daltons of AcChR/toxin site.

[b]Subunit molecular weight (10^3 daltons).

[c]The same specific activity can be obtained from affinity resins B or C (Table 1) with elution by succinylcholine gradients.

[d]Reported as 10^3 daltons/AcCh site.

[e]Suggested by "the integral ratios of amino acid residues derived from the histidine content" [32].

[f]Assay was by affinity labeling with [^{3}H]-MBTA.

The wealth of information about two components in the synapse, AcChR and AcChE, has been especially advanced because of the relative ease by which these components are separated from other membrane proteins by affinity chromatography. The inherent advantages of affinity chromatography will be extremely useful for future advances in neurochemical research and membrane biochemistry. There are many other components involved in synaptic transmission and axonal conduction, which because of their low concentrations would be difficult to purify by ordinary methods. The structure and function of membrane-bound proteins is not yet well understood. Affinity chromatographic techniques applied to these problems should aid future research on these systems.

ACKNOWLEDGMENTS

We thank Dr. Yuan Chao for the synthesis of the affinity ligand for resin B and for helpful consultation.

This work was supported by USPHS Grant No. 10294 and by a grant from the Sloan Foundation.

REFERENCES

[1] R. Axen, J. Porath, and S. Ernback, Nature (London), 214, 1302 (1967).

[2] J. Porath, R. Axen, and S. Ernback, Ibid., 215, 1491 (1967).

[3] J. Porath, J. C. Janson, and T. Loas, J. Chromatogr., 60, 167 (1971).

[4] P. Cuatrecasas, J. Biol. Chem., 245, 3059 (1970).

[5] B. H. J. Hofstee, Anal. Biochem., 52, 430 (1973).

[6] R. D. O'Brien, M. E. Eldefrawi, and A. J. Eldefrawi, Ann. Rev. Pharmacol., 12, 19 (1972).

[7] Z. W. Hall, Ann. Rev. Biochem., 41, 925 (1972).

[8] H. P. Rang, Quarterly Reviews of Biochemistry, 7, 283 (1975).

[9] D. J. Triggle, Neurotransmitter Receptor Interactions, Academic, New York, 1971, p. 344.

[10] C. Y. Lee and C. C. Chang, Mem. Inst. Butantan, Sao Paulo, 33, 555 (1966).

[11] J. A. Cohen and M. G. P. J. Warringa, Biochem. Biophys. Acta, 11, 52 (1953).

[12] M. E. Eldefrawi, A. G. Britten, and R. D. O'Brien, Pesticide Biochem. Physiol., 1, 101 (1971).

[13] D. Mebs, K. Narita, S. Iwanage, Y. Samejima, and C. Y. Lee, Biochem. Biophys. Res. Commun., 44, 711 (1971).

[14] D. G. Clark, D. D. Macmurchie, E. Elliot, R. G. Wolcott, A. M. Landell, and M. A. Raftery, Biochemistry, 11, 1663 (1972).

[15] J. Schmidt and M. A. Raftery, Anal. Biochem., 52, 349 (1973).

[16] D. Nachmansohn, Chemical and Molecular Basis of Nerve Activity, Academic, New York, 1959.

[17] M. A. Raftery, J. Schmidt, R. Vandlen, and T. Moody, Neurochemistry of Cholinergic Receptors (E. de Robertis and J. Schacht, eds.), Raven Press, New York, 1974.

[18] M. A. Raftery, J. Schmidt, and D. G. Clark, Arch. Biochem. Biophys., 152, 882 (1972).

[19] J. Schmidt and M. A. Raftery, Biochemistry, 12, 852 (1973).

[20] J. D. Berman and M. Young, Proc. Natl. Acad. Sci. U. S. A., 68, 395 (1971).

[21] Y. Dudai, I. Silman, N. Kalderon, and S. Blumberg, Biochem. Biophys. Acta, 268, 138 (1972).

[22] J. Schmidt and M. A. Raftery, J. Neurochem., 23, 617 (1974).

[23] P. Cuatrecasas, M. Wilchek, and C. B. Anfinsen, Proc. Natl. Acad. Sci. U. S. A., 61, 636 (1968).

[24] W. H. Hopff, G. Riggio, and P. G. Waser, FEBS Lett., 35, 220 (1973).

[25] M. Martinez-Carrion and M. A. Raftery, Biochem. Biophys. Res. Commun., 55, 1156 (1973).

[26] I. Wilson, M. Hatch, and S. Ginsburg, J. Biol. Chem., 235, 2312 (1960).

[27] M. J. Jung and B. Belleau, Mol. Pharmacol., 6, 589 (1972).

[28] M. A. Raftery, Arch. Biochem. Biophys., 154, 270 (1973).

[29] J.-C. Meunier, M. Huchet, P. Boquet, and J.-P. Changeux, C. R. Acad. Sci., D, 272, 117 (1971).

[30] E. Karlsson, E. Heilbronn, and L. Widlund, FEBS Lett., 28, 107 (1972).

[31] J. Lindstrom and J. Patrick, in Synaptic Transmission and Neuronal Interaction (M. V. L. Bennett, ed.), Raven Press, New York, 1974.

[32] R. Klett, B. Fulpius, D. Cooper, M. Smith, E. Reich, and L. Possani, J. Biol. Chem., 248, 6841 (1973).

[33] M. Eldefrawi and A. Eldefrawi, Arch. Biochem. Biophys., 159, 362 (1973).

[34] G. Biesecker, Biochemistry, 12, 4403 (1973).

[35] J. Axelsson and S. Thesleff, Acta Physiol. Scand., 43, 15 (1958).

[36] J. Patrick, J. Lindstrom, B. Culp, and J. McMillan, Proc. Natl. Acad. Sci. U. S. A., 70, 3334 (1973).

[37] J.-C. Meunier, R. Olsen, A. Menez, J.-L. Morgat, P. Fromageot, A. M. Ronseray, P. Boquet, and J.-P. Changeux, C. R. Acad. Sci., D, 273, 595 (1971).

[38] M. A. Raftery, J. Schmidt, D. G. Clark, and R. G. Wolcott, Biochem. Biophys. Res. Commun., 45, 1622 (1971).

[39] P. Taylor, J. W. Jones, and N. M. Jacobs, Mol. Pharmacol., 10, 78 (1974).

[40] C. Tanford, Y. Nozaki, J. Reynolds, and S. Makino, Biochemistry, 13, 2369 (1974).

[41] S. Makeno, J. A. Reynolds, and C. Tanford, J. Biol. Chem., 248, 4926 (1973).

[42] G. Fairbanks, T. Steck, and D. Wallach, Biochemistry, 10, 2606 (1971).

[43] A. Karlin, Life Sci., 14, 1385 (1974).

[44] M. A. Raftery, J. Schmidt, M. Martinez-Carrion, T. Moody, R. Vandlen, and J. Duguid, J. Supramol. Str., 1, 360 (1973).

[45] E. Neufeld and V. Ginsburg, eds., Methods Enzymol., 8, 93 (1966).

[46] J.-C. Meunier, H. Sugiyama, J. Cartaud, R. Sealock, and J.-P. Changeux, Brain Res., 62, 307 (1973).

[47] T. Moody, J. Schmidt, and M. A. Raftery, Biochem. Biophys. Res. Commun., 53, 761, (1973).

[48] J.-C. Meunier and J.-P. Changeux, FEBS Lett., 32, 143 (1973).

[49] R. Vandlen, T. Claudio, Y. Chao, T. Lee, and M. A. Raftery, Manuscript in Preparation.

[50] J. Schmidt and M. A. Raftery, Anal. Biochem., 52, 349 (1973).

[51] T. Moody and M. A. Raftery, Unpublished.

[52] J. Bode and M. A. Raftery, Unpublished.

[53] A. Karlin, J. Prives, W. Deal, and M. Winnik, J. Mol. Biol., 61, 175 (1971).

[54] A. Karlin, D. Cowburn, Proc. Natl. Acad. Sci. U. S. A., 70 3636 (1973).

[55] A. Karlin, Fed. Proc., 32, 1847 (1973).

[56] S. Parsons, L. Tao, E. Dahlquist, C. Borders, T. Groff, J. Racs, and M. A. Raftery, Biochemistry, 8, 700 (1969).

[57] J. D. Rawn and G. E. Lienhard, Biochem. Biophys. Res. Commun., 56, 654 (1974).

[58] Y. Chao, R. Vandlen, and M. A. Raftery, Biochem. Biophys. Res. Commun., 63, 300 (1975).

[59] E. Heilbronn and Ch. Mattson, J. Neurochem., 22, 315 (1974).

[60] J.-P. Changeux, J.-C. Meunier, R. Olsen, M. Weber, J.-P. Bourgeois, J.-L. Popot, J. Cohen, G. Hazelbauer, and H. Lester, Proceedings of the Symposium on Drug Receptors, MacMillan Co., London, 1973.

[61] F. Hucho and J.-P. Changeux, FEBS Lett., 38, 11 (1973).

[62] G. L. Ellman, K. D. Courtney, V. Andres, and R. M. Featherstone, Biochem. Pharmacol., 7, 88 (1961).

HYDROPHOBIC EFFECTS IN ADSORPTIVE PROTEIN IMMOBILIZATION

B. H. J. HOFSTEE[1]

Biochemistry Division
Palo Alto Medical Research Foundation
Palo Alto, California 94301

INTRODUCTION

In an aqueous environment, noncovalent polar bonding (i.e., through ionic, charge transfer, or hydrogen bonds) may be assumed to be relatively weak because of the strong charge-solvating and hydrogen-bonding ability of water. It has been suggested, therefore, that apolar (hydrophobic) forces may be the most important single factor providing the driving force for noncovalent intermolecular interactions in an aqueous milieu [1, 2]. However, in aqueous solutions electrostatic interaction may become important when the pertaining charges are shielded by accompanying hydrophobic groups from the quenching effect of water [1]. In this connection it may be pointed out that, on account of the low dielectric constant of the hydrophobic interior of a protein, intramolecular interaction of oppositely charged surface groups can be strong indeed [3].

Shielding from the quenching effect of water may also be assumed to occur when the binding of a particular molecule takes place in a complementary cavity of a larger entity, e.g., in the case of the interaction of a substrate with an

---

[1] With the technical assistance of N. Frank Otillio.

enzyme or of the binding of a protein in a cavity on a solid surface (see below). Furthermore, it is possible that the interaction of two oppositely charged molecules, in both of which the charge is shielded by a hydrophobic group, may be enhanced by a mutual reinforcing effect of (long-range) electrostatic and (short-range) hydrophobic forces. From these points of view the two types of interaction may be of equal importance in biochemical processes.

Enhanced binding through the postulated shielding of charges by hydrophobic groups is a phenomenon entirely different from the formation of hydrophobic "bonds." The former remains essentially electrostatic and can be reversed by salt, whereas hydrophobic bond formation may be stabilized by salt, e.g., NaCl in concentrations of 1 to 4 M [4-10]. There is no reason to assume that in the interaction of oppositely charged compounds with relatively large hydrophobic groups the two effects could not occur at the same time.

The study in aqueous solution of hydrophobic bond formation between compounds with large hydrophobic groups has been hampered by the inherently low solubility of such compounds. Chemical studies in general can be properly carried out only with the interacting compounds in a state of molecular dispersion. However, highly hydrophobic and insoluble compounds may be molecularly dispersed in an aqueous milieu by their covalent attachment to an insoluble but hydrophilic (wettable) matrix. Although adsorbents with hydrophobic groups have previously been applied to chromatographic separations (for literature quotations, see Ref. 9), systematic studies of hydrophobic phenomena by means of such adsorbents and initiated by Yon [11] have been made only relatively recently [4-13].

Because of recent advances in the preparation and chemical substitution of polymers, many compounds can be attached to a hydrophilic carrier suitable for chromatographic procedures, e.g., beaded agarose [14-17]. A major part of the work on protein binding thus far has been carried out with agarose substituted via CNBr activation with various ligands carrying primary amino groups. In the final product the amino group retains its basic properties [18] and is positively charged at the pH values at which most of the experiments usually are carried out.

A charge on the adsorbent is necessary for study of combined hydrophobic and electrostatic effects. Furthermore, such a charge may be assumed not to interfere with studies of true hydrophobic bonding which is favored at high salt concentrations where charge effects generally would be quenched. Nevertheless, it is of interest to carry out studies with adsorbents that are uncharged [10, 19] or where the positive charge on the agarose-bound amino group is neutralized by the negative charge of an ionized carboxyl group, as in the case of a substituting amino acid [20, 21].

The current series of investigations [4-8, 20, 21] was originally started on the basis of the conclusion that hydrophobic forces are involved in the interaction of the active site of many enzymes with their substrates or other specific ligands [22]. In view of the generally nonspecific nature of hydrophobic effects per se [23], it was postulated [4] that the apparently accessible hydrophobic sites in the active center would exhibit nonspecific affinity for many other hydrophobic ligands. In the course of these investigations it became apparent, however, that depending on the hydrophobicity of the immobilized ligands most of the proteins that were tested, including BSA,[2] γ-G, and β-LG, were capable of forming

---

[2] The abbreviations used are: BSA, bovine serum albumin; OV, ovalbumin; γ-G, 7S γ-globulin; β-LG, β-lactoglobulin; EG, ethylene glycol; A-C_{1-8}, adsorbents prepared from CNBr-activated agarose (Sepharose 4B) through substituion with an n-alkylamine with a C-chain length varying from C_1 to C_8; A-PBA, agarose substituted via CNBr-activation with 4-phenyl-n-butylamine: XO, xanthine oxidase; LDH, lactate dehydrogenase.

salt-stable hydrophobic bonds with n-alkylamine-substituted agaroses. These findings are in accord with recent observations [21, 24] indicating frequent occurrence of hydrophobic groups on the surface of protein molecules. It appears that the occurrence of such groups is not limited to relatively small ones such as the $-CH_3$ side-chain of alanine, but includes the larger hydrophobic side-chains such as those of phenylalanine and tryptophan as well [24]. In any event, the possibilities for hydrophobic bonding of proteins by immobilized hydrophobic ligands are much more widespread than originally surmised. In fact, it appears now that hydrophobicity is a general characteristic of proteins and at least in the case of enzymes, may play an important role in their attachment to intracellular matrices as well as in their biological function per se. With respect to chromatographic protein purification, hydrophobicity may be of equal importance as properties such as isoelectric point or molecular weight which underlie ion-exchange and exclusion chromatography, respectively [4, 20].

GENERAL EXPERIMENTAL APPROACH AND OBJECTIVES

The investigations described below deal with the binding of highly purified proteins by agarose (Sepharose 4B) substituted with ligands carrying hydrophobic groups. Studies with known proteins are prerequisite to possible applications of hydrophobic phenomena, e.g., in new procedures for the separation of proteins in mixtures of unknown composition [4, 20], or in adsorptive enzyme immobilization for reactor-type experimentation [6]. For this reason, investigations have been made of factors such as the degree of hydrophobicity and the electric charge of the protein and of the adsorbent. Studies have also been made of the apparent "irreversibility" of binding (immobilization) and possible inhomogeneity of the adsorbent binding sites and of the relationship of these factors to the degree of substitution of the adsorbents. Experimental details are described in previous publications [4-8] and in the legends to the figures presented below.

Frequent use is made of the known difference in hydrophobicity of OV and BSA [25]. OV has little or no ability to form hydrophobic bonds stable at high salt concentrations. By contrast, BSA readily forms such bonds. Other proteins that are relatively hydrophobic in this respect are γ-G and β-LG [7,25]. It is a rather simple matter to separate BSA and OV on the basis of hydrophobicity (see Ref. 4), but it is much more difficult to attain good separation of the more hydrophobic proteins from each other. In order to achieve this goal, suitable adsorbents and eluants must be found and the proper experimental conditions established.

On the basis of the finding that certain salts stabilize hydrophobic bonding [26], separation should in principle be obtained by means of a gradient of <u>decreasing</u> concentration of such a salt (e.g., see Ref. 10). However, in order to reverse hydrophobic bonds and also to circumvent possible electrostatic binding at the lower salt concentrations, it would seem logical to apply a gradient of increasing concentrations of a "chaotropic" agent [27]. A difficulty here is the apparent nonuniformity of the binding of a pure protein by a particular adsorbent [4, 7] which also was observed with several ion-exchangers [28-30]. Although this difficulty does not apply to reactor-type experiments in which enzymes are immobilized through adsorption by certain substituted agaroses [6], it must be resolved before application of hydrophobic phenomena can be made to efficient chromatographic protein separation.

Most often the adsorbents used were prepared by substitution of CNBr-activated agarose [15, 16, 31] with n-alkylamines of varying C-chain lengths. As was noted, the bound amino group retains its basic properties. In some experiments the adsorbent was DEAE-agarose (Bio Rad) where the ligand is bound to the agarose via an ether linkage.

A convenient means for determining the degree of substitution of the applied adsorbents has been worked out on the basis of irreversible binding of Ponceau S at low ionic strength [7]. For a homologous series of n-alkylamines (C_1 to C_8) it was found that the binding capacity with respect to saturation by the dye was proportional to the binding capacity for OV under similar conditions. The constancy of the dye:protein ratio indicates that the binding of the dye and of the protein is to the same adsorbent sites and occurs in the same or a similar manner. The great disparity in structure of the dye and of the protein emphasizes the absence of specificity of their interactions with the adsorbents.

Chromatographic analyses of protein mixtures on columns, in particular when carried out with the aid of a concentration gradient of an eluant, are best achieved by continuous application of the eluant solution and, if possible, continuous monitoring of the eluate. However, with one such set-up only one analysis can be made at the time. Chromatographic information on a single pure protein, e.g., the relative effect of a series of eluants, can be obtained much more rapidly by means of a series of identical columns eluted by manual applications of the eluant and manual assay of the eluates, e.g., measurement of protein content by spectrophotometry. A major portion of the data discussed below were obtained through the use of series of 1 ml columns in disposable pipets (Van-Lab) and provided with a small plug of glass wool. The amount of protein in the filtrates was determined from the light absorbance at 280 nm.

DUALITY OF NONSPECIFIC PROTEIN BINDING BY n-ALKYLAMINO-AGAROSES

Previous results [7], compiled in Fig. 1, indicated that the lower members (A-C_{1-6}) of a homologous series of n-alkylamine-substituted agaroses (A-C_{1-8}) strongly bind negatively charged proteins (e.g., OV, BSA, and β-LG) by electrostatic (salt

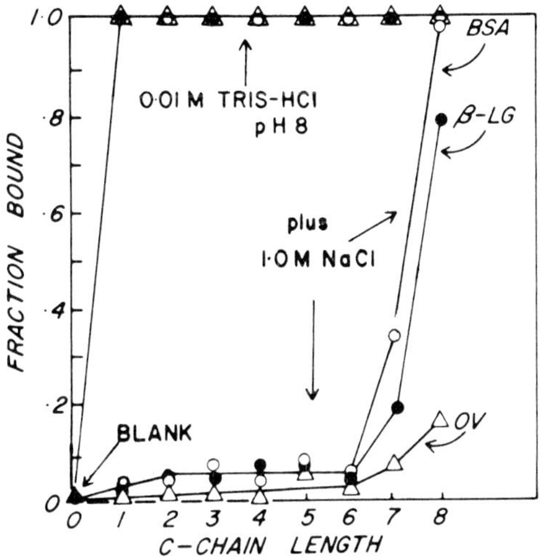

FIG. 1.
Effect of the C-chain length of n-alkylamino-agaroses on their fractional binding of bovine serum albumin (BSA), β-lactoglobulin (β-LG), or ovalbumin (OV) in 0.01 M Tris-HCl buffer, pH 8, in the absence or in the presence of 1.0 M NaCl. Aged CNBr-activated but unsubstituted agarose was used as a blank. Two milligrams of a protein were applied to a 1-ml column of an adsorbent equilibrated at ≈5° with the buffer. The fraction of the protein bound refers to "irreversible" adsorption (see text). (Data compiled from previous results [7] with permission of Publisher.)

reversible) interaction. In addition, the larger n-alkyl groups (>C_6) appear to form hydrophobic (salt stable) "bonds" with a more hydrophobic protein, such as BSA. The following presentation is prompted by the results of further studies on these two aspects of protein binding by substituted agaroses.

Electrostatic (Salt Reversible) Binding

Since the amino groups of the substituting ligands are positively charged under the experimental conditions [18], one might assume that the salt reversible binding of the applied negatively charged proteins by the short chain (<C_6) ligands (Fig. 1) could be the result of the presence of this positive charge. Confirmation of this proposition might be derived from the fact that under the same conditions and in the absence of specific interaction, positively charged proteins are generally not bound by these adsorbents [4-7]. Also, binding occurs as readily with relatively nonhydrophobic OV as with much more hydrophobic BSA of β-LG (see Fig. 1). Furthermore, previous data [7] showed little or no difference in the salt (NaCl) concentration at which OV or BSA are eluted either from C_1 or C_4. In addition, recent observations [32] indicate that the mere treatment of CNBr-activated agarose with ammonia, that is, in the absence of any hydrophobic group, endows the material with a high capacity for electrostatic protein (OV, BSA) binding.

For the interpretation of a possible effect of hydrophobic shielding on electrostatic interactions, it should be taken into consideration that such shielding not necessarily requires the presence of hydrophobic groups on the protein. An electrostatic bond could be shielded by a hydrophobic group in conjunction with the charge on one of the reactants only. As noted above, further protection against quenching by the aqueous medium may occur when the binding is in a cavity on the adsorbent, in particular in the case of a close fit between the protein molecules and the cavity (see also below).

An entirely different interpretation of the effect of hydrophobic groups on electrostatic binding could be based on the recent suggestion by Lewin [33] that the structured water around a hydrophobic group is oriented with the hydrogen outward. This could present opportunities for hydrogen bonding and preferential binding of anions. It should be noted that also in this case binding would not necessarily depend on the hydrophobicity of the protein.

Hydrophobic (Salt Stable) Bonding

It is clear from the data of Fig. 1 that with the applied proteins and under the experimental conditions, hydrophobic (salt-stable) "bonding" occurs primarily when the C-chain length of the adsorbent is $>C_6$. This type of binding is much stronger for BSA and β-LC than for OV. It had previously been shown by McClure and Edelman, through binding studies with naphthalene derivatives [25], that BSA and β-LG are much more hydrophobic than OV, which confirms that the binding depends on hydrophobic interaction between the protein and the ligand.

The data of Fig. 2 show that, whereas at low salt concentration (<0.1 M) BSA is bound by any of the adsorbents tested, at high salt concentration (>1 M) the extent of binding of the protein is greatest for the more hydrophobic ligands, e.g., n-C_8 > n-C_7 > n-C_6. The data suggest, furthermore, that the hydrophobicity of -DEAE is less than that of n-C_6 (see also Ref. 8) and that the hydrophobicity of the PBA-ligand is intermediate between those of the n-C_7 and n-C_8 ligands.[3] Since the extent of binding would also depend on the degree of substitution, it should be noted that the order of the latter (determined by saturation with Ponceau S, see above) was n-C_6 > n-C_7 $\gtrapprox$ DEAE > PBA > n-C_8, which at least for the n-alkyl adsorbents is the opposite of the order of fractional BSA-binding (Fig. 2). Thus binding is determined by the hydrophobicity of the ligand rather than by its concentration on the matrix.

At the highest NaCl concentration (3.3 M) that was applied (Fig. 2), only the PBA, n-C_7, and n-C_8 adsorbents form the salt-stable type of hydrophobic bond with BSA. Even at this high ionic strength the immobilized n-C_6 ligand apparently forms this type of bond only with part of the protein, whereas -DEAE does not form such bonds at all under the applied conditions.

---

[3] This observation is in accord with early findings [55, 56] that, with respect to micelle formation of certain fatty acid esters, the hydrophobicity of the benzene ring corresponds to 3 to 4 straight chain methylene groups [see also Ref. 57].

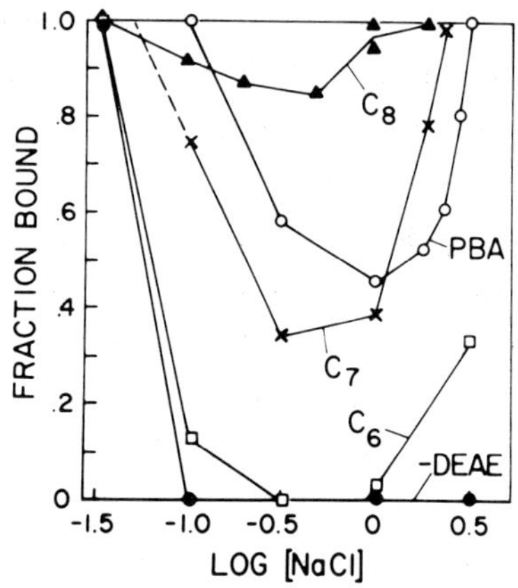

FIG. 2.
Effect of salt (NaCl) concentration on the fractional binding of bovine serum albumin by n-caprylamino(n-C_8)-, 4-phenyl-n-butylamino(PBA)-, n-heptylamino(n-C_7)-, n-hexylamino(n-C_6)-, or diethylaminoethyl(DEAE)-agaroses under the same general experimental conditions as for Fig. 1.

APPARENT "IRREVERSIBILITY" OF BINDING

More than 30 bed volumes of low ionic strength eluant can be passed through a 2-ml DEAE-agarose column loaded with 2 mg OV, without any protein appearing in the filtrate. Elution does not occur unless the ionic strength is raised to a certain level. Earlier results [7] showed that this also holds true for n-alkylamine-substituted agaroses. Furthermore, the data of Fig. 3A indicate that 1 mg OV remains bound on a 2-ml A-PBA column after washing with more than 600 ml (≃ 300 bed volumes) of 0.01 M Tris-HCl buffer, pH 8, over a period of ≃ 15 hr. Also, the elution curve is not significantly affected by the amount of low ionic strength buffer that is washed through the loaded column prior to starting the salt gradient. Furthermore, large differences in the rate at which the salt gradient is applied (from ≃ 10 to ≃ 40 ml/hr, Fig. 3B) have only relatively little effect on the results. Thus under the ambient conditions the strength of binding is tantamount to immobilization of

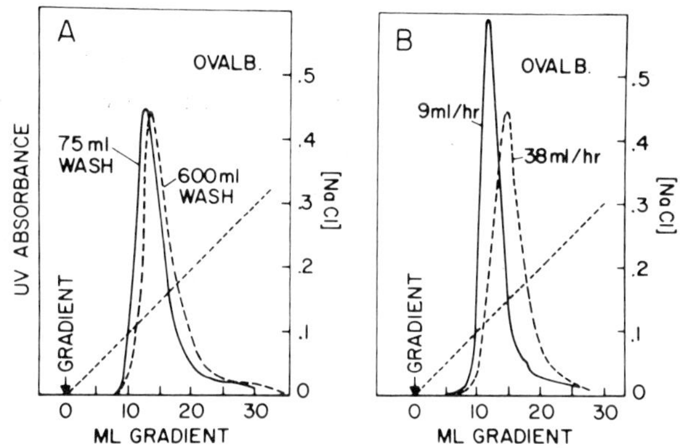

FIG. 3.
Elution pattern (A_{222}) of ovalbumin from 4-phenylbutylamino-agarose. One milligram of the protein was applied to about 2 ml of the adsorbent in a 5-mm wide column equilibrated at ≈5° with 0.01 M Tris-HCl, pH 8, and eluted by means of a linear NaCl gradient in the buffer. (A) Effect of washing of the loaded column with buffer alone prior to applying the NaCl gradient. The washing was either with 75 ml (solid curve) or with 600 ml (dashed curve) of the buffer. The flow rate in each case was ≈40 ml/hr. (B) Effect of the flow rate of the salt gradient either 9 ml/hr (solid curve) or 38 ml/hr (dashed curve).

the protein. This phenomenon has been applied to the immobilization of enzymes in reactor-type experiments (Ref. 6, see also below).

It would seem unlikely that the observed strong binding by mere adsorption could be the result of a single bond per protein molecule, even if the bond were formed through the combined effects of hydrophobic and electrostatic forces. Evidence has been presented that nonspecific binding as opposed to the "one-to-one" specific type of binding, such as that of a substrate by the active center of an enzyme, sharply decreases when the degree of substitution of the adsorbent is lowered to the point where the distance between the ligand molecules is larger than the diameter of the protein molecule [34]. Therefore, as also suggested by others [28, 29] it is assumed that strong nonspecific binding is the result of multiple point attachment of the same protein molecule to several ligand molecules on the matrix. That the number of binding sites involved need not be large is indicated also by the fact that binding of Ponceau S by these adsorbents is "irreversible." The molecule of the dye contains only four negative charges in conjunction with hydrophobic groups. In any event, it appears that the simultaneous interaction of only a few sets of opposite charges may result in strong, often virtually irreversible, interaction.

It should be emphasized that the term "irreversible" is used here in a strictly experimental sense. The term merely indicates that under the applied conditions the protein cannot be washed off the column by the ambient medium without changing the composition of the latter. Evidently, in adsorptive binding an equilibrium must obtain between bound and unbound protein. However, the equilibrium apparently is often in favor of binding, even to the extent that no free protein can be detected by the applied procedures.

FACTORS UNDERLYING APPARENT INHOMOGENEITY OF BINDING OF PURE PROTEINS

Inhomogeneity of Adsorbent Binding Sites

Previous results [7] indicated that the homogeneity of a particular preparation of OV seemingly was not always the same with respect to column chromatography on an adsorbent such as n-butylamino-agarose. The elution pattern depended on the protein:adsorbent ratio (i.e., the protein load) and the degree of substitution. These observations are further substantiated by the results of Fig. 4 where DEAE-agarose was saturated with OV in 0.001 *M* Tris-HCl buffer, pH 8, and washed with the buffer alone until no protein could be detected in the filtrate. The data indicate that a small fraction of the

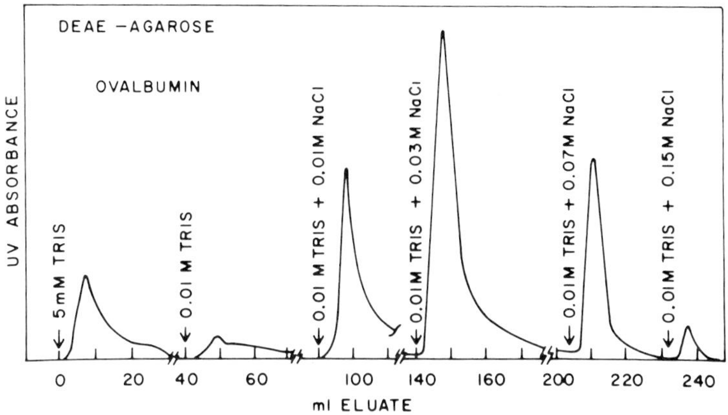

FIG. 4.

Elution patterns (A_{280}) of ovalbumin (5X crystallized and nearly homogeneous by isoelectric focusing) from DEAE-agarose by means of stepwise increase of the salt concentration. A saturating amount of the protein (≈75 mg) was applied to a 5 X 100 mm column of the adsorbent equilibrated at ≈5° with 1.0 m<u>M</u> Tris-HCl buffer, pH 8. About 60 mg remained bound upon subsequent exhaustive washing with the buffer alone. The stepwise applications of eluant solutions of varying composition are indicated by the arrows. (From previously published data by author [8]).

OV bound in 0.001 <u>M</u> Tris can be eluted with 0.005 <u>M</u> buffer, after which the UV absorbance of the filtrate returns to base line. Another small fraction can be eluted with 0.01 <u>M</u> buffer, and more is eluted by stepwise increase of the ionic strength of the eluant until no protein is left on the column. With respect to isoelectric focusing, the 5X crystallized OV used here showed only one major and one minor fraction [35].

It is important to note (Fig. 4) that after each successive step the protein content of the eluate returns to zero. Although this point often is reached very slowly, it indicates that at the pertaining level of the salt concentration the remaining protein is "irreversibly" bound and, under the ambient conditions, in fact, is immobilized on the column. Thus these results confirm previous indications that these as well as other types of adsorbent [see Ref. 28] may be endowed with binding sites having varying degrees of affinity for a particular protein and that, at least with high protein loads, apparent inhomogeneity may indicate inhomogeneity of the adsorbent binding sites rather than of the protein. Similar conclusions have been reached by Hixson and Nishikawa [36] and by Nishikawa et al. [37].

A consequence of the inhomogeneity of binding sites would be that not only the binding capacity of an adsorbent but also the strength of binding of a fixed amount of protein would depend on the number of available binding sites, i.e., on the amount of adsorbent applied and on its degree of substitution. The data of Fig. 5 show that about 65% of the applied amount of BSA is washed off by 0.33 M NaCl from a 1-ml column of n-octylamine-substituted agarose. The degree of substitution of the adsorbent was relatively low and the applied salt concentration unfavorable for binding (see Fig. 2). It is of interest, however, that under the same conditions all of the protein remains irreversibly bound by a larger(5 ml) column. If only one type of binding site were involved, one would expect that the extent of the elution by the same number of bed volumes of the eluant (0.33 M NaCl) would be the same for the two columns. The results clearly show that this is not the case. Apparently only the larger column contains a sufficiently large number of "strong" binding sites to bind all of the protein irreversibly under the applied conditions.

It should be noted that for both columns the major portion of the protein that remains irreversibly bound in 0.33 M NaCl can subsequently be eluted by including ethylene glycol in the eluant (see below). Thus these data further confirm the proposition that nonuniformity of the adsorbent binding sites may result in apparent chromatographic inhomogeneity of a pure protein. It may be assumed that the surface of the adsorbent matrix is irregular and presents cavities of varying size and shape. An extremely tight binding would occur when such a

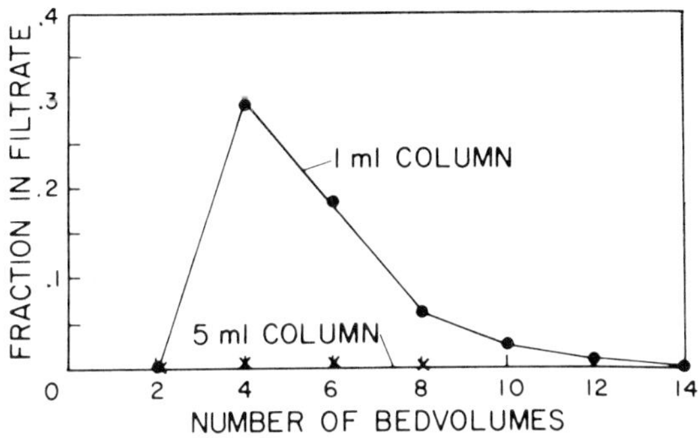

FIG. 5.

Effect of the adsorbent:protein ratio on the strength of binding of bovine serum albumin by n-octylamino-agarose. The two columns (1 and 5 ml, both about 4 cm high), were equilibrated at ≈5° with 0.33 M NaCl in 0.01 M Tris-HCl, pH 8, and each was charged with ≈1.7 mg of the protein, followed by washing with the ambient NaCl-buffer solution. The volumes of the filtrates are expressed as the number of bedvolumes for each column, i.e., as multiples of 1 or 5 ml, respectively.

cavity contains several ligand molecules properly placed and provided with the proper charges and/or hydrophobic groups, in particular when the shape and size of the cavity fits that of the protein molecule (see also Refs. 36 and 37). In this manner simultaneous binding of a protein molecule with several ligand molecules can take place in a three-dimensional fashion.

The data are consistent with the assumption of a wide range of binding sites of varying binding strengths but nevertheless all being able to bind protein "irreversibly" under the proper experimental conditions (see also below).

Simultaneous Salt Reversible (Electrostatic) and Salt Stable (Hydrophobic) Binding.

When 3 mg BSA is applied to a column of 5 ml PBA-agarose, part of the protein is removed in a 0 to 1.0 M NaCl gradient. However, a second fraction can subsequently be eluted by a solution of 50% ethylene glycol in 1 M NaCl. By contrast, under the same conditions all or nearly all of the much less hydrophobic OV can be eluted by salt alone [4]. The fact that elution of the second fraction of BSA requires an agent such as ethylene glycol suggests that hydrophobic bonding is involved [4]. This type of apparent inhomogeneity is not necessarily the result of inhomogeneity of the adsorbent binding sites, which most often manifests itself primarily at much higher protein loads than were applied here. A possible interpretation is as follows: At low ionic strength the negatively charged BSA binds irreversibly through electrostatic interaction as discussed above. As the salt concentration is raised, this type of binding is reversed and the protein begins to be released. However, at the same time "true" hydrophobic binding, which increases with the salt concentration (see Fig. 2), becomes more and more predominant. Thus part of the protein would remain on the column by virtue of true hydrophobic bonding at the higher salt concentrations and can then be removed by an agent that weakens hydrophobic bonds (e.g., ethylene glycol).

ADSORPTIVE IMMOBILIZATION OF ENZYMES

General Observations

The electrostatic and the salt-stable hydrophobic type of protein binding are both of an unspecific nature and most probably involve charged and/or hydrophobic groups that could occur on the molecular surface of any protein [21, 24]. The nonspecificity of the binding is emphasized by the finding [7] that binding by the present adsorbents of a protein such as OV or a dye such as Ponceau S occurs in a similar fashion. From this it follows that adsorptive binding of a biologically active protein, e.g., an enzyme, may occur through a site or sites other than the active center. In fact, the binding of enzymes to intracellular matrices, which does not seem to depend on covalent bonds, must occur in a manner that leaves the active center free to react with substrate. One could even speculate that the in vivo binding of such enzymes occurs through electrostatic and hydrophobic forces similar to the binding of BSA or OV to the present adsorbents. There is considerable evidence that both these forces indeed play a role in the intracellular binding of certain enzymes [2, 38]. Although presumably of the adsorptive type, the binding nevertheless appears to be extremely strong and virtually "irreversible" under intracellular conditions.

Strong adsorptive binding of enzymes to a variety of artificial solid matrices has been observed on many occasions (e.g., for reviews see Refs. 39 and 40). In view of possible leakage problems, the reviewers express some skepticism as to the use of this type of binding in enzyme "reactors." However, thus far attempts at adsorptive binding have been mainly empirical. The more systematic studies discussed above show virtually irreversible binding of enzymes and other proteins by agaroses substituted with primary amines carrying hydrophobic groups. A crucial factor, in addition to the presence of hydrophobic groups, appears to be a high degree of substitution, allowing multiple point attachment of the protein molecule to the supporting matrix (see also below).

Another point that should be taken into consideration here is the evidence that a particular adsorbent appears to be endowed with binding sites that may range from extremely weak to extremely strong. Although under a given set of conditions part of the protein may be reversibly bound, it is often impossible to remove all of the protein, even by exhaustive washing with the ambient medium. Apparently a residue of the protein remains bound to the stronger binding sites. For the less hydrophobic amine-substituted adsorbents this holds true only at relatively low ionic strength, but for adsorbents with the larger hydrophobic groups irreversible binding may occur at any salt concentration (see Fig. 2). In any event if initial leakage occurs under a given set of conditions, this does not necessarily mean that under those conditions eventually all of the protein will be washed off the adsorbent.

It is possible that under certain conditions, e.g., those required for optimum activity of an enzyme, only relatively few sites on the adsorbent bind the protein in an irreversible fashion, i.e., only a small fraction of a given amount of protein may remain bound by a given amount of adsorbent. As shown by Fig. 5 more protein will be bound irreversibly simply by increasing the amount of adsorbent.

Application to Enzyme "Reactors"

On the basis of the above considerations and in view of the nonspecific but strong binding of several proteins by the present type of adsorbent, the binding of a number of enzymes by n-butyl- or n-octylamino-agarose was investigated [6]. It was noted that in 0.01 M Tris-HCl, pH 8, several enzymes with isoelectric points in the acid region (xanthine oxidase, lactate dehydrogenase, DN-ase I, alkaline phosphatase, and urease) are strongly, virtually irreversibly, bound by these two adsorbents. By contrast, RNase, the isoelectric point of which lies at pH values >9, showed little or no binding. In the presence of 1 M NaCl, the acidic enzymes, negatively charged at pH 8, were invariably more strongly bound to A-C_8 than to A-C_4. The binding to A-C_4 of all of the enzymes could be largely reversed by 1 M NaCl, indicating a predominantly electrostatic type of interaction. On the other hand, binding to A-C_8 could most often only partly be reversed by 1 M NaCl. Little or no reversal by salt occurred for the case of DNase, although most of the enzyme could be eluted from A-C_8 by a combination of 1 M NaCl and 50% ethylene glycol. This indicates the involvement of the salt-stable hydrophobic type of binding.

The major finding of these investigations [6] was that for all of these enzymes, conditions can be found under which virtually irreversible binding (immobilization) occurs by one or the other n-alkyl amino-agarose, often even at high ionic strength. It was also found that in most, if not all, cases the adsorbent-bound enzyme retained at least part of its activity, confirming the supposition that binding does not necessarily involve the active center. In fact, the high frequency of this noninvolvement for the five arbitrarily chosen enzymes could mean that these enzymes are endowed with binding sites, "nonfunctional" with respect to their catalytic action, but necessary for their proper attachment to intracellular matrices. In any event, protein binding by adsorbents of this type may present not only models for protein binding in vivo, but also may provide opportunities for the immobilization of enzymes in "reactor"-type experiments.

Reactor-type (long-term) experiments with several enzymes were carried out with 1 to 2 mg of, most often, highly purified enzyme preparations adsorbed on 1 to 2 ml columns of either A-C_4 or A-C_8, equilibrated with 0.01 M Tris-HCl buffer of pH 8. After exhaustive washing of the column, a substrate solution was filtered through and the presence of bound active enzyme detected by the continuous appearance of product in the filtrate during extended time periods [6].

For xanthine oxidase the substrate solution was 0.1 mM xanthine in 0.01 M Tris-HCl, pH 8, which was applied at ≃ 5° to the C_4-adsorbent containing the enzyme. It was found through measurement of the increase in absorbance at 290 nm that after establishment of a steady state,at least 90% of the substrate was converted continuously into uric acid over a time period of several days and the passage of about 500 bed volumes of the xanthine solution. Similar results were obtained with LDH and with urease. It can be seen (Fig. 6) that in the presence of 0.01 M Tris-HCl, XO, LDH, and DNase are also immobilized on DEAE-agarose. As was found with the A-C_4 adsorbent, in each case most of the enzyme protein could be eluted by 1 M NaCl.

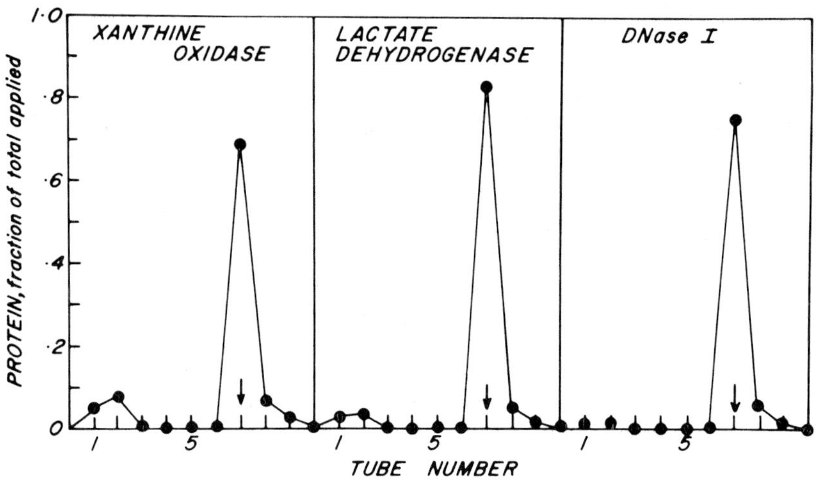

FIG. 6.
"Irreversible" adsorption (immobilization) of various enzymes by DEAE-agarose. Of the order of 1 mg of XO (Miles Seravac 36/611), LDH (Miles-Seravac 36-301), or DNase (Worthington, DSV, Standard), dissolved in 2 ml of 0.01 M Tris-HCl, pH 8, was applied at ≈5° to a 1-ml (8.7 X 16 mm) column of the adsorbent which then was washed 5 times with a 2-ml portion of the buffer and subsequently 4 times with 2 ml buffer containing 1.0 M NaCl (beginning at the arrow). The protein contents of the filtrates were determined from the light absorbance at 280 nm. (From previously published results by author [8]).

The results of a reactor-type experiment with 0.5 to 1 mg of XO on a 0.5 x 10 cm cooled (≈ 5° C) DEAE-agarose column were similar to the ones with A-C_4. With moderate flow rates (e.g., 25 ml/hr), no decrease in activity was noted after more than 50 hr of continuous operation of the column during which period more than 1500 ml (>750 bed volumes) of substrate solution (0.1 mM xanthine in 0.01 M Tris-HCl, pH 8) were passed through. At this time and as determined from the UV absorbance, more than 90% of the amount of enzyme protein originally bound could be recovered by elution with 1 M NaCl. However, after this treatment the column still was enzymatically active. The data of Fig. 6 indicate that a small part of the enzyme remains bound in 1 M NaCl, presumably due to the above discussed inhomogeneity of the adsorbent binding sites. In any event, it would seem that under the applied conditions little or no enzyme protein was lost during more than 2 days of continuous operation of the reactor.

Under conditions favorable for activity, i.e., in the presence of 0.01 M $MgCl_2$ and at pH 5, DNase (Worthington, DSV, Standard) remains bound to A-C_8 in toto and in part also to A-C_4 [6]. However, reactor-type experiments with high molecular DNA as the substrate were hampered by binding of the negatively charged nucleic acid by the positively charged adsorbent.

In the case of alkaline phosphatase (Worthington PC), the major portion of ≃ 1 mg of the enzyme preparation was bound by 1 ml of the A-C_4 as well as by the A-C_8 adsorbent in 0.01 M Tris HCl, pH 8, in the absence of additions [6]. On the other hand, most of the enzyme protein was washed off these columns by 0.01 M glycine buffer, pH 9.5, containing 0.01 M $MgCl_2$, a medium favorable for activity of this enzyme. Nevertheless, after continued washing with the glycine-$MgCl_2$ mixture until no enzyme could be detected in the filtrate, the column remained active as was indicated by the hydrolysis of o-carboxyphenyl phosphate, measured through an increase in the absorbancy of the filtrate at 300 nm [41]. Although the efficiencies gradually decreased, activity was still observed after passage of ≃ 100 and more than 300 bed volumes of glycine-$MgCl_2$ through the A-C_4 and A-C_8 columns, respectively, during a period of 16 hr at room temperature.

The strong binding of only a relatively small amount of alkaline phosphatase under conditions favorable for activity (i.e., at pH 9.5 and in the presence of $MgCl_2$) may be ascribed to the presence of only a limited number of sufficiently strong binding sites under those conditions. Apparently other sites are present that bind the enzyme strongly at pH 8 and in the absence of $MgCl_2$. In any event, the presumed inhomogeneity of the binding sites (see above), which may be a drawback for chromatographic protein separation [4], is not necessarily a disadvantage in the use of these adsorbents for enzyme reactors. Provided the rate of the enzyme reaction is sufficiently large, no extremely large columns would be required to provide a sufficient number of "strong" binding sites in order to immobilize the relatively small amount of enzyme that would be needed. Furthermore, in view of recent developments, in particular those on the stabilization of hydrophobic binding at high concentrations of "structure forming" salts, the capacities for irreversible protein binding of an adsorbent such as A-C_8 might be considerably increased. Under such conditions, binding is also independent of the overall charge of the protein [20, 21].

Immobilization of enzymes by adsorptive means not only is less cumbersome, but also is less likely to result in irreversible alteration of the enzyme structure than may be assumed to be the case during chemical manipulation needed for covalent binding. Also, depending on the hydrophobicity of the adsorbent, the enzyme can be recovered by changing the composition of the medium.

The above noted unexpectedly high frequency of retention of activity, shown by a series of arbitrarily chosen enzymes after binding by the adsorbent, may be explained as follows: In view of the nonspecificity of the binding of protein by the present adsorbents, it can be expected that not all of the protein molecules are bound in the same manner. Binding may be assumed to be possible through several ionic and/or hydrophobic groups located in different areas on the protein surface, including the active center of an enzyme. However, if binding would occur through the active center, the presence of substrate would tend to prevent the latter mode of binding and the enzyme would tend to become attached through a site other than its active center, i.e., in a fashion not susceptible to reversal by the substrate. Thus in the presence of substrate the enzyme would be bound preferentially in a manner in which the active center is available for interaction with the substrate. If this supposition is correct, adsorptive binding would have an additional advantage over covalent binding, because in the latter case binding involving the active center would inactivate the enzyme irreversibly.

PROTEIN SEPARATION BY HYDROPHOBIC ADSORPTION CHROMATOGRAPHY

The principle of separation of proteins on the basis of differences in their ability to form hydrophobic bonds is demonstrated by the data of Fig. 1. It can be seen that at pH 8 and in the presence of 1 *M* NaCl, all of the BSA is bound by A-C_8, whereas little OV is bound under such conditions. In view of the fact that the isoelectric points of the proteins do not greatly differ, separation of BSA and OV would be difficult to achieve by electrochemical procedures alone. Thus the hydrophobic factor presents an additional parameter for separation of proteins with similar electrochemical properties. The same would apply to proteins with similar molecular weights which would make them difficult to separate by exclusion chromatography.

It has been tentatively suggested [4] that the rather strong binding, e.g., of BSA by A-C_8, might be the result of cooperation between electrostatic and hydrophobic binding. The major point of evidence for this supposition was that binding of BSA to A-C_8 in 0.01 *M* Tris was reversed by the combination of 1 *M* salt and 50% EG but not by the salt alone or by the EG alone. However, from the above observations it is now clear that in the presence of 1 *M* NaCl alone strong hydrophobic binding is possible, whereas in 0.01 *M* Tris-HCl and with EG alone strong electrostatic binding takes place.

The results of Fig. 2 show overlap of the regions of the salt concentration in which the electrostatic and the true hydrophobic type of binding predominate. As a result, all or most of the BSA remains bound to A-C_8 at any salt concentration, although in an intermediate range of the latter the binding appears relatively weak. These findings do not preclude elution by means of a decreasing salt concentration gradient, especially when protein release is enhanced by the presence of an agent such as EG. Under such conditions all of the protein may be released before the onset of electrostatic binding, which does not begin until the salt concentration has dropped to relatively low values ($\overset{\approx}{<}$0.3 M).

Electrostatic effects can also be circumvented by employing neutral adsorbents. These can be obtained through the introduction of a negative charge that neutralizes the positive charge on the amino group, e.g., the ionized carboxyl group in the case of amino acids [20, 21, 42]. For the same purpose, adsorbents free of charge have also been prepared [10, 19]. However, the positively charged amine-substituted agaroses, which have been used for studying electrostatic effects at low ionic strength, can be used for "true" hydrophobic protein binding, merely by carrying out the procedures in the presence of a "structure forming" salt in a concentration that not only quenches electrostatic effects but at the same time enhances hydrophobic bonding. This has recently been applied to a procedure for protein separation by means of a hydrophobicity gradient of a series of connected columns of agaroses substituted with ligands of increasing hydrophobicities and equilibrated with ≃ 3 M NaCl. The protein mixture is washed-in with the salt solution and the subsequently disconnected columns are eluted with 50% ethylene glycol in 1 M NaCl [20]. Use of a hydrophobicity gradient has the advantage that a particular protein is bound by the adsorbent that provides the minimum degree of binding strength that is required. Thus elution is likely to be achieved by means of a relatively mild eluant [see Ref. 20].

It can be seen (Fig. 7) that the fraction of a protein (BSA) that can be eluted by EG decreases with increasing

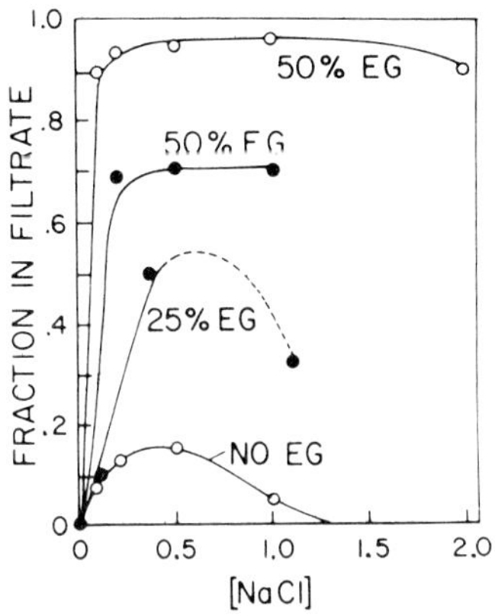

FIG. 7.
Effect of the concentrations of salt (NaCl) and of ethylene glycol (EG) on the elution of bovine serum albumin from preparations of n-octylamino-agaroses with different ligand contents. The experimental conditions were generally the same as those for Figs. 1 and 2. The open and closed circles refer to adsorbent preparations with different degrees of substitution (Ponceau values of about 0.47 and 0.72 mg/ml adsorbent respectively, see text).

degree of substitution. This could be interpreted on the basis of the assumption that with increasing degree of substitution additional opportunities for multiple point (three-dimensional) attachment become available, resulting in the formation of "stronger" binding sites (see above). Thus lowering the degree of substitution would facilitate elution, and under such conditions an eluant such as EG may completely counteract hydrophobic binding at high salt concentrations. The same result is obtained by lowering the hydrophobicity of the ligand instead of its concentration. In concentrations as high as 50%, EG seems to have no irreversible effect on most proteins [43]. Further details of systematic investigations of this type and of actual applications to protein separation will be presented elsewhere [44].

NONSPECIFIC VERSUS SPECIFIC PROTEIN BINDING

Studies of the nonspecific hydrophobic effects discussed above are useful not only with respect to their potential applications to protein immobilization or separation, but also with respect to the possible interference of these nonspecific effects with the isolation of biologically active proteins by biospecific affinity chromatography.

It was shown [4] that an adsorbent "designed" for the specific adsorption of a particular enzyme may bind a wide variety of unrelated proteins (including enzymes) in a nonspecific manner involving hydrophobic effects in conjunction with electrostatic forces. Subsequently, it was suggested [e.g., Refs. 5, 12, and 45] that such nonspecific interaction could occur through the n-alkylamine "arm" by which the ligand usually was attached to the matrix. However, nonspecific binding also could occur through hydrophobic and/or charged groups on the ligand itself, even when the latter is bound by an inert arm to an inert matrix. This type of binding is schematically depicted in the upper and lower parts of Fig. 8. Since nonspecific one-to-one binding may be assumed to be generally weaker than specific binding, the latter is favored when the distance between the matrix-bound ligand molecules is larger than the diameter of the enzyme molecule (Fig. 8, middle part). This condition prevents multiple point nonspecific attachment which, by contrast, may be assumed generally to be much stronger than the one-to-one specific type of binding. Figure 8 also attempts to show that the mode of multiple point binding tends to be more dependent on the locale of the binding site and, in contrast to the one-to-one type, is not necessarily the same for all molecules of a particular protein. This could be the basis of the above discussed false chromatographic inhomogeneity of a protein.

Specific interaction, e.g., that between an enzyme and a substrate or substrate analog, presumably involves complementarity

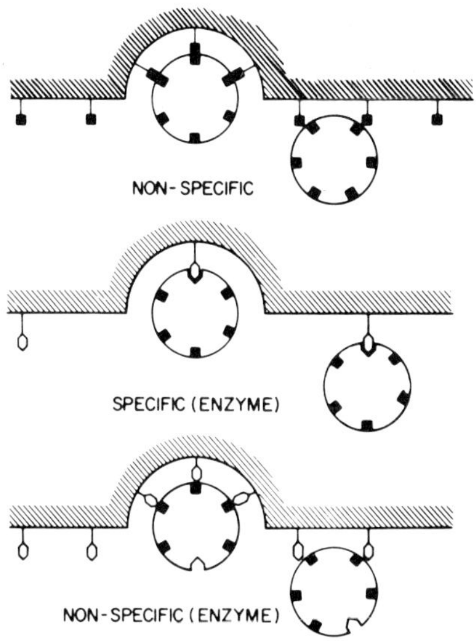

FIG. 8.
Schematic presentation of nonspecific multiple point protein binding by an immobilized ligand (upper part), as compared to the specific "one-to-one" binding of an enzyme with an immobilized substrate analog (middle part). Nonspecific multiple point binding of an enzyme with retention of activity is depicted in the lower part. An attempt is also made to show different modes of nonspecific binding of two molecules of the same protein. The nonspecific type of binding is assumed to take place through electrostatic-hydrophobic interaction of the immobilized ligand (possibly the substrate analog in the case of an enzyme) with corresponding groups of the protein which not necessarily are located in the "active center." "Specific" binding presumably depends on complementarity of the molecular contours of the ligand and the active center of the enzyme, but is assumed often to be weaker than nonspecific multiple-point binding.

(i.e., fit) of molecular contours of the ligand and the active center of the enzyme. For instance, α-chymotrypsin activity appears to depend on the interaction of the aromatic moiety of the substrate (usually a derivative of tyrosine, phenylalanine, or tryptophan) with the "tosyl hole" of the enzyme active center [46]. There may also be assumed to be complementarity with respect to bond formation, including the formation of electrostatic and hydrophobic bonds. The same could apply to interaction of an antibody with the antigen or hapten [47]. It should be emphasized that for an enzyme at least, specificity of interaction not necessarily relates primarily to strength of binding, but rather to the imposition of a particular orientation with respect to the substrate molecule as a whole as well as with respect to its intramolecular conformation [e.g., see Ref. 22].

One may assume that the binding of chymotrypsin by PBA-agarose [4, 48] is due to the fact that the ligand is a substrate analog. In this case both the protein and the adsorbent carry a positive overall charge and binding occurs despite apparent electrostatic repulsion. Chymotrypsinogen, also positively charged, showed little affinity for PBA-agarose [4]. Furthermore, the enzyme and other positively charged proteins, including chymotrypsinogen, lysozyme, and RNase, were not strongly bound by n-alkylamine-substituted agaroses [5], presumably because of the absence of specificity. On the other hand, and as already noted above, at relatively low ionic strength (0.01 to 0.05) the positively charged alkylamine-substituted agaroses generally strongly bound all negatively charged protein species that were tested, including a number of enzymes and other unrelated proteins such as BSA, OV, β-LG, and γ-G [4-7]. The latter unspecific type of binding seems to depend on the overall charge of the protein.

As suggested previously [7], in addition to lowering the ligand concentration on the matrix (Fig. 8), the absence of a stringent dependence on overall charge in the case of specific affinity might present another opportunity to circumvent nonspecific binding during attempts to isolate proteins by means of biospecific affinity. A procedure that may have more or less general application is as follows: The protein mixture containing the enzyme is filtered over an ion exchanger that carries the same overall charge as the enzyme to be isolated and that does not have a specific affinity for the latter, e.g., DEAE-agarose in the case of chymotrypsin. The filtrate, which in this case contains only positively charged protein species, then is applied to an adsorbent with a positively charged ligand that binds the enzyme specifically. The contaminating positively charged protein species generally will be repelled by the adsorbent but the enzyme would be specifically bound, despite its positive charge. After washing the adsorbent-bound enzyme with the buffer solution alone, elution may be achieved, e.g., by changing the ionic conditions [see Ref. 4]. For a negatively charged enzyme a negatively charged ion exchanger and a negatively charged specific adsorbent would be used.

POSSIBLE IMPLICATIONS AND FUTURE APPLICATIONS OF HYDROPHOBIC EFFECTS

Hydrophobic affinity per se or in combination with electrostatic effects is relevant not merely to the immobilization or the separation and purification of proteins as discussed above. As exemplified by the virtually irreversible adsorption of a dye such as Ponceau S by the applied adsorbents, this type of interaction may be assumed to be involved in the binding of relatively low molecular compounds as well.

The general importance of hydrophobicity in drug action is indicated by the work of Hansch and co-workers [see Ref. 49]. The implication of hydrophobic effects in narcosis, as expressed in the Meyer-Overton theory, has been known since the turn of the century. A tentative listing and discussion of other biochemical processes in which hydrophobic effects may play a role and of their possible future applications, some of which are as yet highly conjectural, is presented as follows.

The Interaction of Enzymes with Substrates or Effectors (Cofactors, Inhibitors, Drugs)

It has been shown that small changes in the hydrophobic moiety of a substrate may have a large effect on the activity of esterolytic enzymes (for a discussion, see Ref. 22). Since a change in hydrophobicity of the substrate most often affects the V_m of the reaction, but not necessarily results in a change in K_m, it was concluded that at least in such cases the effect is on the orientation and conformation of the substrate molecule in the enzyme-substrate complex rather than on the enzyme-substrate affinity.

Hydrophobic effects are not limited to esterolytic enzymes and seem to play a role in many enzymatic reactions including certain redox systems and several other types [see Ref. 50]. Exact studies on the requirements with respect to size and shape of the hydrophobic moiety of the substrate or cofactor might aid in the "design" of inhibitors and could contribute to the elucidation of the mechanism of action of enzymes in general. Such inhibitors (e.g., substrate analogs) covalently bound to an insoluble matrix can be of use in the isolation of the enzyme from crude extracts by biospecific affinity chromatography. A typical example is the strong binding of α-chymotrypsin by 4-phenyl-butylamino-agarose [48].

Interaction of Antibodies and Antigens

Similar to the case of enzyme-substrate interactions, complex formation between antibodies and antigens (or haptens) often seems to involve hydrophobic effects [47]. It is of interest in this respect that γ-G is strongly bound by some of the hydrophobic adsorbents, e.g., PBA-agarose [4], used in the present investigations. Rabbit immunogloblin is bound by L-phenylalanine-Sepharose at high salt concentrations, which also suggests hydrophobic binding [42].

Binding of Metabolites (and Drugs) by Serum Albumin as a Transport Protein

Extensive studies have been made on the binding of various hydrophobic ligands by serum albumin [e.g., Refs. 23, 51, 52].

The binding of long-chain fatty acids to serum albumin indicates the presence of at least 8 hydrophobic binding sites per protein molecule [53]. The hydrophobicity of serum albumin is also indicated by fluorescence studies with derivatives of naphthalene sulfonate [25]. The hydrophobic character of BSA relative to certain other proteins, e.g., OV, is confirmed by the present results (see Fig. 1). Thus it may be assumed that hydrophobic forces play an important role in the binding by BSA of metabolites and drugs, which also often are endowed with hydrophobic properties.

Association of Enzyme Subunits

Although as yet conjectural, it would seem reasonable to assume that hydrophobic forces also might be involved in the association of enzyme subunits that seem to occur through noncovalent binding [e.g., see Ref. 2].

Immobilization of Enzymes in vivo

As pointed out above, there is a strong possibility that the immobilization of certain enzymes on intracellular matrices, which also seems to occur without covalent binding, is through hydrophobic interaction, presumably in cooperation with electrostatic forces. Since the enzymes are active, binding must be through sites that are not in the (substrate-binding) active center. Thus the adsorptive immobilization of various enzymes on n-alkylamino-agaroses, which often occurs without blocking the active center [6], might present a "model" for enzyme immobilization in vivo. Such studies would be relevant to the possibility that intracellular enzymes may be endowed with binding sites away from the active center and for the purpose of "specific" attachment of the enzymes to the corresponding sites on the proper membrane.

Structure of Cell Membranes

There seems to be no doubt that hydrophobic phenomena play a role in maintaining the structure of cell membranes (for a recent review of the involvement of hydrophobic forces, see Ref. 2). The great interest in this field is, in part, due to the realization that most intracellular biochemical processes appear to occur on the surface of membranes, which is entirely different from the conditions in classical "solution" chemistry.

Formation of Protein-Nucleic Acid Complexes

Previous findings [54] indicate that positively charged proteins tend to enter into complex formation with nucleic acids. For instance, at low ionic strength chymotrypsin forms a stoichiometric complex with native double-stranded calf thymus DNA. The complex, for which the protein: DNA ratio is 20:1 by weight, sediments by ultracentrifugation as a stable enzymatically active unit. Electron micrograms suggested that the DNA is coated with a monomolecular layer of protein [54]. Although the complex dissociates at relatively low salt concentrations, indicating a primarily electrostatic type of binding involving the charged phosphate groups, recent suggestions as to the possible hydrophobic properties of nucleic acids [e.g., see Ref. 1], could mean that hydrophobic forces also are involved.

The greater general stability of viruses as compared to the artificial complexes could merely be the result of more pronounced hydrophobic properties of the virus proteins and a stronger hydrophobic intermolecular interaction of the latter and/or between the protein and the nucleic acid bases.

Possible Slow Release of Topically Applied Drugs

The observation that, depending on the experimental conditions, a dye such as Ponceau S may be extremely strongly bound by the present type of adsorbent suggests that this may also be the case for certain drugs. One could assume that the rate of release of such a drug from the complex with the adsorbent could be regulated (e.g., by changing the relative hydrophobicity of the ligand) and adapted to certain environmental conditions. Therefore, it is reasonable to assume that adsorbents could be designed that would bind the drug and when applied topically release it at a desired rate under the conditions at the site of application. The adsorbent-bound drug should be contained in a suitable permeable bag. Such a device might be inserted directly into a body cavity or be implanted subcutaneously.

SUMMARY

Evidence has been presented for the duality of nonspecific protein binding by agaroses substituted with primary amines carrying hydrophobic groups. It appears that the binding by these positively charged adsorbents may occur 1) through a more or less general salt-reversible electrostatic effect or 2) through hydrophobic salt stable "bonding," the latter occurring predominantly in the case of the more hydrophobic ligands interacting with the more hydrophobic proteins, e.g., bovine serum albumin as opposed to ovalbumin.

Further evidence is presented that in contrast to electrostatic binding "true" hydrophobic bonding is stabilized with increasing NaCl concentrations. It appears that this is one of the reasons that a highly purified protein may seem chromatographically inhomogeneous when attempts are made at elution by means of a salt gradient of increasing concentration. "Irreversible" protein binding, in conjunction with nonuniformity of adsorbent binding sites, may constitute another factor underlying the often observed and presumably false chromatographic inhomogeneity of a pure protein.

Nonspecific irreversible binding, which is ascribed to simultaneous multiple point attachment of a protein molecule to two or more adsorbent binding sites, may occur with any highly substituted adsorbent that carries charged and/or hydrophobic groups. The latter even could be provided by a ligand attached with an inert spacer to an inert matrix. Although generally a drawback for protein separation by column chromatography, nonspecific irreversible binding is the basis for adsorptive protein immobilization, e.g., for the purpose of reactor-type experiments in the case of enzymes.

Previous results with enzymes immobilized by adsorption on n-alkylamine-substituted agaroses have been augmented with DEAE-agarose as the adsorbent. A number of arbitrarily chosen enzymes with isoelectric points in the acid region were investigated and in each case at least part of the enzyme remained irreversibly bound under conditions favorable for activity.

In view of the apparent nonuniformity of binding sites, postulated to range from extremely "weak" to extremely "strong," initial "leakage" of an enzyme from a column does not necessarily mean that all of the enzyme will be released upon continued washing with the ambient medium.

ACKNOWLEDGMENTS

This work has been supported by U. S. Public Health Service Grant RR-05513, the Harvey Bassett Clarke Foundation and by the Santa Clara County United Fund.

REFERENCES

[1] W. P. Jencks, Catalysis in Chemistry and Enzymology, McGraw-Hill, New York, 1969.

[2] C. Tanford, The Hydrophobic Effect: Formation of Micelles and Biological Membranes, Wiley, New York, 1973.

[3] H. F. Epstein, J. Theor. Biol., 31, 69 (1971).

[4] B. H. J. Hofstee, Anal. Biochem. 52, 430 (1973).

[5] B. H. J. Hofstee, Biochem. Biophys. Res. Commun. 50, 751 (1973).

[6] B. H. J. Hofstee, Ibid., 53, 1137 (1973).

[7] B. H. J. Hofstee, in Immobilized Biochemicals and Affinity Chromatography (R. B. Dunlap, ed.), Plenum, New York, 1974, pp. 43-59.

[8] B. H. J. Hofstee, Polym. Prepr., Am. Chem. Soc., Div. Polym. Chem., 15, 311 (1974).

[9] S. Hjertén, J. Chromatogr., 87, 325 (1973).

[10] J. Porath, L. Sundberg, N. Fornstedt, and I. Olsson, Nature (London), 245, 465 (1973).

[11] R. J. Yon, Biochem. J., 126 765 (1972).

[12] Z. Er-el, Y. Zaidenzaig, and S. Shaltiel, Biochem. Biophys. Res. Commun., 49, 383 (1972).

[13] S. Shaltiel and Z. Er-el, Proc. Natl. Acad. Sci., U. S. A., 70, 778 (1973).

[14] S. Hjertén, Biochem. Biophys. Acta, 79, 393 (1964).

[15] R. Axén, J. Porath, and S. Ernbäck, Nature (London), 214, 1302 (1967).

[16] J. Porath, R. Axén, and S. Ernbäck, Ibid., 215, 1491 (1967).

[17] P. Cuatracasas and I. Parikh, Biochemistry, 11, 2291 (1972).

[18] J. Porath, Nature (London), 218, 834 (1968).

[19] R. Jost, T. Miron, and M. Wilcheck, Biochem. Biophys. Acta, 362, 75 (1974).

[20] B. H. J. Hofstee, Prep. Biochem., 5, 7 (1975).

[21] B. H. J. Hofstee, Biochem. Biophys. Res. Commun., 63, 618 (1975).

[22] B. H. J. Hofstee, Nature (London), 213, 42 (1967).

[23] C. Hansch, and W. J. Dunn, J. Pharm. Sci., 61, 1 (1972).

[24] I. M. Klotz, Arch. Biochem. Biophys., 138, 704 (1970).

[25] W. O. McClure and G. M. Edelman, Biochemistry, 5, 1908 (1966).

[26] P. H. von Hippel and T. Schleich, in Structure and Stability of Biological Macromolecules (S. N. Timasheff and G. D. Fasman, eds.), Dekker, New York, 1969, pp. 417-574.

[27] K. Hamaguchi and E. P. Geiduschek, J. Am. Chem. Soc., 84, 1329 (1962).

[28] H. A. Sober, R. W. Hartley, W. R. Carrol, and E. A. Peterson, in The Proteins, Vol. 3, 2nd ed. (H. Neurath, ed.), Academic, New York, 1965 pp. 1-97.

[29] S. R. Himmelhoch, Methods Enzymol., 22, 273 (1971).

[30] K. E. Hopper, Biochem. Biophys. Acta, 293, 364 (1973).

[31] P. Cuatrecasas, J. Biol. Chem., 245, 3059 (1970).

[32] B. H. J. Hofstee, in Enzyme Engineering, Vol. 3 (E. K. Pye, ed.) Plenum, New York, In Preparation.

[33] S. Lewin, Displacement of Water and Its Control of Biochemical Reactions, Academic, London, 1974.

[34] J. Schmidt and M. A. Raftery, Biochemistry, 12, 852 (1973).

[35] D. Bobb and B. H. J. Hofstee, Anal. Biochem., 40, 209 (1971).

[36] H. F. Hixson and A. H. Nishikawa, Arch. Biochem. Biophys., 154, 501 (1973).

[37] A. H. Nishikawa, P. Bailon, and A. H. Ramel, Abstr. SERACS Meeting, Charleston, South Carolina, November 1973.

[38] G. Vanderkooi and D. E. Green, Proc. Natl. Acad. Sci. U.S.A., 66, 615 (1970).

[39] I. H. Silman and E. Katchalski, Ann. Rev. Biochem., 35, 873 (1966).

[40] O. R. Zaborsky, Immobilized Enzymes, CRC Press, Cleveland, 1973.

[41] B. H. J. Hofstee, Arch. Biochem. Biophys., 51, 139 (1954).

[42] G. J. Doellgast, V. A. Memoli, A. G. Plaut, and W. H. Fishman, Fed. Proc., 33, 1561 (1974), Abstracts.

[43] C. Tanford, Adv. Protein Chem., 23, 209 (1968).

[44] B. H. J. Hofstee, in Methods of Protein Separation , Vol. 2, N. Catsimpoolas, ed.), Plenum, In Preparation.

[45] P. O'Carra, S. Barry, and T. Griffin, Biochem. Soc. Trans., 1, 289 (1973).

[46] T. A. Steitz, R. Henderson, and D. M. Blow, J. Mol. Biol., 46, 337 (1969).

[47] S. J. Singer, in The Proteins, Vol. 3, 2nd ed. (H. Neurath, ed.), Academic, New York, 1965, pp. 269-357.

[48] K. J. Stevenson and A. Landman, Can. J. Biochem., 49, 119 (1971).

[49] Biological Correlations - The Hansch Approach, Adv. Chem. Series 114, (W. Van Valkenburg, ed.), Am. Chem. Soc., Washington, D. C., 1972.

[50] B. R. Baker, Design of Active-Site-Directed Irreversible Enzyme Inhibitors, Wiley, New York, 1967.

[51] P. D. Boyer, F. G. Lum, A. Ballou, J. M. Luck, and R. G. Rice, J. Biol. Chem., 162, 181 (1946).

[52] J. D. Teresi, J. Am. Chem. Soc., 72, 3972 (1950).

[53] J. E. Fletcher, A. A. Spector, and J. D. Ashbrook, Biochemistry, 10, 3229 (1971).

[54] B. H. J. Hofstee, Biochem. Biophys. Acta, 55, 440 (1962).

[55] R. G. Paquette, E. C. Lingafelter, and H. V. Tartar, J. Am. Chem. Soc., 65 686 (1943).

[56] B. H. J. Hofstee, Arch. Biochem. Biophys., 78, 188 (1958).

[57] C. Tanford, J. Am. Chem. Soc., 84, 4240 (1962).

DESIGN PARAMETERS IN AFFINITY CHROMATOGRAPHY

A. H. Nishikawa, P. Bailon, and A. H. Ramel

Chemical Research Department
Hoffmann-La Roche Inc.
Nutley, New Jersey 07110

INTRODUCTION

Well over 400 papers have been published in the past few years which describe the application of the affinity method to the purification of biopolymers. It is fair to say that most are concerned only with solving a unique problem at hand, be it to purify a given enzyme, antigen, or whatever. Very likely the majority of the scientists with the isolation or purification problem at hand are not in a situation to consider more than a few variations on the several stock "recipes" in the literature. If only some of the tricks should happen work, then the results may be reported as a partial purification of the material. If the attempts do not work outright, of course they do not see print. It is not clear that the failures are understood or why only partial purification was achieved.

Yet as early as 1967 a set of guidelines was published which pointed out the criteria important to the preparation of affinity sorbents. In his book entitled Design of Active-site-directed Irreversible Enzyme Inhibitors, the late Baker [1] included a chapter (the last) on "Enzyme-Specific Columns." Here was spelled out the need for inert, hydrophilic insoluble carriers, as well as the importance of knowing what part of the ligand molecule could be used for immobilization ("bulk tolerance") without losing affinity for enzyme. Baker also suggested the need for obtaining an optimum density of functional groups in the carrier matrix so that efficient binding would result.

While Baker's suggestions went largely unnoticed, a paper from a group at the N.I.H. proclaimed "affinity chromatography" [2] and introduced the notion of a spacer-arm (we prefer the term "leash") molecule. This was recommended because, when the ligand was attached to the insoluble carrier via a leash or tether moiety, its interaction with active-site of an enzyme was optimized. Enzymes whose active-sites are located in deep clefts had steric problems in binding to ligands which were directly attached to insoluble surfaces. The leash is not a universal requirement however, and may even cause nonspecific binding problems [3], as we shall see.

Another important parameter is gel porosity. Cuatrecasas has observed that the affinity binding of an enzyme with a molecular weight of 17,000 daltons (staphylococcal nuclease) required the use of 4% agarose gel, which has a molecular weight exclusion limit of 15 million daltons [4]. The need for this high porosity is not yet clearly understood.

Even with knowledge and application of the preceding criteria, success has not always presented itself in the preparation of an affinity sorbent. Thus there must be other factors which must be considered. One which is receiving more attention recently is the adventitiously introduced ionic groups in the affinity gel which result in nonspecific binding of proteins. A second is the nonspecific hydrophobic interactions due to leash (and ligand) groups introduced into gels. And a third is the efficacy of affinity binding as affected by the concentration of ligands which have been bonded into the sorbent [5-7]. These topics will be the central concern of this paper. By proper evaluation and understanding of these parameters, we should then be better able to tackle the yet uncharted affinity purification problems.

IONIC EFFECTS

Because proteins are polyelectrolytes, one can expect ionic functional groups on any chromatographic sorbent to affect their elution behavior. These interactions are, of course, central to the more established technique of chromatography on ion-exchange celluloses. In affinity chromatography the presence of ionic groups in the ligand, leash, or matrix itself may give rise to problems of nonspecific binding.

Leftover Leashes

One source of extraneous ionic groups in the matrix can be the leftover charges due to incomplete attachment of ligands onto leashes. Such a situation is shown in FIG. 1. In the upper

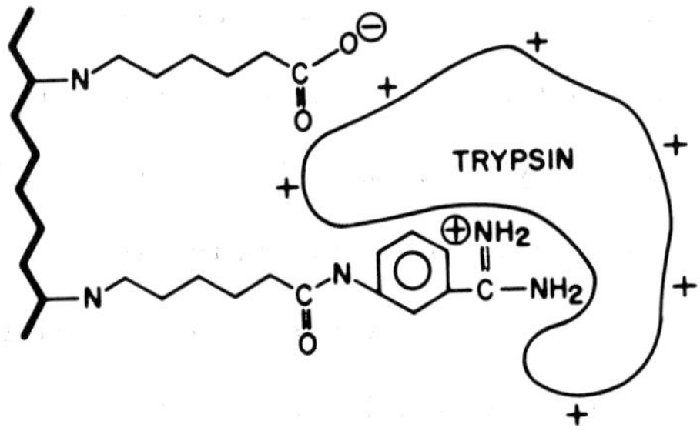

FIG. 1

part of the diagram, a carboxylate anion due to unreacted caproic leash moiety is forming a salt with the cations on the surface of the enzyme. In the lower portion we see the expected active site interaction of the benzamidine ligand. The nonspecific interactions due to the caproic anions can be substantial if they are present in large amounts-perhaps as little as 10% of all of the functional groups. These problems were encountered by Hixson and Nishikawa [5] in their trypsin and thrombin studies.

They therefore used the percolation of chymotrypsin through affinity gel columns as a test for functional specificity. As can be seen in FIG. 2, chymotrypsin and trypsin share similar physical properties. Since they are both polycations at pH 8, we can expect both proteins to be bound or retarded in flow rate when passing through a gel containing many free caproate anions. However, a properly prepared benzamidine gel exhibits none of these anions. Thus, owing to the strict and different substrate specificities of the enzymes, the trypsin will be adsorbed to

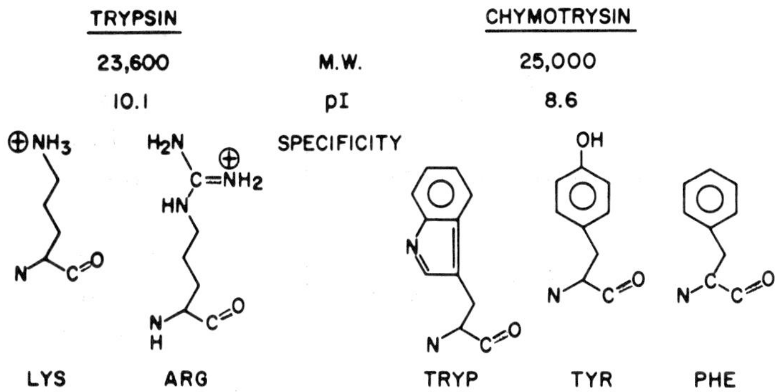

FIG. 2

the properly prepared gel while the chymotrypsin will not.

To avoid having these incompletely reacted leashes in the gel, Wilchek [8] has encouraged the strategy of attaching the ligand to the leash moiety first, then introducing the ensemble into the carrier gel. However, for many workers such a course requires much more skill and work in organic chemical synthesis compared to the effort of doing a sequence of reactions on an insoluble gel. FIG. 3 shows a comparison. The <u>in vitro</u> approach requires a protecting group on a bifunctional ligand (Cw) which is coupled to a protected leash moiety (yB). The protecting group (y) is removed from the leash which is then coupled to a second protected leash moiety (zA). The leash (z) and ligand (w) protecting groups are removed and the ensemble attached to the gel (G-x). Each of these steps (coupling as well as de-protecting) requires purification of the product before the next step is assumed. Depending on the complexity of the ligand-leash

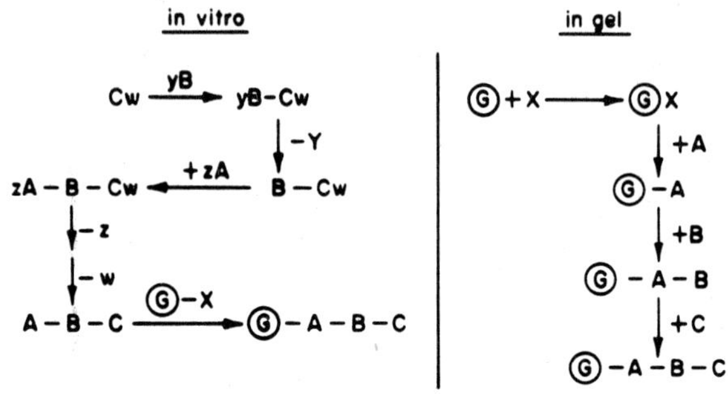

FIG. 3 Strategy for affinity gel preparation.

structure, the <u>in vitro</u> route may take several days to complete synthesis.

By comparison the <u>in-gel</u> approach is much simpler in that excess or unreacted reagents from a given reaction step are easily washed out of the gel particle and the following step is readily assumed. Such has been the attraction of Merrifield's solid-phase peptide synthesis (see J. M. Stewart's article elsewhere in this volume). Because covalent coupling to a gel results in the insolubilization of a compound, that compound may often be bifunctional and used without a protecting group for the second functionality. To avoid the problems of leaving unreacted leash moieties within a gel, care must be taken to drive each reaction step to 100% completion. This is usually done by using large excesses in reagent. Aliquots of gels should be taken after each step and tested by some appropriate means for completeness of reaction.

Leash Linkage

Another source of unwanted charges may arise from the linkage connecting leash to gel. Here we focus our attention on the most frequently used reaction of cyanogen bromide on agarose gel.

This reaction, which was developed by Porath and co-workers [9, 10], has been used extensively to couple organic amines to polysaccharide carriers. Several structures have been proposed to describe the final mode of linking of the amine to the polysaccharide when cyanogen bromide is used. Of these, the isourea linkage has the potential for generating a positively charged group. The reaction pathway outlined in FIG. 4 indicates how this is possible. In agarose, the 1,3-propanediol structure is most likely involved in the primary reaction with BrCN. The formation of the cyclic imidocarbonate intermediate has been strongly supported by model compound studies using methyl 4,6-O-benzylidene-α-D-glucopyranoside [11] and trans-1,2-cyclohexanediol [12]. In our own studies (Nishikawa, unpublished observations) it appears that BrCN does not react readily with isolated hydroxyl groups to form even the unstable cyanate.

The formation of an isourea linkage is supported by the fact that ammonia is not released during reaction with the organic amine. Comparison of titration data with elemental nitrogen analysis shows that more nitrogen is obtained in the gel derivative than can be accounted for by titratable functional groups [13].

A complete titration curve also shows that the isourea linkage is associated with a high pK_a as seen in FIG. 5. There

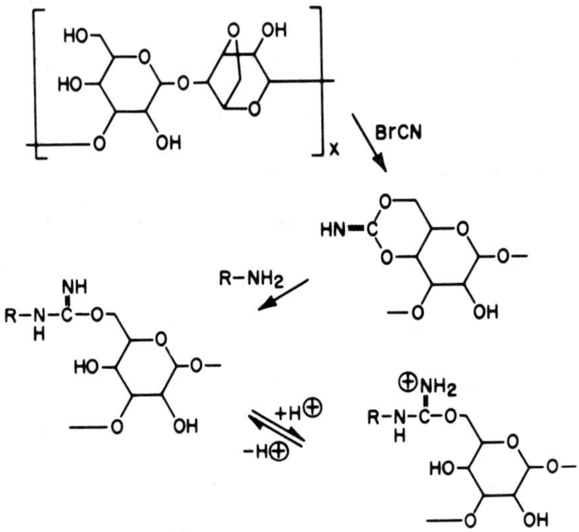

FIG. 4 Reaction of agarose with cyanogen bromide.

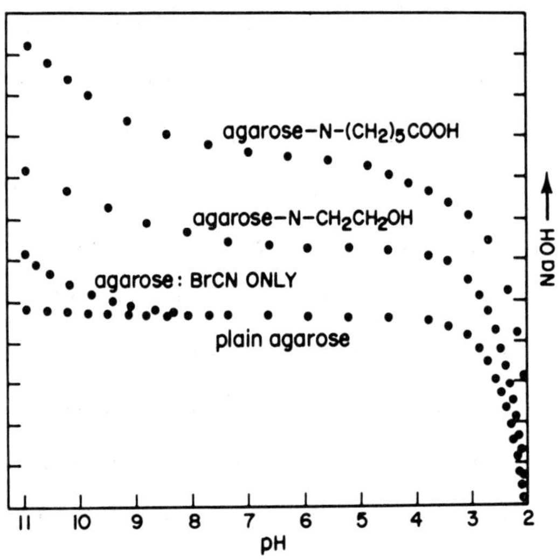

FIG. 5 Titrimetric analysis of agarose gels.

is a distinct change in the uptake of sodium hydroxide at high pH in the gel samples bearing ethanolamine and 6-aminohexanoic acid adducts as compared to plain agarose or BrCN activated agarose. Wilchek recently also reported a similar observation and assigned a value of 10.4 for the pK_a of an isourea resulting from a primary alkylamine [8]. To avoid these adventitiously generated cationic groups, Wilchek found effective the use of acylhydrazides, such as adipic dihydrazide, instead of alkylamines. As indicated in TABLE 1, the pK_a for the acylhydrazide nitrogen is very low, and the pK_a for the resulting isourea linkage would be similarly low. Hence at physiological pH a leash structure derived from acylhydrazides would be nonprotonated.

TABLE 1

Leash Functions in Cyanogen Bromide Coupling

Structure	pK_a
$-CH_2-CH_2-C(=O)-NH-NH_2$	~4.0
$-CH_2-NH-C(=O)-CH_2-NH_2$	~8.2
$-CH_2-CH_2-CH_2-NH_2$	~10.6

Work is in progress in our laboratories on evaluating the isourea linkages derived from these hydrazides as well as on leash structures which are derived without the use of cyanogen bromide.

It will be instructive to consider presently some preliminary findings in our laboratory which demonstrate the interfering effects of unwanted ionic groups on affinity chromatography. In our program to purify galactosidase from Aspergillus niger, we started with an examination of the work of Steers et al. [14] who purified the β-galactosidase from Escherichia coli on sorbents like those shown on FIG. 6. Gel B was found to retard the flow of a band of enzyme percolating through the column. Gel C in pH 7 buffer bound the enzyme so tightly that a substrate solution would not desorb it from the gel. However, the enzyme could be eluted with a pH 10 buffer. In view of the fact that the enzyme had in the past been purified on DEAE-cellulose [15] and that the

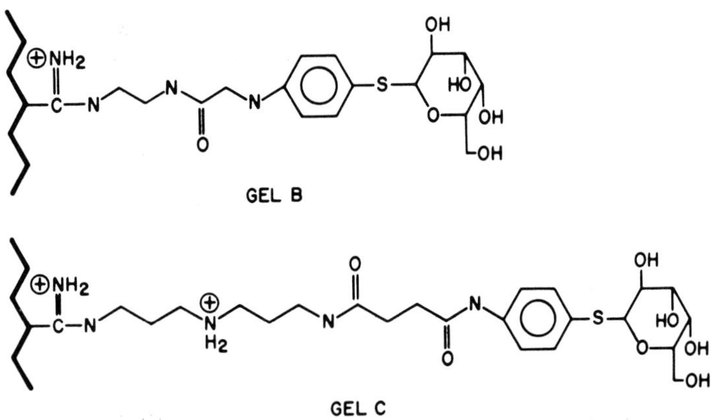

FIG. 6 Affinity gels for β-galactosidase [14].

pK of isourea is near 10, it seemed to us that the binding of enzyme by Gel C could be explained by strong ionic interactions.

We have confirmed the observations of Steers and co-workers in our own lab. Furthermore we have examined appropriate control adsorbents and the results are summarized on Table 2. The number of net cationic charges per ligand or leash site has been calculated taking into account the contribution of the isourea linkage to agarose (A). It is evident from the adsorption pattern of enzyme onto the different gels that the polyanionic character of the protein dominates and directs the binding to sorbent sites. Of particular interest is the sorbent where the leash structure is attached to the agarose via the acylhydrazide (the fourth gel in Table 2). Where there is no neighboring

TABLE 2

Affinity Sorbents for β-Galactosidase

sorbent	charge/site	adsorbs
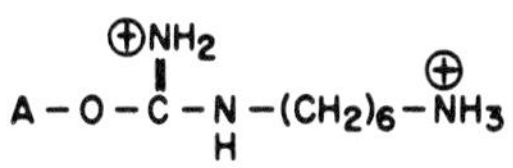	++	YES
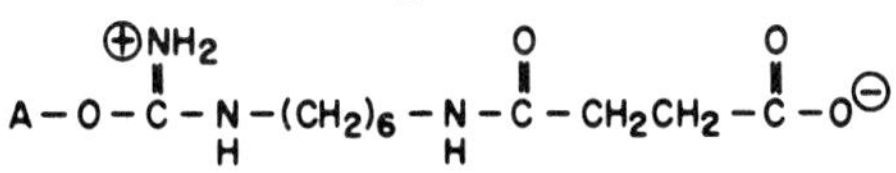	0	NO
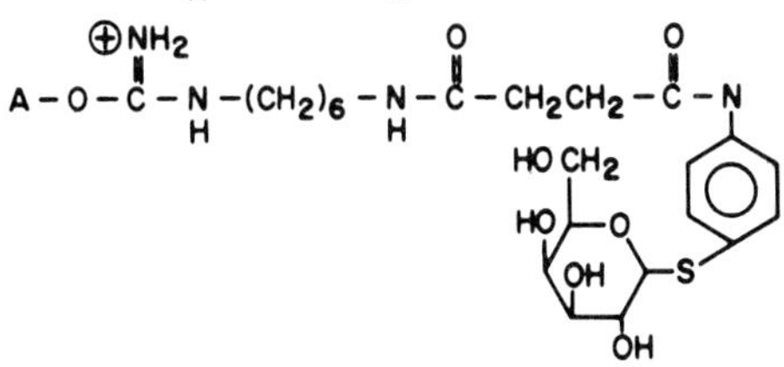	+	YES
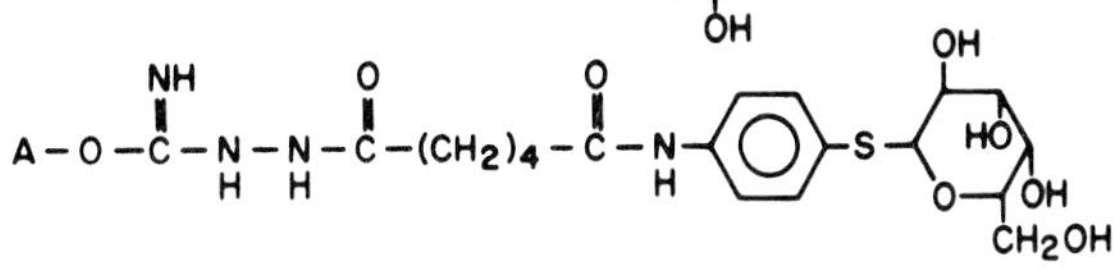	0	NO

cationic group, the phenyl-thiogalactopyranoside is indeed a weak ligand as is suggested by the inhibition constants, ~5 m*M* [8]. Full detailed studies on this problem will be published elsewhere [16].

Our studies on the use of the p-aminophenyl-thio-galactopyranoside as the ligand for β-galactosidase purification suggest that where solutions of high salt concentration or with markedly different pH are used to elute or desorb the enzyme, the process most likely involves nonspecific ionic binding. This is most strongly supported by the data in Table 3 where the K_i of the phenyl-thiogalactopyranoside is little affected by a tenfold change in salt concentration. In this connection the observations of Robinson et al. [17] on their scale-up study of

TABLE 3

Inhibition Constants for β-Galactosidase (*E. coli*)

Compound		K_i
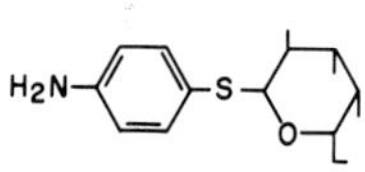		1.2×10^{-3} *M*
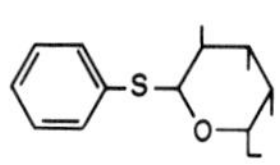	0.02 *M* NaCl	0.4×10^{-3} *M*
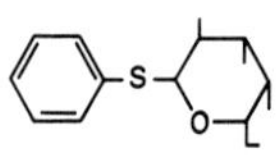	0.25 *M* NaCl	0.3×10^{-3} *M*

the affinity system reported by Steers et al. [14] is particularly interesting. They observed 6000 units of enzyme to bind to their sorbent in 0.01 ionic strength buffer; however, this value fell to 350 units in 0.05 ionic strength buffer. Since the binding (inhibitor) characteristics of the ligand itself are not sensitive to salt in the medium, it would appear that Robinson et al. [17] as well as Steers et al. [14] were dealing largely with non-specific ionic adsorptions to the gel. O'Carra [3] and coworkers [18] have recently also come to the same conclusion regarding the β-galactosidase purification by affinity chromatography. Their work, however, led them to conclude that hydrophobic interactions were primarily responsible for the nonspecific binding. Recent findings in our lab suggest that hydrophobic effects are prominent in this system (as will be discussed in a following section of this paper). Full details are presented elsewhere [16].

A final example of nonspecific binding due to possible ionic groups in the leash moiety is the one observed by Blumberg et al. [19] in their study of the affinity purification of lectins for L-fucose. FIG. 7 shows the structure of the affinity sorbent containing L-fucosylamine which bound the lectins. The

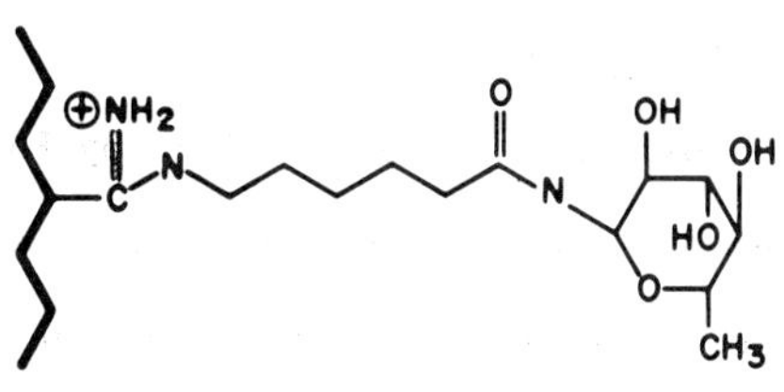

FIG. 7 Affinity gel for fucose lectin [19].

same sorbent was also observed to bind the β-galactosidase from *E. coli*. The lectins could be subsequently desorbed by solutions of L-fucose, but the enzyme could not. The latter could be eluted out by buffers containing high concentrations of salt.

Ionic Ligands

Ligands which are ionic can also give rise to ion-exchange problems in the sorbent. With the m-aminobenzamidine ligand for trypsin (see FIG. 1), Hixson and Nishikawa [5] found that the inclusion of 0.5 *M* KCl in the buffer medium enhanced the specificity of the binding. Another example is that reported by O'Carra and Barry [20], who studied the use of immobilized oxamate for affinity purification of lactate dehydrogenase. The inclusion of 0.5 *M* NaCl in the loading and washing buffer eliminated non-specific ionic binding to gel. However, high ionic strenth is not a universal cure as shown by acetylcholinesterase in its binding to sorbents containing quaternary ammonium ion ligands. Dudai et al. [21] and Schmidt and Raftery [22] have observed that high salt weakens the binding avidity of the enzyme to the ligand. Where ionic ligands must be used, experience suggests that the inclusion of salt (at least 0.15 *M*) in the buffering medium be investigated to optimize specificity of binding.

General Comments on Ionic Effects

Although it has been pointed out earlier by Porath [10] and others [23] that the cyanogen bromide coupling of amines to agarose can give rise to isourea linkages which are adventitiously accompanied by cationic charges, this as a problem source has been largely ignored. Recognizing this, one can, upon reviewing a number of affinity chromatography papers, understand why nonspecific binding problems have been encountered. Furthermore, it is apparent that the claims for specific affinity in a number of systems are invalid. At best, binding in these cases seems to be due to a combination of effects--ligand specificity reinforced by ionic binding.

In view of the several possibilities for nonspecific adsorption due to ionic effects, it is undesirable to effect elution (desorption) of an enzyme by increases in salt concentration in the buffer. Since only meager data are available on salt effects on affinity binding or on K_i of competitive inhibitors, it should be demonstrated for the enzyme at hand that high salt concentrations do indeed affect these parameters. As we have shown above for the phenyl-thiogalactopyranoside, there is no justification here to use salt gradients for elution of β-galactosidase from an affinity sorbent containing this ligand. Similarly, Barry and O'Carra have shown that the K_m of NADH to lactate dehydrogenase is little effected by KCl from 0 to 0.5 M [24]. Hence, where these enzymes are desorbed by salt gradients from gels containing immobilized NADH, one is not observing true affinity chromatography.

Similarly the use of pH-shifts in the eluting medium should be studiously avoided. Compared to salt gradients, this procedure provides for greater prospects of obtaining nonspecifically bound proteins. As Blumberg et al. [19] and others have pointed out, the only reliable indication of successful affinity chromatography should be the demonstration that desorption of an affinity bound enzyme can be effected by a solution of specific inhibitor or substrate. Once this has been verified for a particular system, it may then be simpler or more economical to use salt gradients or pH shifts to effect desorption provided that appropriate control tests are made for these conditions.

HYDROPHOBIC EFFECTS

In addition to being complex polyelectrolytes, proteins also possess many structural elements which arise from intra- as well as intermolecular hydrophobic interactions. Generally these hydrophobic elements are buried in the interior of the protein structure; however, there may be crevices and/or pockets near the surface which are important in the structural integrity and/or the functioning of the protein. It is possible that the interactions of a protein with certain hydrophobic surfaces result in an deleterious and irreversible adhesion.

A number of recent reports [25-27] have drawn attention to the significant effects of hydrophobic interactions in affinity chromatography. The term"hydrophobic chromatography" has been coined and a systematic method for taking advantage of this effect has been set forth by Shaltiel and Er-el [29]. But here we shall focus on the "side effects" or perhaps the unwanted aspects of hydrophobic interactions.

Matrix

With regard to the carrier gel or matrix, enzyme chemists have long observed that proteins are very frequently adsorbed nonspecifically and irreversibly to hydrophobic surfaces such as polystyrene beads. Ion-exchange resins of polyacrylic acid/ ethylene copolymers have seen some utility in the chromatographic purification of "compact" low molecular weight proteins such as RNAse A or cytochrome C. But the examples are few.

For affinity work, the hydrophilic structure of the polysaccharide, agarose has afforded a relatively inert, non-interactive support material and thus has been the carrier matrix of choice. Cellulose is another polysaccharide that has been used for affinity sorbents. While it has seen less service than agarose in recent years, immunochemists have used it widely for their immunoadsorbents, especially in the years preceeding 1968 (see review [30]).

While nonspecific adsorption problems due to hydrophobic effects have been largely avoided through the use of polysaccharide carriers, they have been exacerbated by the chemical functionalizations which have introduced the often hydrophobic leash and ligand structures.

Leash

The indiscriminate use of long hydrophobic leash structures in affinity gels have been called into question by O'Carra and co-workers [18]. In their study of the affinity purification of β-galactosidase as described by Steers et al. [14], they prepared a gel with the structure shown in FIG. 8. The length of the leash structure from the isourea linkage to the end of the phenyl ring was comparable to that shown for Gel C in FIG. 6.

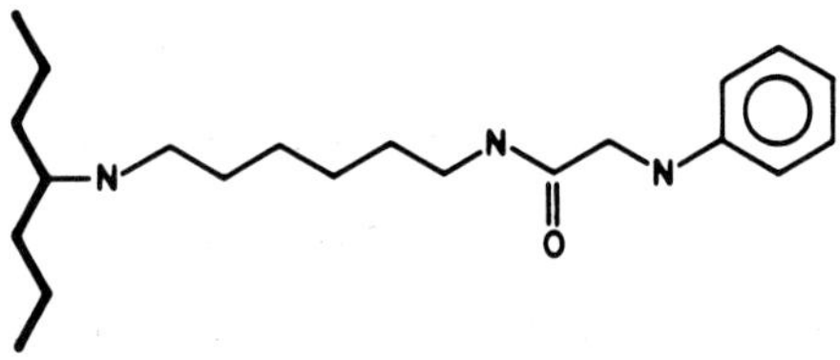

FIG. 8 Hydrophobic affinity leash [18].

Despite the absence of the galactopyranose ligand, the β-galactosidase is bound to the gel so firmly that even a 0.5 M KCl solution did not desorb the enzyme. Because of the aromatic structure involved, O'Carra [3] placed the nonspecific adsorption problems at the doorstep of hydrophobic forces. In view of the analysis of the isourea linkage in the first section of this paper, we lean to the interpretation that the cationic charge on the isourea linkage reinforces the hydrophobic interactions due to the N-phenylglycyl moiety in O'Carra's leash example.

A similar interpretation may be placed on the linear aliphatic structures developed by Shaltiel and co-workers [27, 29] for their hydrophobic chromatography. The fact that concentration gradients of salts (which were not necessarily chaotropic) could desorb proteins from these hydrophobic gels suggests a significant interplay of ionic forces here. By attaching hydrophobic structures to agarose gels by nonionic linkages, we have observed "true" hydrophobic binding. The results of these studies will be reported elsewhere.

<u>Ligands</u>

Hydrophobic ligands, especially aromatic ones, can also yield nonspecific binding of proteins. These ligands often are not very different from the hydrophobic leash structures discussed in the preceding section. The work of Stevenson and Landman [31] is an interesting case in point.

In their work to purify chymotrypsins from a variety of sources, these authors selected 4-phenylbutyl amine (PBA) as the ligand. Since chymotrypsins recognize the phenyl ring, such a ligand was a class-specific one. As a control test for the specificity of their gel preparation, Stevenson and Landman prepared a sample of active-site inhibited chymotrypsin as shown schematically in FIG. 9. In elution diagram A we see the behavior of native chymotrypsin on their affinity gel. The enzyme binds the phenylbutane structure and is desorbed by 0.1 M acetic

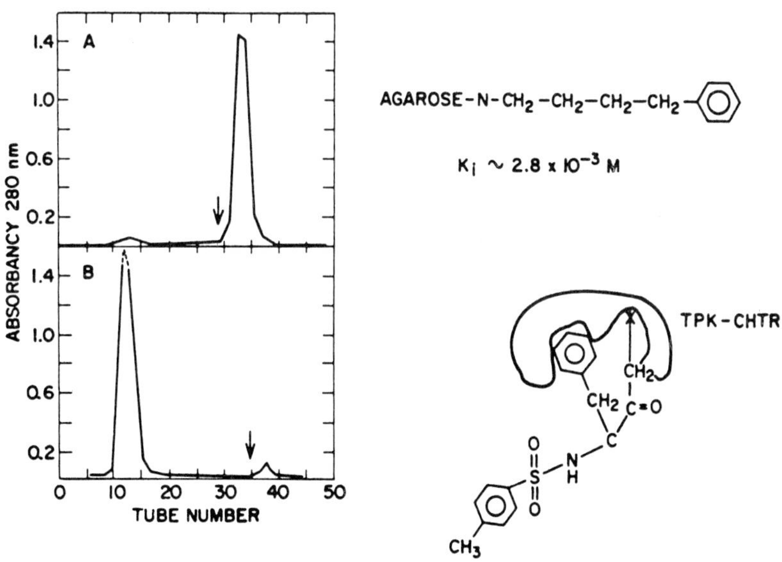

FIG. 9 [31].

acid (vertical arrow). In diagram B the active-site inhibited enzyme does not bind to the gel and is washed straight through the column. A change of the solvent to acetic acid yields a very small amount of protein [presumably due to incomplete reaction of the enzyme with tosylphenylalanylchloroketone (TPCK)]. Since chymotrypsin is a polycation, the binding of the enzyme to these gels containing 4-phenylbutylamine via the cyanogen bromide reaction most likely represented true affinity binding.

Even with seemingly unequivocal data indicating true affinity binding, the agarose gel containing PBA was shown later to exhibit nonspecific protein binding. Hofstee, as shown in FIG. 10, clearly demonstrated that such a gel has a strong avidity for a number of acidic proteins like ovalbumin or

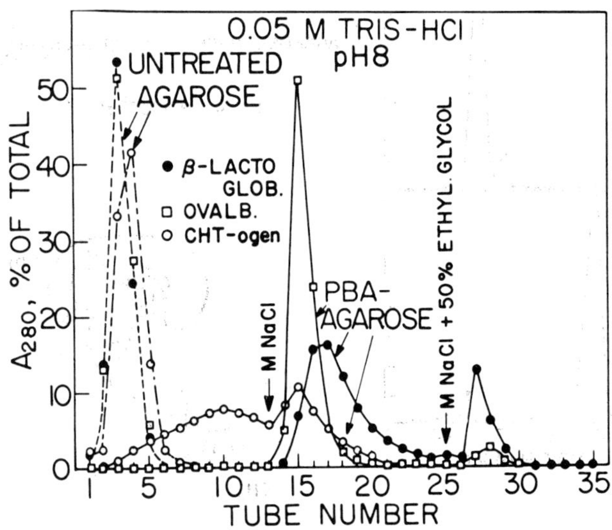

FIG. 10 [25].

β-lactoglobulin [24]. The retarded elution pattern for chymotrypsinogen (a polycation) suggested some ability for this zymogen to recognize the ligand.

We have also encountered hydrophobic problems. In our studies on the affinity purification of thrombin, the specific sorbent reported by Thompson and Davie [32] was studied and some preliminary results are shown on FIG. 11. Our gel contained p-chlorobenzylamine (pCBA) ligand at 13 μeq/gel compared to 21 μeq/ml gel reported by Thompson and Davie for their 4% agarose preparation. These relatively high ligand concentrations were necessitated by the relatively modest K_i (0.22 m$\underline{M}$) of the p-chlorobenzylamine to thrombin. In view of Hofstee's work we expected that a sorbent bearing this functional group would bind

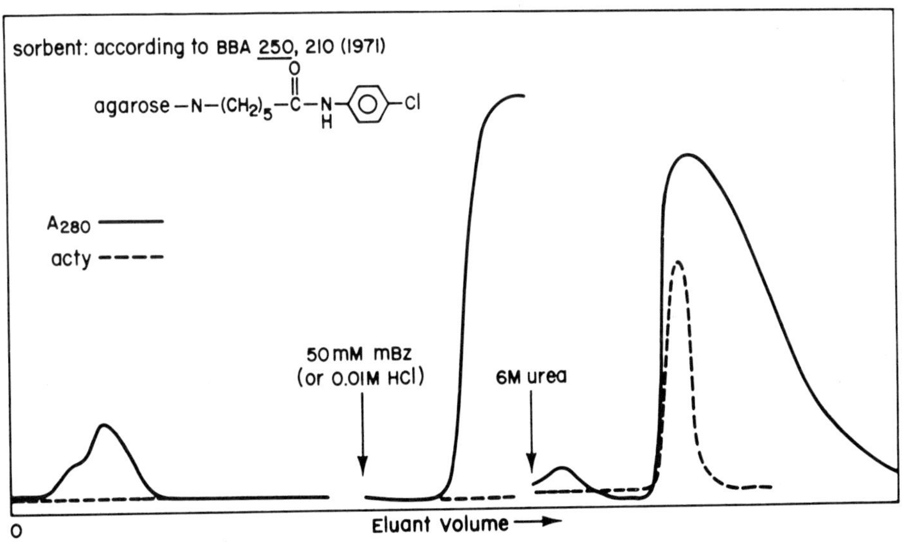

FIG. 11 Hydrophobic adsorption of trypsin on sorbent for thrombin.

proteins nonspecifically, and the experiments shown in FIG. 11 bear this out.

Both trypsin and chymotrypsin were bound tightly to this sorbent. Only degraded proteins were eluted directly from the column (as seen on the left portion of FIG. 11). A solution of 0.01 M HCl, which would normally suffice to remove these proteins (if they are bound primarily by ionic binding),was ineffective. Similarly ineffective was a solution of m-aminobenzamidine, the specific inhibitor for trypsin. Finally (as seen in the right portion of FIG. 11), a solution of 6 M urea was able to wash these proteins out of the gel. This, of course, strongly implicated hydrophobic binding as the predominant force in the adsorption.

In other experiments, egg white lysozyme was observed to bind to this gel at pH 8.1. However, 80% of this protein could be washed out with 10 mM HCl and the rest subsequently by 6 M urea. Bovine serum albumin, which is a polyanion at pH 7 and also very lipophilic, was bound tightly to the pCBA gel. It was not eluted with 0.01 M HCl, but 6 M urea effected its removal. These results also implicated strong hydrophobic binding of proteins to this gel.

General Comments on Ionic and Hydrophobic Interactions

A most significant combination of effects on affinity binding is that resulting from simultaneous ionic and hydrophobic interactions. As Hofstee [25], Hjerten [28], and others have pointed out, the evidence so far on "hydrophobic chromatography" suggests that it is complicated by ionic binding as well. Since hydrophobic interactions are entropy driven and dependent on the presence of water (which primarily participates in hydrogen bonding), it is not surprising that ionic and hydrophobic interactions can mutually reinforce each other so that the resultant effect is much greater than the sum of the two. In FIG. 12 we symbolize the overlapping relationships of hydrophobic with both hydrogen bonding and ionic interactions. From this perspective we would expect that whenever there is immobilized a discrete combination of functional groups bearing these interactive capabilities, the result will be a strongly adsorptive surface. The increase or decrease in specificity will depend on the proximity to each other of the fixed groups as well as the differences in chemical topology of the proteins in the mixture exposed to the sorbent. In a recent example, Chu and Chaykin found significant improvement in the affinity purification of aldehyde oxidase if the sorbent bearing the ligand also included a positively charged nitrogen atom in the leash structure approximately 7 Å from the ligand [33].

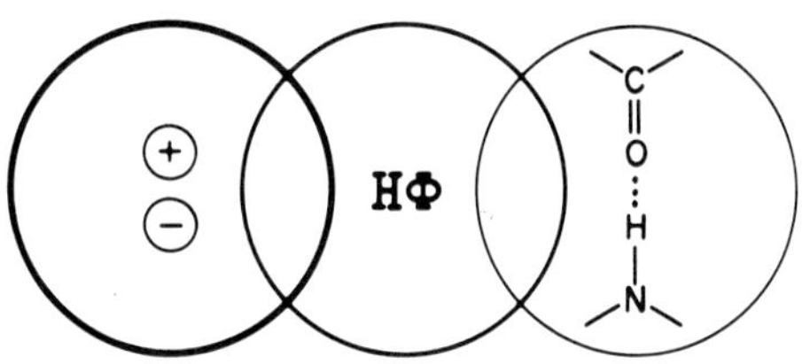

FIG. 12 Interacting forces in affinity adsorption.

This deliberate manipulation of the binding (or inhibition) constant of a ligand-leash structure, of course, is reminiscent of the attempts by enzyme chemists to improve substrate or inhibitor binding by chemical alterations. In addition to increasing knowledge of structure-activity relationships, such efforts should enhance the probabilities for converting ligands with marginal binding capabilities into more useful ones. Much work lies ahead in this area.

LEASH AND LIGAND CONCENTRATION EFFECTS

Perhaps the most neglected of the important parameters in the design of affinity sorbents is the concentration of ligand (with or without leash) in the gel. Relatively few reports have included data of sufficient quality to document its importance [5-7]. Furthermore, until recently a convenient means for preparing absorbent gels with defined amounts of functional groups has not been available [13].

While Baker had earlier suggested a need to determine the optimum density of functional groups in an affinity sorbent [1], in 1971 Hixson and Nishikawa [34] presented a model with equations for predicting useful affinity binding conditions. This was used to predict the ligand density needed to bind thrombin, which was verified by experimental work [5]. More recently Graves and co-workers [35] have reported a mathematical model which might be used to predict success or failure of affinity binding. Wankat has also proposed an analytical model [36], as have O'Carra [3] and Dean and co-workers [37].

Limits of Binding

Binding capacities and ligand concentrations might be expected to have a significant relationship to each other, and knowing this relationship should facilitate sorbent gel preparation. One approach is to study the adsorption limits of affinity gels. Nishikawa et al. [38] recently presented a Langmuir-isotherm description for affinity adsorption which included a derivation of

$$[EL] = \frac{K_b[E][L_o]}{1 + K_b[E]} \tag{1}$$

The complex [EL] is in units of moles of enzyme per unit volume of gel, and is clearly limited by the concentration of binding site, $[L_o]$, built into the gel. [E] is the concentration of enzyme in contact with sorbent gel at equilibrium and is different from $[E_{tot}]$ which is the concentration of enzyme placed in contact with the gel at the beginning of the experiment. K_b is the binding constant in units of liters/mole.

The hyperbolic function represented by Eq. (1) generates the expected sigmoidal curve when [EL] is plotted against the log of the enzyme concentration. Figure 13 shows a family of curves for a model system in which a binding constant of 1000 is assumed. As the $[L_o]$ is increased with the different gel preparations, we see a corresponding increase in the saturation limits of the bound enzyme complex [El]. However, the 50%

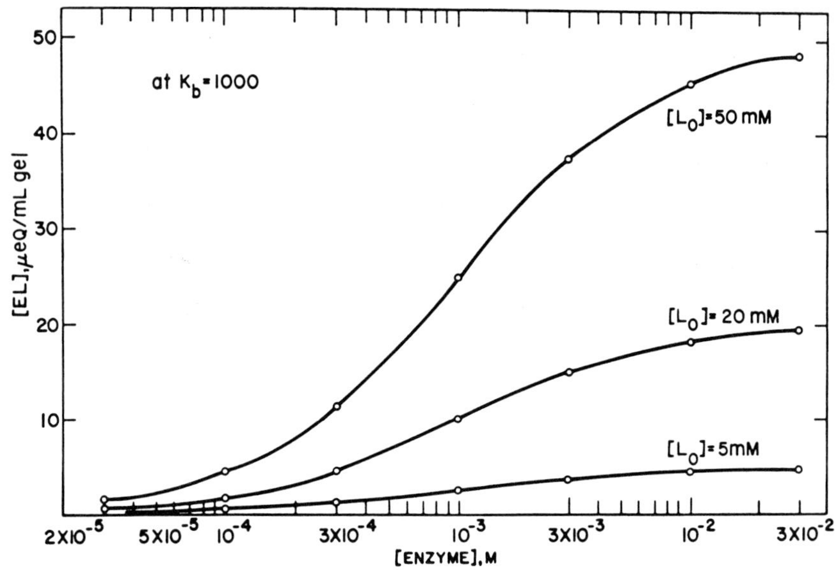

FIG. 13 Enzyme binding to affinity sorbent.

binding concentration for the enzyme is fixed at 1 m*M* in all cases. We would expect that this value is determined by the K_b. In terms of K_d (which is $1/K_b$) equal to 1 m*M*, it would appear that saturation binding would be seen at 100 m*M* in [E]. It is interesting to compare the model curves with that observed for a real system. FIG. 14 shows data obtained from trypsin absorption to agarose gel containing m-aminobenzamidine [38]. In contrast to the ideal model, the concentration of enzyme-ligand complex at saturation is much smaller than that expected from the concentration of $[L_o]$. The apparent or "effective" ligand concentration within the gel matrix is roughly 1/100th of the chemically determined value. Outside of mentioning

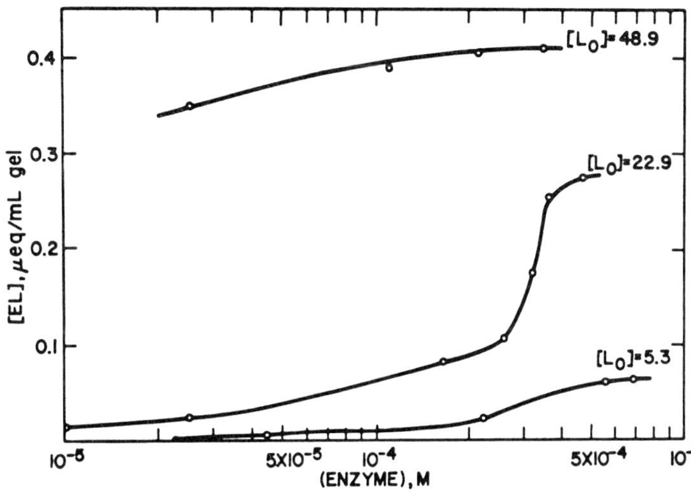

FIG. 14

possible steric problems posed by the agarose polymer to the accessibility of ligands, we presently have no explanation for this low availability. Previously Hixson and Nishikawa [5] also found similarly low values from pulse-loading chromatographic experiments. Low effective ligand concentrations have been observed by Dean and co-workers [39] in sorbents for NAD^+-dependent dehydrogenases. Similarly, Whitney has reported that the effective ligand concentration in affinity sorbents for carbonic anhydrase is about 1% of the chemically determined concentration [40].

In comparing the curve shapes in FIG. 14 to those in FIG. 13, we see only very rough resemblances. Even after examining several other such curves, we are not able to discern a coherent pattern to these shapes. No doubt subtle variations in the chemical reactions on these gels are a contributing factor to the variety of absorption curves.

Extent of Binding

The avidity of binding to a sorbent can be evaluated by the use of Eq. (2), which can be derived by a route analogous to that for Eq. (1) [38]:

$$\frac{[EL]}{[E_o]} = \frac{K_b[L]}{1 + K_b[L]} = \theta \qquad (2)$$

(fraction of enzyme bond to sorbent). Here we treat the ligand concentration as the variable for some fixed amount of enzyme input, $[E_o]$. If we assume the same system considered in FIG. 13 at $[L] = 50$ m*M*, we find $\theta = 0.98$. That is, 98% of the enzyme will be bound to a gel which contains 50 μeq ligand/ml gel. At $[L] = 10$ m*M*, we find that 91% of the enzyme is bound. These calculations suggest that if a ligand has $K_i = 5$ m*M*, then a gel must contain at least 50 μeq ligand/ml gel to effect good binding of enzyme. Note that with this model we have not even taken into account the markedly low availability of a ligand that is immobilized into a gel.

With ligands exhibiting K_i larger than 1 mM, the problem is that of coupling ligands into a gel to high densities. Here there may be a practical limit. We have observed, as shown in Table 4, that the coupling of organic amines to agarose gel beads with cyanogen bromide appears to have distinct limits [13]. We have calculated the maximum hydroxyl concentration in the various gels from the density and the mean residue weight of the fundamental disaccharide unit. If we assume that only the 1,3-diol structure (see FIG. 4) is involved in the coupling, then the maximum possible functional group concentrations would be 0.073, 0.146, and 0.219 M, respectively, for the 2, 4, and 6% gels. However, the observed maxima for covalent bonding of leash functions in these gels are considerably lower. While the reasons for this are not clear, one might expect that the gel structure of agarose, which arises from inter- and intrachain double helices involving hydrogen-bonding [41], lowers the availability of hydroxyl groups for chemical reaction.

While the exclusion molecular weights for each of the gels listed in Table 4 relate to gel porosity, we have no idea of the _distribution_ of hole sizes in any of the gels. This information would be useful in the preparation of affinity gels for enzymes of

TABLE 4

Agarose Parameters

Gel density (%)	Max[ROH] (M)	MW exclusion limit (daltons)	$[L]_{limiting}$ (μeq/ml)
2	0.292	50,000,000	~5
4	0.583	15,000,000	15-20
6	0.875	5,000,000	50-60

very high molecular weight. Implicit in the requirement for high porosity [4] is the notion that the apparent ligand concentration in a gel would vary according to the hydrodynamic volume of the enzyme being tested. For a gel with a given $[L_o]$, the $[L]^{eff}$ would be higher when measured by a low molecular weight enzyme than when tested by one with a high molecular weight. The lack of quantitative data on gel porosity presently precludes answers to questions such as: Can an enzyme of 300,000 daltons and possessing a single active-site be expected to bind to a 2% agarose gel containing 5 μeq ligand/ml with K_i = 100 μ<u>M</u>?

<u>In-gel parameters by Klotz plots</u>

In preceding sections we have dealt with values for K_i (or K_d) which have been determined for the soluble inhibitor or ligand. The values for [L] have been determined chemically on the bulk of the gel at hand. For better understanding of the affinity process as well as to better predict by the design the utility of a particular sorbent, it would be useful to determine the two parameters in situ or in gel. Where a given sorbent does not seem to bind an enzyme very well, such an analysis might help to explain why. With affinity gels that work, the in-gel analysis of K_i and [L] is a worthwhile documentation that the system is functioning properly.

Our analytical approach was suggested by the work of Hirose and Kano [42] who analyzed the partitioning of soluble ligands between adsorption to a protein and permeation of gel beads. FIG. 15 shows the manipulation of the equation describing enzyme

$$R = \frac{[EL]}{[E]} = \frac{K_b[L_o]}{1+K_b[E]}$$

as reciprocals

$$\frac{1}{R} = \frac{1+K_b[E]}{K_b[L_o]} = \frac{1}{K_b[L_o]} + \frac{[E]}{[L_o]}$$

COMPARE TO KLOTZ PLOT:

$$\frac{1}{R} = \frac{K}{N}\,\frac{1}{[D_o]} + \frac{1}{N}$$

FIG. 15 [42, 43].

adsorption to affinity gel. We define the ratio R = [EL]/[E]. A plot of 1/R vs [E] should give us a straight line with slope = $1/[L_o]$, and the ordinate intercept should be $1/K_b[L_o]$. In form, our equations and plots closely resemble those originated by Klotz in 1946 [43]. FIG. 16 is an example of the analysis applied to a trypsin affinity gel where $[L_o]$ = 19.2 μeq/ml gel of 4% agarose. Because most enzymologists usually deal with dissociation constants, we have handled our calculations in terms of K_d which is equal to the reciprocal of K_b and has units of mole/liter. The results show an affinity gel dissociation constant of 1.4×10^{-5} $\underline{M}$ and an effective in-gel ligand concentration of 0.44 μeq/ml gel. As in the saturation binding experiments (see the section entitled "Limits of Binding"), we again observe the effective ligand concentration to be but a small fraction of the chemically determined values.

Table 5 lists the results of this analysis applied to several different gel preparations. Generally there is a correlation between the $[L_o]$ and $[L_o]^{eff}$. However, the values for the K_d^{app}

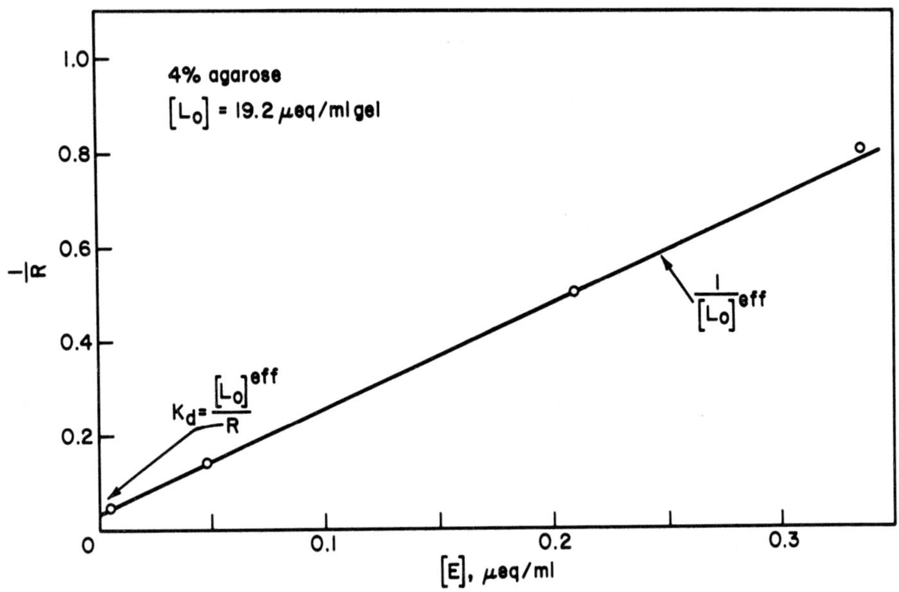

FIG. 16 Trypsin binding to mBz gel.

appear to be somewhat erratic when comparing different gel preparations. We have no ready explanations at present for this large variation. The best binding gel binding gel ([L] = 48.9) shows a K_d^{app} value approaching that of the free solution inhibition constant (= 8 X 10^{-6} M) for m-aminobenzamidine. But mostly the values are in the range of 50 to 100 μM. Clearly, more preparations must be examined by this technique to ascertain the significance of these variations.

Elution Ratio

In addition to static adsorption methods for measuring affinity processes, it is also of interest to have chromatographic parameters which are measured under dynamic (operating) conditions

TABLE 5

In-gel Binding Parameters for Trypsin Sorbents

Agarose (%)	$[L_o]$ (μeq/ml)	$[L_o]^{eff}$ (μeq/ml)	K_d^{app} (m$\underline{M}$)
6	48.9	0.41	.0049
6	22.9	0.19	.086
6	11.1	0.095	.143
6	4.9	0.125	.36
4	19.2	0.44	.014

in a packed column. One attempt to relate enzyme binding to chromatographic parameters was reported in 1972 by Hixson and Nishikawa [44]. Another approach is outlined below starting from an expression used in adsorption chromatography [46] for the retention volume, which we will call the elution volume:

$$V_e = V_o + K_p V_g$$

The partition coefficient, Kp, is the ratio (adsorbed enzyme)/(unadsorbed enzyme); V_o is the volume of the mobile phase; and V_g is the volume of the sorbent. For affinity sorbents derived from agarose, there is present a gel-filtration effect on the enzyme in addition to affinity sorption. Thus

$$V_e = V_o + K_{av} V_i + K_p V_g$$

where K_{av} and V_i are distribution constant and imbibed volume, respectively (standard terms in gel-filtration theory). With 4%

agarose gel, $K_{av} = 1$ for an enzyme-like trypsin. Thus the "straight through" elution volume of the protein (without any affinity sorption) is $V_s = V_o + V_i$. Now $K_p = [L]/K_d$, terms which are defined in the adsorption isotherm model. On making the substitutions and factoring, we obtain the expression for the <u>elution ratio</u>:

$$\frac{V_e}{V_s} = 1 + \frac{[L]}{K_d} \frac{V_g}{V_s} \tag{3}$$

Experimentally, we have observed that the gel bed volume, V_g, is approximately equal to V_s. The effective ligand concentration, however, is only about 1% of the chemically determined value, as we noted in a preceding section. Thus the selection of V_g as well as [L] presents a problem. For the sake of simplicity, the calculation of the elution ratio in the experiments described in FIGS. 17 through 20 have assumed that V_g and V_s are equal. Furthermore, the calculations in FIGS. 18 through 20 have used affinity parameters which have been obtained from Klotz plots.

Since the elution ratio is readily measured experimentally, we have a feasible way of assessing the ratio of affinity parameters, [L]/Kd, the value of which can be also obtained by the Klotz plot. In FIG. 17 is shown the pH dependence of the elution ratio as well as the K_d and the $[L_o]^{eff}$ for a single affinity gel preparation. This sorbent for trypsin was made from 6% agarose and contained 19.1 μeq of m-aminobenzamidine/ml gel.

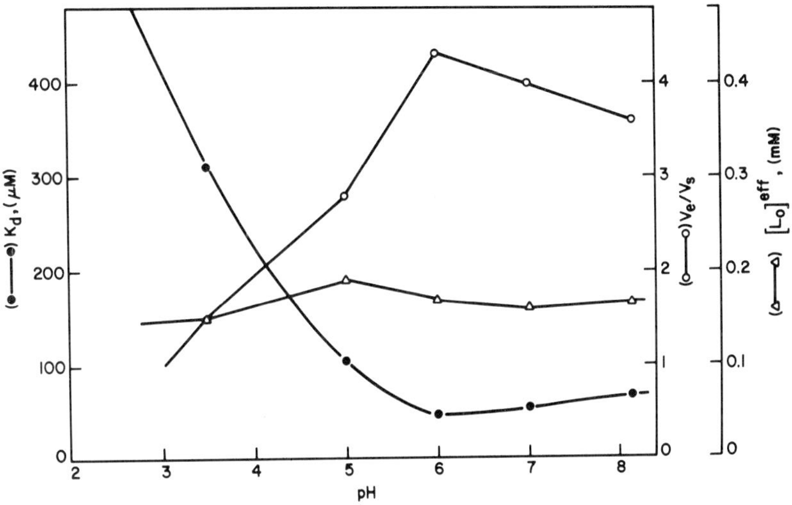

FIG. 17 Trypsin sorbent: In-gel binding parameters vs pH.

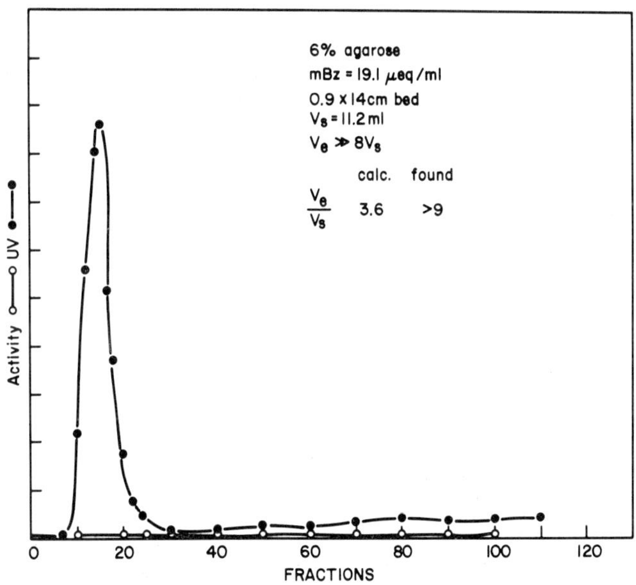

FIG. 18

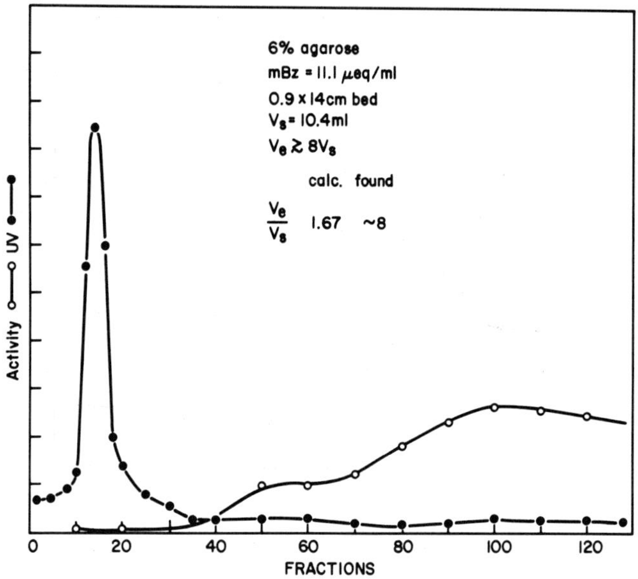
6% agarose
mBz = 11.1 μeq/ml
0.9 x 14cm bed
V_s = 10.4ml
$V_e \gtrsim 8V_s$
calc. found
$\frac{V_e}{V_s}$ 1.67 ~8
Activity ○—○ UV ●—●
0
20
40
60
80
100
120
FRACTIONS

FIG. 19

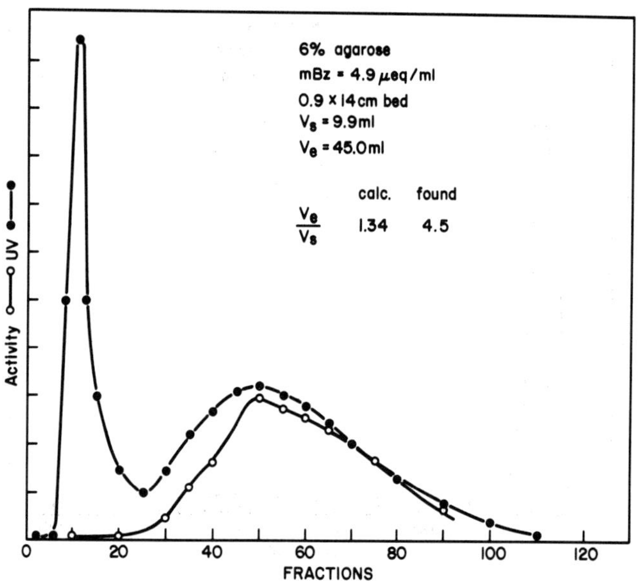
6% agarose
mBz = 4.9 μeq/ml
0.9 x 14cm bed
V_s = 9.9ml
V_e = 45.0ml
calc. found
$\frac{V_e}{V_s}$ 1.34 4.5
Activity ○—○ UV ●—●
0
20
40
60
80
100
120
FRACTIONS

FIG. 20

At lower pH the elution ratio declines due to poorer binding of the ligand by the enzyme, which is also indicated by an increase in the K_d at low pH. And as it should be, the effective ligand concentration remains the same throughout the pH range tested. Thus even though a coherent pattern in the variation of these parameters is not evident when comparing a variety of sorbent preparations, such as in Table 5 when the analyses are applied to a single gel preparation, meaningful information is revealed.

The elution ratio is further examined in a series of chromatograms presented in FIGS. 18 through 20, where each was obtained with a different gel preparation. At the higher ligand concentration the enzyme binds well, and correspondingly the elution ratio is very large. At lower ligand concentrations the enzyme binds more weakly to the sorbent and this is accompanied by a correspondingly lower elution ratio. In these examples there is a striking difference between calculated values and the experimentally observed data. The observed values are roughly 3.5 to 4.5 times greater than the calculated ones. While not completely ruled out, the high ionic strength (0.5 M) of the buffers used should diminish the possible Donnan-equilibrium effects [45] due to the immobilized benzamidine. Regardless of the present reasons for the quantitative discrepancies, the determination of the elution ratio seems to be a useful measure for predicting chromatographic utility of an affinity gel. The ratio value is fairly revealing for a very poorly binding gel as well as for the very strongly binding one. Certainly more work is needed to assess its usefulness on gels with marginal performance characteristics.

General Comments

It is interesting to note that the Langmuir adsorption isotherm model is a reasonable description of the affinity process. The chromatogram in FIG. 20 agrees with the shape predicted for a convex isotherm from adsorption theory [46] as shown in FIG. 21.

To summarize the various considerations that we have given to the quantitative parameters in affinity chromatography, we offer some calculated predictions in Table 6. The values for the limiting K_d's have been obtained using Eq. (2) and (3). The effective ligand concentration has been estimated as 1% of the chemically determined ligand concentration. The apparent ligand concentration has been estimated as 3.5X the effective [L]. The K_d values are somewhat lower than 1 mM and differ between static and dynamic binding experiments. These values are predicted limits for "true" affinity binding. Adventitious ionic and

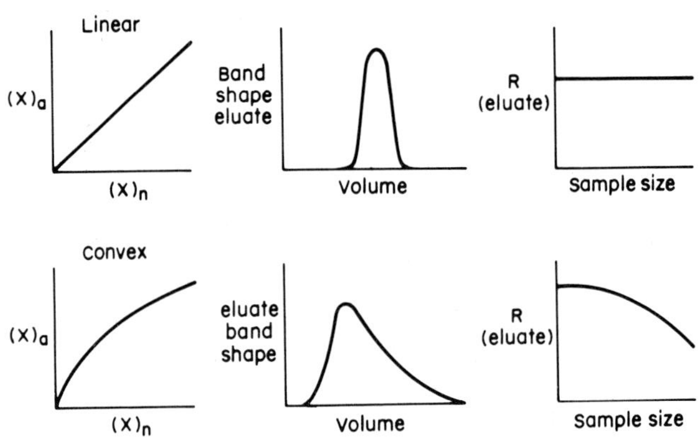

FIG. 21 Isotherm type and chromatographic parameters [46].

TABLE 6

Calculated Limits for K_d of Useful Ligands

Agarose (%)	$[L]^{eff}$ a (mM)	$[L]^{app}$ b (mM)	K_d^{stat} c (μM)	K_d^{dyn} d (μM)
6	0.6	2.1	67	230
4	0.2	0.7	22	77
2	0.05	0.175	5.6	19

a $[L]^{lim}$ X 0.01 (see Table 4 data).
b $[L]^{eff}$ X 3.5.
c Calculated where $\theta > 0.9$.
d Calculated where $\frac{Ve}{Vs} > 10$.

hydrophobic effects could raise or lower these values. With ligands whose K_d values are greater than those predicted in Table 6, useful sorbents may still be obtained. We expect that differential migrations on these sorbents could still offer a practical isolation of the desired protein.

For the near future in affinity chromatography we expect that more quantitative data will be forthcoming so that the design of affinity sorbents can be put on a more rational basis. We anticipate that new hydrophobic gel derivatives which are free of undesired ionic and hydrophobic groups will become available. The synthesis of new hydrophobic polymers, such as polyhydroxy ethyl methacrylate, will be welcome, especially in macroreticular form, since the need continues for mechanically stable carriers. These developments then should make large-scale and commercial applications of affinity chromatography a practical endeavor.

REFERENCES

[1] B. R. Baker, Design of Active-site-Directed Irreversible Enzyme Inhibitors, Wiley, New York, 1967, p. 301.

[2] P. Cuatrecasas, M. Wilchek, and C. B. Anfinsen, Proc. Natl. Sci. U.S.A., 61, 636 (1968).

[3] P. O'Carra, in Industrial Aspects of Biochemistry (B. Spencer, ed.), Federation of European Biochemical Societies, Amsterdam, North-Holland Publ. Co., 1974.

[4] P. Cuatrecasas, J. Biol. Chem., 245, 3059 (1970).

[5] H. F. Hixson, Jr., and A. H. Nishikawa, Arch. Biochem. Biophys., 154, 501 (1973).

[6] M. Robert-Gero and J-P. Waller, Eur. J. Biochem., 31, 315 (1972).

[7] T. L. Rosenberry, H. W. Chang, and Y. T. Chen, J. Biol. Chem., 247, 1555 (1972).

[8] M. Wilchek, Adv. Exp. Med. Biol., 42, 15 (1974).

[9] R. Axen, J. Porath, and S. Ernback, Nature (London), 214, 1302 (1967).

[10] J. Porath, Ibid., 218, 834 (1968).

[11] L. Ahrgren, L. Kagedal, and S. Akerstrom, Acta. Chem. Scand., 26, 285 (1972).

[12] G. J. Bartling, H. D. Brown, L. J. Forrester, M. T. Koes, A. N. Mather, and R. O. Stasiw, Biotech. Bioeng., 14, 1039 (1972).

[13] A. H. Nishikawa and P. Bailon, Anal. Biochem., 64, 268 (1975).

[14] E. Steers, Jr., P. Cuatrecasas, and H. B. Pollard, J. Biol. Chem., 246, 196 (1971).

[15] G. R. Craven, E. Steers, Jr., and C. B. Anfinsen, Ibid., 240, 2468 (1965).

[16] A. H. Nishikawa and P. Bailon, Arch. Biochem. Biophys. 168, 576 (1975).

[17] P. J. Robinson, P. Dunnill, and M. D. Lilly, Biochim. Biophys. Acta, 285, 28(1972).

[18] P. O'Carra, S. Barry, and T. Griffin, Biochem. Soc. Trans., 1, 289 (1973).

[19] S. Blumberg, J. Hildesheim, J. Yariv, and K. J. Wilson, Biochim. Biophys. Acta, 264, 171 (1972).

[20] P. O'Carra and S. Barry, FEBS Lett., 21, 281 (1972).

[21] Y. Dudai, I. Silman, N. Kalderson, and S. Blumberg, Biochim. Biophys. Acta, 268, 138 (1972).

[22] J. Schmidt and M. A. Raftery, Biochemistry, 12, 852 (1972).

[23] P. Cuatrecasas and C. B. Anfinsen, Methods Enzymol., 34, 345 (1971).

[24] S. Barry and P. O'Carra, Biochem. J., 135, 595 (1973).

[25] B. H. J. Hofstee, Anal. Biochem., 52, 430 (1973).

[26] R. J. Yon, Biochem. J., 126 765 (1972).

[27] Z. Er-el, Y. Zaidenzaig, and S. Shaltiel, Biochem. Biophys. Res. Commun., 49, 383 (1972).

[28] S. Hjerten, J. Chromatogr, 87, 325 (1973).

[29] S. Shaltiel and Z. Er-el, Proc. Natl. Acad. Sci. U.S.A., 70, 778 (1973).

[30] I. Silman and E. Katchalski, Ann. Rev. Biochem., 35, 873 (1966).

[31] K. Stevenson and A. Landman, Can. J. Biochem., 49, 119 (1971).

[32] A. R. Thompson and E. W. Davie, Biochim. Biophys. Acta, 250, 210 (1971).

[33] A. Chu and S. Chaykin, Adv. Exp. Med. Biol., 42, 165 (1974).

[34] H. F. Hixson, Jr., and A. H. Nishikawa, Fed. Proc., 30, 1078 (1971), Abstract.

[35] D. J. Graves and Y-T. Wu, Methods Enzymol., 34, 140 (1974).

[42] M. Hirose and Y. Kano, Biochim. Biophys. Acta, 251, 376 (1971).

[43] I. M. Klotz, J. Am. Chem. Soc., 68, 1486 (1946).

[44] H. F. Hixson and A. H. Nishikawa, A.I.Ch.E. Meeting, New York, November 1972.

[45] L. W. Nichol, M. Janado, and D. J. Winzor, Biochem. J., 133, 15 (1973).

[46] L. R. Snyder, Principles of Adsorption Chromatography, Dekker, New York, 1968, Chap. 2.

[36] P. C. Wankat, Anal. Chem., 46, 1400 (1974).

[37] C. R. Lowe, M. J. Harvey, and P. D. G. Dean, Eur. J. Biochem., 42, 1 (1974).

[38] A. H. Nishikawa, P. Bailon, and A. H. Ramel, Adv. Exp. Med. Biol., 42, 33 (1974).

[39] M. J. Harvey, C. R. Lowe, D. B. Craven, and P. D. G. Dean, Eur. J. Biochem., 41, 335 (1974).

[40] P. L. Whitney, Anal. Biochem., 57, 467 (1974).

[41] D. A. Rees, Chem Ind. 1972(16), 630 (August 19).

SYMPOSIUM ON POLYMER GRAFTS IN BIOCHEMISTRY

INTRODUCTION TO SECTION II--BIOACTIVE MATERIALS

The first section in the symposium dealt with one important class of biologically active polymers, those used for affinity chromatography. This second section surveys a number of other highly significant and scientifically intriguing areas in the broad field of biologically active polymeric materials. These areas include:

1. Chemical modification and immobilization of enzymes, antigens, and hormones.

2. Localization and controlled release from polymeric compositions of bioactive agents such as drugs, hormones, or pesticides.

3. Synthesis of biopolymers such polypeptides through monomer sequence development on synthetic polymer substrates (e.g., the Merrifield-type preparation of polypeptides).

4. Synthesis of artificial cells or membranes which mimic natural biological systems.

5. Polymeric compositions which have a detoxifying function by sequestering or chelating toxic agents.

These papers in Section II touch upon many aspects of these areas and suggest new opportunities at the interface between polymer science and biological science. It was, in fact, a primary objective in structuring this symposium to expose the "traditional" polymer scientist to some of these exciting research opportunities and thereby provoke more attention to some of the key problems which now beg solution. By way of further introduction to this section, brief comments concerning each of the papers follow.

Immobilized Proteins and Peptides by Falb and Grode surveys the preparation and application of enzymes and other natural biopolymers when attached to insoluble polymeric substrates. There are numerous implications for the use of such materials in catalysis, therapy, analytical and diagnostic chemistry, production and purification of biological materials, and the fundamental understanding of biological functions. Production of these biomaterials, is of course, at the heart of the affinity chromatography techniques as well. Despite the considerable progress reported since the pioneering work of E. Katchalsky, one can judge that this facet of biomaterials research is yet in its infancy and developing rapidly.

The Controlled Release of Bioactive Compounds for Polymeric Systems by Allan, Friedhoff, McConnell, and Powell concerns a vitally important and burgeoning facet of the biomaterials field. Ability to localize and control the release of biologically active agents through chemical attachment or admixture with polymeric hosts affords opportunities for controlling the application of useful hazardous or toxic materials. In the case of pesticides which are so important to agricultural development, but often dangerous in widespread application, this approach to control will undoubtedly achieve a major step forward in the future utilization of pesticides. In fact, the microencapsulation of pesticides has recently won final approval from the EPA for commercial marketing. Pennwalt Corp. has introduced a product, Penncap-N, for use on cotton and alfalfa which utilizes a cross-linked polyamide to microencapsulate a methyl parathion insecticide for control of activity.

Another important example is the controlled release of drugs or hormones and the localization in the body of substances which are harmful when introduced systematically. The recent development by Alza Corp. of a contraceptive device (the "Progestasert" System) based on controlled release of contraceptive hormones from a plastic uterine insert is likely to become a widespread practical achievement in this field of research. FDA approval has been obtained recently for this system.

A treatment for glaucoma also developed by Alza Corp. is yet another current example of the possibilities inherent in precisely controlling the release of a biological agent from a polymeric composition. The system uses a laminated polymeric wafer containing pilocarpine. The FDA has also approved the marketing of this "Ocusert" material.

Enzymes Immobilized by Microencapsulation within Spherical Ultrathin Polymeric Membranes by Chang represents the first step toward the preparation of cellular structures which exhibit biological activity and which may function either in-vivo or in-vitro to achieve enzymatic or detoxifying functions or to introduce and localize hormonal or immunochemical activity. The preparation and the study of physical behavior of polymeric membranes which is inherent in this research is also, of course, of broad implication to the understanding of biological functions.

Solid Phase Peptide Synthesis by Stewart reviews the development and recent progress in the synthesis of polypeptides, particularly by the Merrifield technique, which has resulted in automated instrumentation for the synthesis of complex polypeptides. The achievement of such syntheses by Merrifield and co-workers has also demonstrated one crucially important point in clarifying property-structure relationships for biological materials. This is the fact that once the enzymatic polypeptide chain structure is achieved through synthesis, its molecular conformation, which is so important to its function, is a natural energetic consequence of its molecular structure. The opportunity afforded by the solid phase peptide synthesis technique to produce new polypeptides as well as duplicate natural polypeptide structures has opened up a new era in the field of biopolymers as well as in molecular biophysics and biochemistry. One possible defect of this technique, however, is the occasional "mistake" in sequence development. The use of a homogeneous solution attachment method which aboids this problem is under investigation by Patchornik and co-workers at the Weizmann Institute.

The reported isolation of polypeptides from the brain of animals after particular learning or stress situations by Unger at Baylor and Routtenberg, et al. at Northwestern creates the prospect, both intriguing and frightening, that there are chemical "codes" to behavior and that our ability to synthesize and measure biopolymers with greater facility today may produce a new behavioral science with a strong biophysical and biochemical basis.

Rigid Support Materials for the Immobilization of Enzymes by Royer, Green, and Sinha is devoted to a more detailed consideration of enzyme immobilization and serves to provide an insight into the complexity of the chemistry and physics associated with attaching a biologically active agent such as an enzyme to a polymeric substrate.

The last paper, Polymer-Drug Grafts for Iron Chelation by Ramirez and Andrade, looks at a specific system for detoxification; iron removal from the circulatory system. Although this paper focuses on a particular polymeric system for metal chelation detoxification, the broad ranging implications of this approach are obvious for the removal of other metallic as well as organic and metabolic toxic agents which occur by accidental ingestion or by biological malfunction.

Finally, some further comments are appropriate. Other important facets of the field embraced by the synthesis and properties of polymeric agents possessing biological activity which are not so clearly represented in this symposium include:

1. Development of "artificial organs" utilizing combinations of polymeric materials having selective surface and permeability characteristics with living cells as, for example, reported recently by Dr. Wolf of the New York Hospital-Cornell Medical Center with liver cells grown on hollow fibers or the work of Dr. Chick at Harvard using pancreas cells also grown on hollow fibers.

2. Synthesis and characterization of polymers which intervene in the body's natural immunochemistry to enhance resistance to viral attack, carcinogenesis, etc. This is an important future thrust in the field of bioactive polymers. An example is the study

of pyran polymers and polysaccharides which have been demonstrated to affect Interferon stimulation. Understanding Interferon induction and the study of property-molecular structure relationships in polymers which affect the human immunochemical system are clearly some of the more exciting areas for further study and progress.

This symposium specifically did not concern itself with polymeric Biocompatible Materials which represents a major area of investigation of both scientific and practical importance. Indeed, the issues of biocompatibility and bioactivity are necessarily intertwined in the design and development of materials having improved biological function as well as for the important prosthetic applications of polymers. The chemistry and physics of polymer biocompatibility is receiving rapidly increasing attention. Although not represented in this symposium because of other recent surveys of the field, biocompatibility deserves emphasis as one of the high priority areas of biopolymer research.

I am indebted to the authors for their fine contributions, their help and cooperation, and especially their patience in developing the symposium and this collection of expanded papers. Very special thanks are due Dr. Harry Hixson my cochairman in this adventure.

Eugene P. Goldberg

Department of Materials Science
and Engineering
University of Florida
Gainesville, Florida 32611
September 1, 1975

IMMOBILIZED PROTEINS AND PEPTIDES

R. D. Falb and G. A. Grode

Department of Biology and Medical Sciences
Battelle-Columbus Laboratories
Columbus, Ohio 43201

In recent years a great deal has been learned about the structure and function of proteins and about the roles they play in the complex process of a living organism. The classical approach of the biochemist in studying a given protein is to free it from the cellular species. The properties and structure of the protein in solution are then examined and analyzed. While this approach is logical and has yielded much information about proteins, one should always keep in mind that in nature a great many proteins function not in solution, but at an interface or within solid state assemblages in cells. The protein removed from a surface is thus often not in its natural environment and can display an altered reactivity and stability.

The attachment of proteins and peptides to solids has progressed rapidly in recent years from a laboratory curiosity to a widely useful technique in both laboratory research as well as for medical and industrial applications. Proteins and enzymes immobilized in various ways now find increasing use in giving valuable insight into basic interactions of proteins and, in addition, have many applications in medicine, clinical analysis, and synthetic chemistry. Many different types of proteins including enzymes, antibodies, enzyme inhibitors, proteinaceous antigens, and peptide hormones have been shown to possess biological activity. More than 50 enzymes have been attached to polymers to make preparations which have many applications because their insolubility makes them easily separable from reaction mixtures and generally renders them more stable than the corresponding soluble polymers. Solid-phase antibodies now find use in diagnostic methods and also are valuable laboratory tools for purification of antigens. Conversely, immobilized antigens can also be used for analytical and preparative purposes. In general, the basic rational for the attachment of a protein or peptide to a solid phase is twofold: (1) to provide a method for conveniently manipulating the protein while making use of its biological properties, or (2) altering or localizing the behavior of the biopolymer.

The field of immobilized proteins requires the skills of both polymer chemists and biochemists. In attaching proteins to the surfaces of polymers, the most common approach is to form a covalent bond between a reactive group on the surface of the polymer and a group on the protein. A very large number of synthetic polymers including polystyrene, ethylene-maleic anhydride, phenolic resins, polyacrylamide, nylon, silicone rubber, polyethylene, polyaminoacids, and polyhydroxyethyl acrylamide have been used as matrices for protein attachment. In general, the polymer is derivatized to contain an active species such as an aliphatic or aromatic amino, hydroxyl, carboxylic, or sulfhydryl group. The proteins contain amino, carboxyl, sulfhydryl, hydroxyphenyl, and hydroxyl groups. By the appropriate choice of activating of coupling agents, these moieties can be joined to appropriate groups on the polymers. For example, one of the most common methods for protein attachment involves utilization of the protein's free amino groups which act as nucleophiles to attack surface-bound activated carboxyl groups (i.e., acid halides, acid azides), activated aliphatic or aromatic halides, and cyclic iminocarbonate esters. In coupling reactions, care must be taken to avoid inactivation of the protein and thus one must carefully select appropriate solvent systems and utilize reactive moieties which do not block the active site of the protein molecule.

Research activity in methods of attachment and applications for immobilized proteins and peptides has been very active in the last 5 years. In the field of immobilized enzymes alone, a literature survey conducted by the authors disclosed over 400 references from June 1973 until the end of 1974. Much of the work on immobilized enzymes is conducted in industrial laboratories and is not available in the open literature. It is beyond the scope of this review to discuss all of the methods available for protein immobilization. The reader is referred to a monography by Zaborsky [1] for a comprehensive review of the methods of preparation and properties of immobilized enzymes. Numerous reviews important to surveying this field have been written and are referenced [2-18].

When the protein has been attached, the composite has many interesting and even exciting properties. Applications for immobilized enzymes are being rapidly developed for food and beverage processing, synthesis of drugs and fine chemicals, and analysis of constituents of body fluids. For synthetic purposes, the polymer-enzyme adduct may be regarded as a heterogeneous catalyst and placed in a column, in a suspension, or coated on the walls of reaction vessels and conduits.

The attachment of a protein to a solid surface may be done in such a way to ensure that the resulting adduct preserves the biological activity of the protein. This requires that the active site of the protein be accessible to the surrounding milieu and not be buried in a pore or blocked by some other component of the surface. It also requires that the reactive groups that comprise the active site of the protein not be modified chemically during the attachment reaction. Another feature which is desirable in most cases is that the protein be irreversibly bonded to the solid phase under the conditions in which the adduct will function. For example, the accuracy of a solid-phase antibody used in an immunoassay system is severely compromised if the antibody desorbs from the solid. An immobilized enzyme system will lose activity over a time period, and the product could be contaminated by the enzyme if the enzyme is not irreversibly attached to the solid phase.

GENERAL METHODS OF IMMOBILIZATION

Proteins have been attached to solid phases in a variety of configurations:

(1) Covalent bonding

(2) Physical adsorption

(3) Cross-linking at solid surfaces

(4) Cross-linking with difunctional reagents

(5) Inclusion in a gel phase

(6) Encapsulation

Because of the desirability of irreversible attachment, most of the work done with protein immobilization has been with covalent bonding. For this purpose, a very large number of synthetic and natural polymers and inorganic materials have been used. Generally, covalent attachment proceeds through a functional group on the surface reacting with a reactive site on the protein, usually a carboxyl, amino, hydroxyl, sulfhydryl, or hydroxyphenyl group. The surfaces to which the proteins are to be attached can be modified in numerous ways to obtain the desired functional groups. Thus a very large number of approaches can usually be used in the attachment of a given protein to a surface. The choice of the method of attachment depends upon the reaction conditions which can be tolerated by the protein, the ease of the attachment sequence, and the availability of reactive groups on the protein for attachment. With respect to the latter point, one usually tries to avoid reaction with a group at the active site which is essential for biological activity. Thus, if an amino group is an essential element of the protein's active site, attachment reactions which involve other groups on the enzyme are usually utilized for the attachment or the amine groups are reversibly blocked during the attachment sequence.

As mentioned previously, one of the primary requirements for utilizing proteins bonded to solids is that the active site be accessible. To achieve this, sometimes it is necessary to place the attached protein a certain distance away from the solid surface.

This is usually done by means of a tether linkage, one end of which is attached to the surface and the other end of which is attached to the protein. Tethers of varying lengths have been used in the attachment of proteins, and very often the length of the tether has a profound effect on the biological activity of the proteins.

The desirable amount of protein per given volume of solid varies according to the intended use of the solid-phase adduct. Generally for immobilized enzymes, one wishes to have as large an amount of active enzyme per given weight as possible. The same is true for solid-phase antibodies and antigens which are used for adsorption and purification processes. If an adduct has a relatively low amount of protein per given weight, a column of this material used for synthetic or preparative purposes will be unnecessarily large and the process will thus be inefficient. As a rule of thumb, it is desirable to have at least 100 units of enzyme activity per gram of solid to avoid excessive bulkiness. (A unit of enzyme activity is defined as 1 micromole converted per minute.)

In cases where the solid-phase proteins are used for analytical purposes, the most important consideration is usually the reproducibility of protein binding per given unit surface area. For example, in solid-phase radioimmunoassay the antigen being assayed is usually present in very small amounts. The antibody used to bind the antigen at the surface needs to be present in corresponding amounts and thus the surface coverage of the antibody is quite small. One needs to know very accurately the amount of antibody per given area or the test will not be accurate.

A number of solid-phase materials specifically prepared for protein attachment are now available commercially. These include derivatized porous glass beads, various types of derivatized cellulose, Sephadex and agarose particles, and acrylamide gels. One of the most commonly used methods at present involves attachment to materials containing adjacent hydroxyl groups such as cellulose, Sephadex, or agarose by use of cyanogen bromide activation. This method has been used for a very large number of different proteins and appears to have good general applicability. In the attachment of protein to any solid, an important considration is the amount of active protein which can be recovered in the solid-phase adduct. Large amounts of biological activity are frequently lost in the attachment sequence, and the process can be thus unfeasible economically. The cyanogen bromide process usually gives good retention of biological activity in the adduct.

Immobilized enzymes are now being used in the synthesis and resolution of amino acids, and contemplated for use in starch saccarification, beer deproteinization, fruit juice clarification, preparation of invert sugars, and carbohydrate synthesis. For the last several years our laboratory has been engaged in research on the use of immobilized enzymes for the synthesis of carbohydrates in space cabin environments [19,20]. The utilization of enzymes for many of the above applications has generated a need for polymer materials which are relatively inexpensive, can contain a high amount of enzyme, and are amenable for use as column packings, vessel surfaces, fibers, etc.

Immobilization of enzymes may also provide a method of efficiently utilizing the inherent high specificity of many enzymes for the sensitive detection of constituents of body tissue and fluids. An enzyme may be attached to the surface of a pH or ion-specific electrode and catalyze a reaction with a specific substrate which can be detected by that electrode.

The immobilization of antigens and antibodies has many possible practical consequences. Antigens attached to solid particles can be used for the preparation and purification of purified antibodies. Conversely, immobilized antibodies can be used to purify antigens. Such techniques will have future potential in removing undesirable antigens and antibodies in patients with certain diseases or possibly patients with transplanted organs. Immobilized antibodies also find use in clinical chemistry where they can be used to detect very small amounts of a given substance in blood, urine, or other body tissues. For example, our laboratory has been working on solid-phase assay methods for insulin and a gonadotrophic hormone which are based on an immobilized antibody. In the assay for insulin, antibody to insulin is covalently attached to the surface of a 20 μl glass capillary. The sample containing insulin is spiked with a measured amount of radiolabeled insulin and drawn into the capillary. The insulin originally present in the sample competes with the added radiolabeled insulin for antibody sites present on the surface of the capillary. By measuring the radioactivity of the capillaries after rinsing, one can determine the amount of insulin in the sample. This assay is sensitive to a few picograms of insulin. At present, however, the poor reproducibility of the assay resulting from varying amounts of attached antibody limits the usefulness of the assay.

In addition to enzymes, antibodies, and antigens, peptide hormones have been attached to polymers with retention of biological activity. For example, insulin has been immobilized and this adduct has been shown to be effective in stimulating glucose transport in cell membranes [21]. While the results of this work are now being disputed, they do point out the possibility of therapeutic effects of various immobilized hormones and enzymes.

It is not intended that this brief discussion be construed in any sense as a thorough and complete survey of this rapidly burgeoning field with its many exciting scientific and technological thrusts. It is our objective, however, to present at least a cross section of recent significant developments. Hopefully this will be useful to those new to the field as well as those active in immobilized protein research. The following highlights are presented in chart form for clarity and to minimize verbiage. In addition to these highlights, the Bibliography includes many additional references which are separated into a few major categories for clarity.

CHART I. Immobilization of cells in a collagen membrane. [W. R. Vieth, S. S. Wang, and R. Saini, Biotechnol. Bioeng., 15, 565 (1973).]

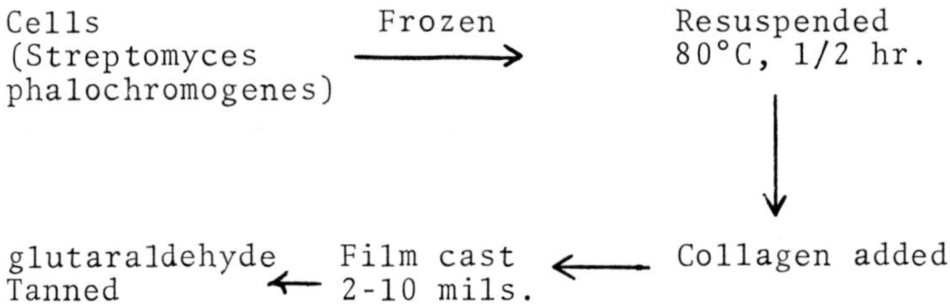

(131 μ/g glucose isomerase activity)

CHART II. Polyacrylamide/glutaraldehyde beads. [P. D. Weston and S. Avrameas, Biochem. Biophys. Res. Commun., 45, 1974 (1971).]

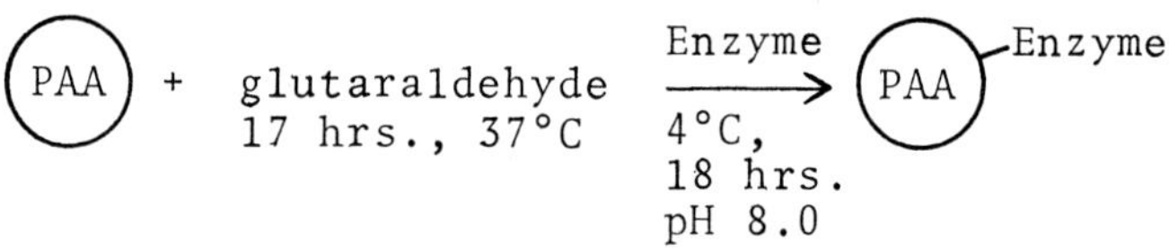

	mg. enzyme/g. polymer	% activity
Acid Phosphatase	18	81
Ribonuclease	20	30
Glucose Oxidase	7	56
Trypsin	18	58
Chymotrypsin	19	65

CHART III. Attachment to poly(allylcarbonate). [J. F. Kennedy, S. A. Barker, and A. Rosevear, J. Chem. Soc., 1971, 2468.]

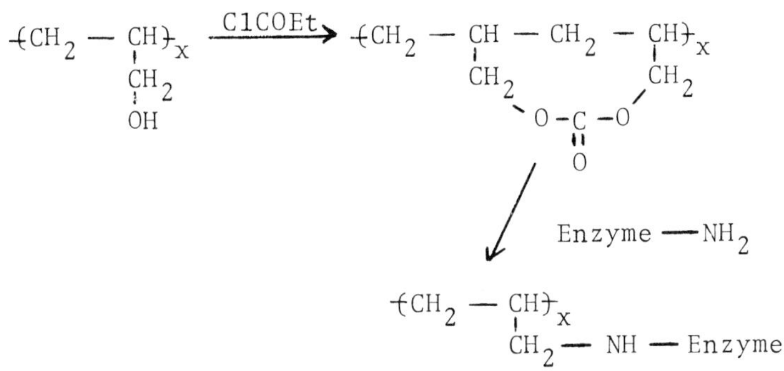

Enzyme	Concentration	Activity
α-Amylase:	2-6 μ/g polymer	2%
β-D-Glucosidase:	6-12 μ/g polymer	30%
Trypsin:	0.25-0.95 μ/g polymer	8%

CHART IV. Magnetic supports for enzymes. [P. J. Robinson, P. Dunhill, and M. D. Lilly, Biotechnol. Bioeng., 15, 603 (1973).]

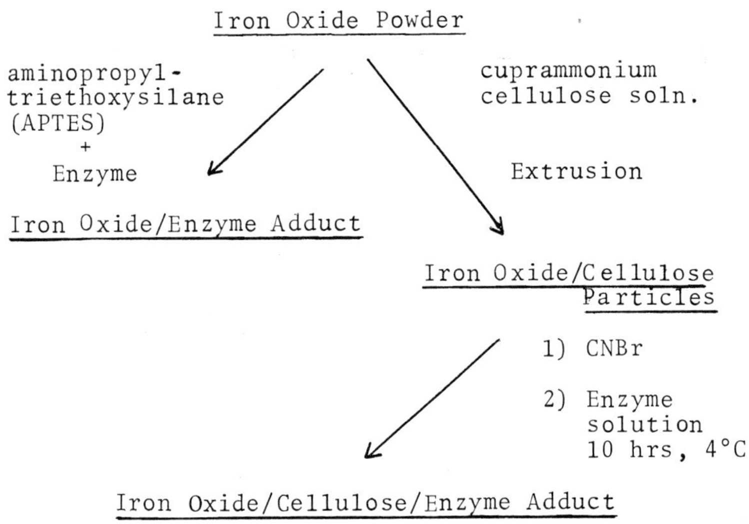

Enzyme	Cellulose	Silane
β-Galactosidase	900 μ/g	1825 μ/g
α-Chymotrypsin	309 μ/g	800 μ/g

CHART V. Enzyme coupling via metal salts. [A. N. Emery et al., Chem. Eng. (London), 1972, 71.]

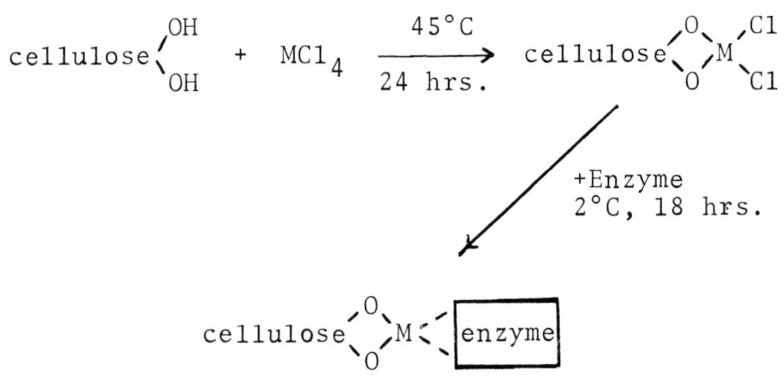

(Metal Salts: $TiCl_4$, $SnCl_4$, $SnCl_2$, $ZrCl_4$, VCl_3, $FeCl_2$)

CHART VI. Enzyme coupling via metal salts.

Enzymes Bound:		
	Amyloglucosidase	730-2780 μ/g
	Invertase	13-147 mg/g
	Trypsin	44-73% activity
	Amylase	
	Urease	
	Uricase	
	Glucose Oxidase	

Supports:	
	Borosilicate Glass
	Soda Glass Beads
	Filter Paper
	Cellulose
	Nylon-66 Fibers
	Brewers Yeast
	B. Subtilis
	E. Coli

CHART VII. Polyacryloyl-4-and-5-aminosalicylic acid. [J. F. Kennedy and J. Epton, Carbohydr. Res., 27, 11 (1973).]

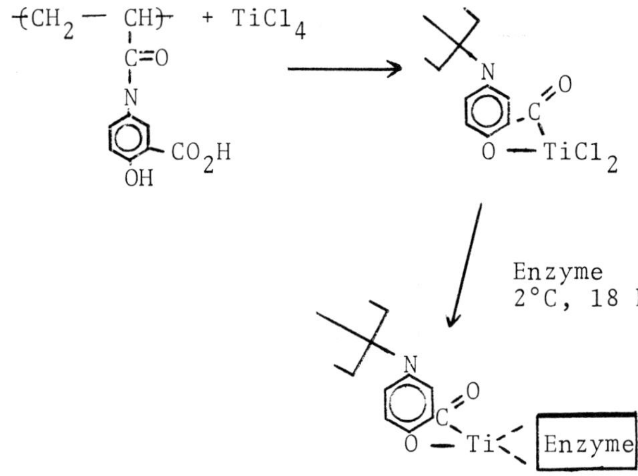

α-Amylase:	2270-4470 μ/g	displaced rapidly
Glucoamylase:	104-469 μ/g	stable 24 hrs./37°C

CHART VIII. Immobilization in nonaqueous solvents. [G. J. Bartling, H. D. Brown, and S. K. Chattopadhyay, Nature, 243, 342 (1973).]

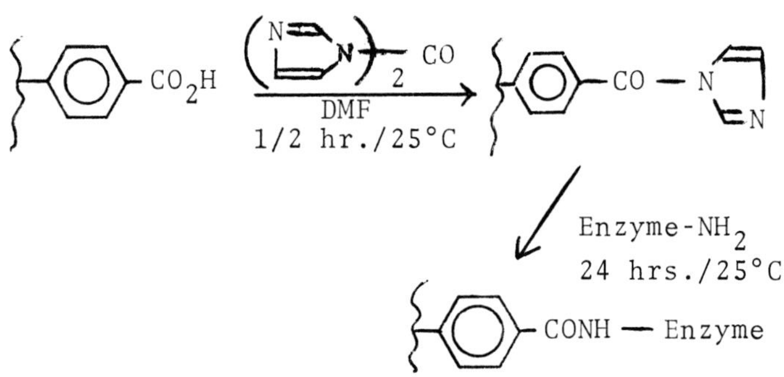

(Lysozyme: 25-35 mg./g. polymer, 40% activity)

CHART IX. Immobilization on phenolic resins. [A. C. Olson and W. L. Stanley, *J. Agric. Food Chem.*, **21**, 441 (1973); W. L. Stanley and R. Palter, *Biotechnol. Bioeng.*, **15**, 597 (1973).]

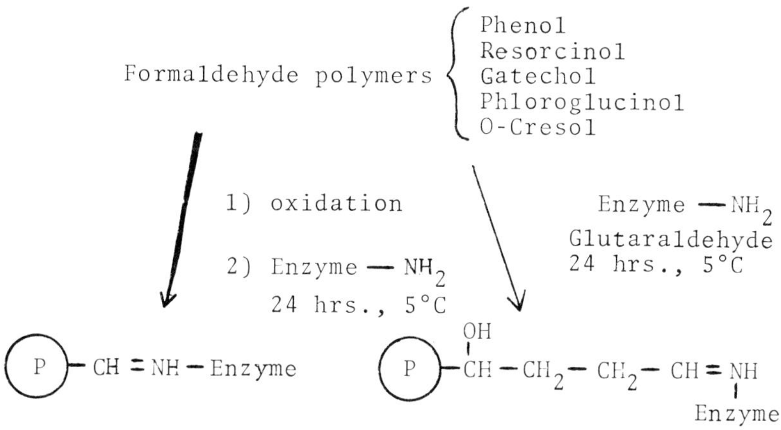

	μ/g. resin
Lactase	200
Invertase	1200
Glucoamylase	17
α-Chymotrypsin	75
Pronase	33

CHART X. Fiber-entrapped enzymes. [D. Dinelli, Process Biochem., 1972, 9.]

Enzyme solution ↘ Emulsion $\xrightarrow[\text{Coagulation Bath}]{\text{Spinneret}}$ Fiber/Enzyme
Polymer solution ↗

Polymers:

Cellulose Triacetate
Cellulose Diacetate
Ethyl Cellulose
γ-Methylpolyglutamate

Properties

Invertase: 160-1500 mg./g. polymer, 60-20% activity
1.75×10^5 μ/g polymer, 90% activity after 3 yrs. at 25°C

Penicillin Acylase
β-Tyrosinase
Tryptophan Synthetase
Glucose Oxidase
β-Galactosidase
Catalase

CHART XI. Other methods of enzyme insolubilization. [G. Broun, D. Thomas, G. Gellf, D. Doniurado, A. M. Berjonneau, and C. Guillon, Biotechnol. Bioeng., 15, 359 (1973).]

Reticulation/Inclusion

Enzyme + Glutaraldehyde $\xrightarrow[\text{pH 6.8}]{\text{4°C, 10 hrs.}}$ "Reticulated" Enzyme (soluble oligomers)

$\swarrow$ acrylamide
agarose, 30 min/4°C

Reticulated Enzyme Gel

(Example: urease 7.2 μ/ml. gel)

Crosslinking with Inactive Proteins

Enzyme + Protein (albumin, fibrinogen, hemoglobin) } solution $\xrightarrow[\text{4°C, 24 hrs.}]{\text{Glutaraldehyde}}$ Crosslinked Film

(Example: Calatase, 400 μ/ml gel)

CHART XII. Porous protein particles, Ibid.

Enzyme + Protein } solution —Glutaraldehyde / -30°C, warmed to 4°C, 4 hrs.→ "Spongelike" Protein Copolymer

↓ Lyophilization grinding

Enzyme/Protein powder

(Example: Urease, 240 μ/ml gel)

Enzymes

Glucose Oxidase	Asparaginase
Urate Oxidase	Triose-P-Isomerase
L-Aminoacid Oxidase	Carbonic Anhydrase
Xanthine Oxidase	Tyrosine Decarboxylase
Catalase	Phenyl Alanine Decarboxylase
Peroxidase	α-Amylase
Lactate Dehydrogenase	β-Galactosidase
Alcohol Dehydrogenase	Trypsin
Hexokinase	Urease
Ribonuclease	

CHART XIII. Cellulose carbonate carriers. [S. A. Barker, S. H. Doss, C. J. Grow, and J. F. Kennedy, Carbohydr. Res., 20, 1 (1971).]

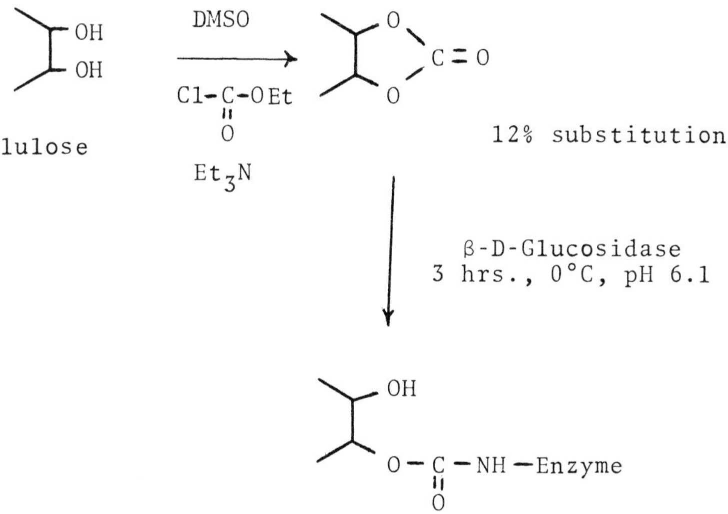

2% protein, 10 μ/gm. carrier

75% retention of activity at 37°C, 50 hrs.

CHART XIV. Microencapsulation via interfacial polymerization. [J.C.W. Ostergaard and S. C. Marting, Biotechnol. Bioeng., 15, 561 (1973).]

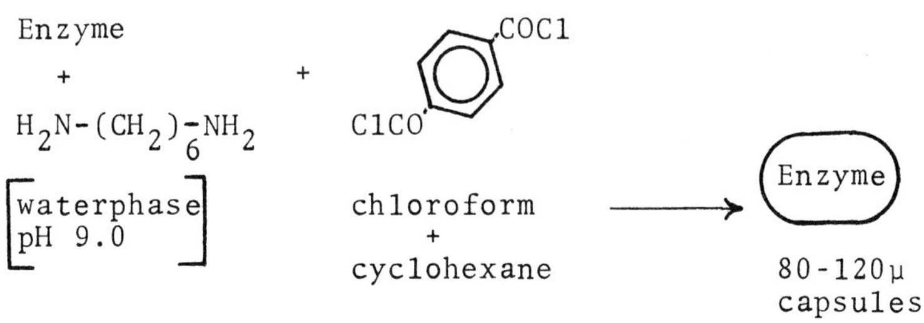

β-Galactosidase: 800 μ/g capsules

28% activity after 18 days/37°C

CHART XV. Encapsulation of enzyme extract. M. Kitajima and A. Kondo, Bull. Chem. Soc. Japan, 44, 3201 (1971).

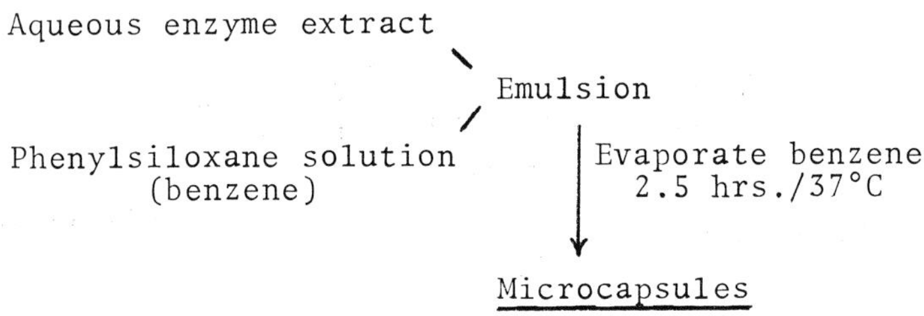

Yeast Extract: Alcohol production, 0.16-0.84 μmole/min g.

Muscle Extract: Lactic Acid production, 0.016 μmole/min g.

CHART XVI. Glass attachment by two different methods. [J. E. Dixon, F. E. Stolzenbach, J. A. Berenson, and N. O. Kaplan, Biochem. Biophys. Res. Commun., 52, 905 (1973).]

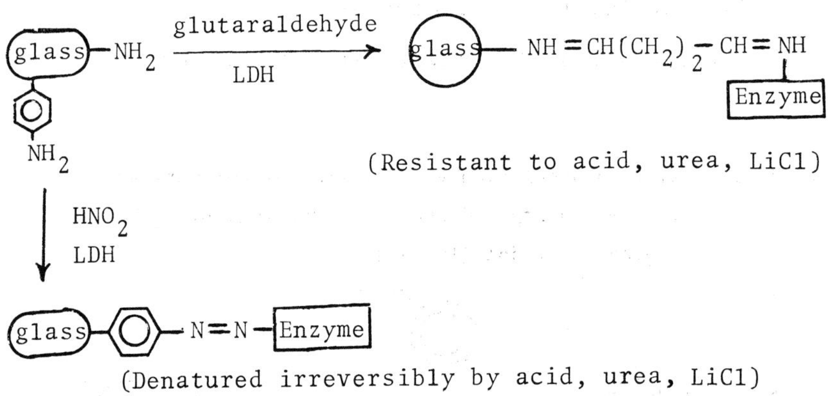

CHART XVII. Entrapment in radiation cross-linked PVA. [H. Maeda, H. Suzuki, and A. Yamauchi, Biotechnol. Bioeng., 15, 607 (1973).]

$$\begin{matrix}\text{Glucoamylase} \\ + \\ \text{Invertase}\end{matrix} + \text{PVA} \xrightarrow[\substack{4\text{-}7 \times 10^6 \text{ rad.} \\ \gamma\text{-ray}}]{11\% \text{ soln.}} \text{Gel}$$

- 9-24% retention of enzyme activity
- no enzyme leakage at higher crosslinking

CHART XVIII. Soluble support materials. [J. R. Wykes, P. Dunhill, and M. D. Lilly, Biochim. Biophys. Acta, 250, 522 (1971).]

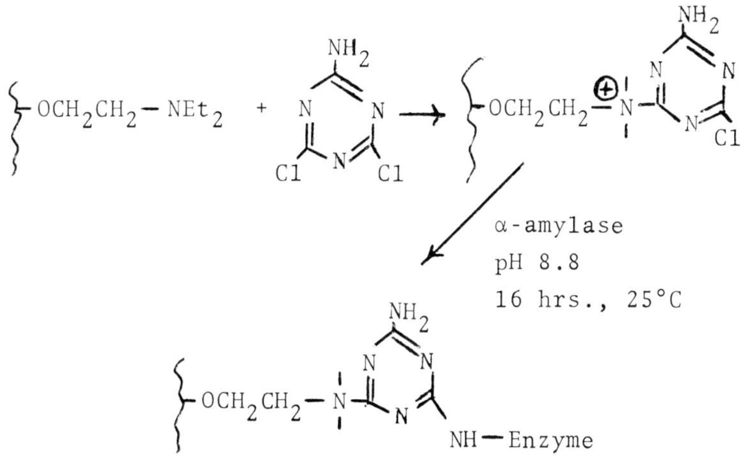

- Protein Content: 0.6-6.1%
- Activity Retention: 25-67%
- Increased Stability at 70°C: 68% ret. after 70 hrs.
- Separation from products by ultracentrifugation.

REFERENCES

[1] O. Zaborsky, Immobilized Enzymes, Chemical Rubber Co., Cleveland, 1973.

[2] I. H. Silman and E. Katchalski, "Water Insoluble Derivatives of Enzymes, Antigens, and Antibodies," Ann. Rev. Biochem., 35, 873 (1966).

[3] G. J. H. Melrose, "Insolubilized Enzymes: Biochemical Applications of Synthetic Polymers," Rev. Pure Appl. Chem., 21, 83 (1971).

[4] G. R. Stark, Biochemical Aspects of Reactions on Solid Supporters, Academic, New York, 1971.

[5] R. D. Falb, "Immobilized Enzymes: An Overview," in Enzyme Engineering, (L. Wingard, ed.), Wiley-Interscience, New York, 1972, p. 177.

[6] R. Goldman, L. Goldstein, and E. Katchalski, "Water-Insoluble Enzyme Derivatives and Artificial Enzyme Membranes," in Biochemical Aspects of Reaction Solid Supports (G. R. Stark, ed.), Academic, New York, 1971, pp. 1-78.

[7] K. Mosbach, "Enzymes and General Biology. Special Reference to Enzyme Synthesis, Purification, and Reactions on Matrixes," J. Dent. Res., 51(2)(Pt.1), 217-223 (1972).

[8] S. J. Updike, "Genetic Engineering, Enzyme Immobilization, and Transplantation," Am. J. Pharm. Educ., 36(5), 718-722 (1972).

[9] A. Wiseman, "Industrial Enzyme Stabilization," Process Biochem., 8(8), 14-15 (1973).

[10] G. P. Royer, P. Andrews, and R. Uy, "Support Materials for Immobilized Enzymes," Enzyme Technol. Dig., 1(3), 99-138 (1973).

[11] M. S. Chang, "Immobilization of Enzymes, Adsorbents, or Both within Semipermeable Microcapsules (Artificial Cells) for Clinical and Experimental Treatment of Metabolite-Related Disorders," Birth Defects, Orig. Artic, Ser., 9(2), 66-76 (1973).

[12] F. Wold, "Chemical Modification of Enzymes," Ibid., 9(2), 46-54 (1973).

[13] H. H. Weetall, "Enzymes Linked by Covalent Bonds to Carriers. Preparation, Properties, and Possibilities of Use," CZ-Chem.-Tech., 1(11), 501-507 (1972).

[14] H. H. Weetall, "Immobilized Enzymes. Applications to Foods and Beverages," Food Prod. Dev., 7(3), 46, 49-50, 52 (1973); 7(4), 94, 96, 98, 100 (1973).

[15] W. Cooreman, A. Lauwers, and S. Scharpe, "Solid-Phase Enzymology," Farm. Tijdschr Belg., 50(3), 170-193 (1973).

[16] W. R. Vieth and K. Venkatasubramanian, "Enzyme Engineering. II. Materials for Immobilized Enzyme Reactions," Chem. Technol., 4(1), 47-55 (1974).

[17] G. Broun, D. Thomas, and M. C. Tran, "Perspectives Offered by the Incorporation of Enzymes into Polymer-Based Membranes," Actual. Nephrol. Hop. Necker, 1972, 113-118.

[18] H. H. Weetall, "Immobilized Enzyme and Its Use in Food and Beverage Industry," Chem. Ztg., 97(11), 611-619 (1973).

[19] D. L. Marshall, "ATP Regeneration Using Immobilized Carbamyl Phosphatase," Biotechnol. Bioeng., 15, 447 (1973).

[20] D. L. Marshall, "Polysaccharide Synthesis on Immobilized Phosphorylase," Carbohydr. Res., 25, 489 (1972).

[21] P. Custrecasas, "Interaction of Insulin with the Cell Membrane: The Primary Action of Insulin," Proc. Natl. Acad. Sci., U.S.A., 63, 450 (1969).

BIBLIOGRAPHY

COVALENTLY BOUND PROTEINS

Chan, W. W. C., "Protein Subunits. V. Specific Interaction Between Matrix-Bound Subunits of Aldolase and Soluble Aldolase Subunits," Can. J. Biochem., 51(9), 1240-1247 (1973).

Cherkasov, I. A., and N. A. Kravchenko, "Structurally Directed Addition of Lysozyme to a Polyacrylamide Support," Izv. Akad. Nauk SSR, Ser. Khim., 1972(10), 2374 (in Russian).

Coughlan, M., and D. B. Johnson, "Preparation and Properties of Immobilized Xanthine Oxidase," Biochim. Biophys. Acta, 320(2), 200-204 (1973).

Flemming, Ch., A. Gabert, and P. Roth, "Synthesis and Properties of Carrier Fixed Enzymes. II. Covalent Coupling of Trypsin to Dialdehyde Epidex, Dialdehyde Sepharose, and 3-Chloro-2-Hydroxypropyl Epidex," Acta Biol. Med. Ger., 31(3), 365-373 (1973).

Forgione, P. S., and R. A. Polistina, "Water-Insoluble Carrier Bound Enzymes," U.S. Patent 3,766,013 (October 16, 1973).

Goldstein, L., "Immobilized Enzymes. Coupling of Biologically Active Proteins to Ethylene-Maleic Anhydride Copolymers of Different Hydride Content," Anal. Biochem., 50(1), 40-46 (1972).

Goldstein, L., "Immobilized Enzymes. Synthesis of a New Type of Polyanionic and Polycationic Resins and Their Utilization for the Preparation of Water-Insoluble Enzyme Derivatives," Biochim. Biophys. Acta, 315(1), 1-17 (1973).

Goldstein, L., "New Polyamine Carrier for the Immobilization of Proteins. Water-Insoluble Derivatives of Pepsin and Trypsin," Ibid., 327(1), 132-137 (1973).

Gray, C. J., C. M. Livingstone, C. M. Jones, and S. A. Barker, "New and Convenient Method of Enzyme Insolubilization Using Diazotized m-Diaminobenzene," Ibid., 341(2), 457-464 (1974).

Katchalski, E., L. Goldstein, Y. Levin, and S. Blumberg, "Water-Insoluble Enzyme Derivatives," U.S. Patent 3,706,633 (December 19, 1972).

Levin, Y., S. Blumberg, E. Katchalski, and L. Goldstein, "Insoluble Polymer-Enzyme Products," U.S. Patent 3,650,900 (March 21, 1972).

Manecke, G., "Immobilization of Enzymes by Various Synthetic Polymers," Biotechnol. Bioeng. Symp., 3, 185-187 (1972).

Mehltretter, C. L., and F. B. Weakley, "Active Water-Insoluble Enzymes," U.S. Patent 3,745,088 (July 10, 1973).

Valaris, M., and W. J. Harper, "Effect of Carboxymethylcellulose on the Proteolysis of Alpha$_5$-Casein by Immobilized Trypsin," J. Food Sci., 38(3), 481-483 (1973).

Vretbald, P., and R. Axen, "Use of Isocyanides for the Immobilization of Biological Molecules," Acta Chem. Scand., 27(8) 2769-2780 (1973).

Weakley, F. B., and C. L. Mehltretter, "Binding of Papain to Dialdehyde Starch," Biotechnol. Bioeng., 15(6), 1189-1192 (1973).

TRAPPED OR SORBED PROTEINS

Broun, G., D. Thomas, G. Gellf, D. Domurado, A. M. Berjonneau, and G. Guillon, "New Methods for Binding Enzyme Molecules Into a Water-Insoluble Matrix. Properties After Insolubilization," Biotechnol. Bioeng., 15(2), 359-375 (1973).

Chibata, I., T. Tosa, and T. Mori, "Insoluble Enzyme Preparations in Polyacrylamide Gel and Microcapsules thus Formed," Fr. Demande 2,157,982 (July 13, 1973).

Hofstee, B. H. J., and N. F. Otillio, "Immobilization of Enzymes through Noncovalent Binding to Substituted Agaroses," Biochem. Biophys. Res. Commun., 53(4), 1137-1144 (1973).

Johansson, A., "Matrix-Bound Enzymes Studied by Flow Microcalorimetry," Protides Biol. Fluids, Proc. Colloq., 20, 567-570 (1972).

Kennedy, J. F., and C. E. Doyle, "Active, Water-Insoluble Derivatives of D-Glucose Oxidase and Alginic Acid, Chitin, and Celite," Carbohydr. Res., 28(1), 89-92 (1973).

Kosugi, Y., and H. Suzuki, "Fixation of Cell-Bound Lipase and Properties of the Fixed Lipase as an Immobilized Enzyme," Hakko Kagaku Zasshi, 51(12), 895-903 (1973).

Park, Y. K., "Enzymic Properties of a Fungal Amyloglucosidase-Resin Complex," Ibid., 52(2), 140-142 (1974).

WATER-SOLUBLE ENZYME GRAFTS

Hixson, H. F., "Water-Soluble Enzyme-Polymer Grafts. Thermal Stabilization of Glucose Oxidase," Biotechnol. Bioeng., 15(5), 1011-1016 (1973).

ENZYME MEMBRANES

Research Corp., "Protein-Enzyme Membranes," Fr. Demande 2,165,832 (September 14, 1973).

Vieth, W. R., S. S. Wang, S. G. Gilbert, and R. Saini, "Immobilized Enzyme-Membrane Complexes by Electrocodeposition," U.S. Patent 3,758,396 (September 11, 1973).

PREPARATIONS USING IRRADIATION

Maeda, H., H. Suzuki, and A. Yamauchi, "Preparation of Immobilized Enzymes by Electron-Beam Irradiation," Biotechnol. Bioeng., 15(4), 827-829 (1973).

Maeda, H., H. Suzuki, and A. Yamauchi, "Preparation of Immobilized Enzymes using Poly(vinyl Alcohol) (with γ-Ray Irradiation)," Ibid., 15(3), 607-610 (1973)

Maeda, H., A. Yamauchi, and H. Suzuki, "Preparation of Immobilized Enzymes γ-Ray Irradiation," Biochim. Biophys. Acta, 315(1), 18-21 (1973).

Walton, H. M., and J. E. Eastman, "Insolubilized Amylases (Using Photopolymerization)," Biotechnol. Bioeng., 15(5) 951-962 (1973).

IMMOBILIZATION ON GLASS

DeJong, P. J., and P. L. Kumler, "Preparation of Immobilized Enzymes, and Determination of Their pH Activity Profile," Integrated Organic-Biology Experiment," J. Chem. Educ., 51(3), 200-201 (1974).

Herring, W. M., R. L. Laurence, and J. R. Kittrell, "Immobilization of Glucose Oxidase on Nickel-Silica Alumina," Biotechnol. Bioeng., 14(6), 975-984 (1972).

Messing, R. A., "Immobilized Papain," Ger. Offen. 2,339,805 (February 14, 1974).

Messing, R. A., and H. H. Weetall, "Silane Coupler for Stabilized Enzyme Preparations," Ger. Offen. 2,036,499 (February 3, 1972).

Monsan, P., and G. Durand, "Enzymes Immobilized by Binding on Supports," French Patent 2,133,370 (December 29, 1972).

Royer, G. P., and J. P. Andrews, "Immobilized Leucine Aminopeptidase Applications in Protein Chemistry," J. Macromol. Sci.-Chem., 7(5), 1167-1179 (1973).

Weetall, H. H., N. D. Havewala, W. H. Pitcher, Jr., C. C. Detar, W. P. Vann, P. Williams, and S. Yaverbaum, "Preparation of Immobilized Lactase and Its Use in the Enzymic Hydrolysis of Acid Whey," Biotechnol. Bioeng., 16(3), 295-313 (1974).

PROPERTIES OF IMMOBILIZED PROTEINS

Brockman, H. L., J. H. Lwa, and F. J. Kezdy, "Catalysis by Adsorbed Enzymes. Hydrolysis of Tripropionin by Pancreatic Lipase Adsorbed to Siliconized Glass Beads," J. Biol. Chem., 248(14), 4965-4970 (1973).

Ford, J. R., R. P. Chambers, and W. Cohen, "Active-Site Titration Method for Immobilized Trypsin," Biochim. Biophys. Acta, 309(1), 175-180 (1973).

Gabel, D., and J. Porath, "Molecular Properties of Immobilized Proteins," Biochem. J., 127(2), 13P-14P (1972).

Gestrelius, S., B. Mattiasson, and K. Mosbach, "Regulation of the Activity of Immobilized Enzymes. Microenvironmental Effects of Enzyme-Generated pH Changes," Eur. J. Biochem., 36(1), 89-96 (1973).

Henry, S., J. Koczan, and T. Richardson, "Bactericidal Effectiveness of Immobilized Peroxidases," Biotechnol. Bioeng., 16(2), 289-291 (1974).

Horigone, T., H. Kasai, and T. Okuyama, "Stability of Taka-amylase A Immobilized on Various Sizes of Matrix," J. Biochem. (Tokyo), 75(2), 299-307 (1974).

Mort, J. S., D. K. K. Chong, and W. W. C. Chan, "Continuous Spectrophotometric Assay of a Sepharose-Bound Enzyme and Its Use to Study Kinetics of Coupling of the Enzyme to Sepharose," Anal. Biochem., 52(1), 162-168 (1973).

Ogiso, T., M. Sugiura, and Y. Kato, "Enzymes, LXX. Bile-sensitive Lipase. X. Preparation and Properties of Carrier-Bound Mucor Lipase," Chem. Pharm. Bull., 20(12) 2542-2550 (1972).

Regan, D. L., P. Dunnill, and M. Lilly, "Immobilized Enzyme Reaction Stability. Attrition of the Support Material," Biotechnol. Bioeng., 16(3), 333-343 (1974).

Royer, G. P., R. Uu, "Evidence for the Induction of a Conformational Change of Bovine Trypsin by a Specific Substrate at pH 8.0," J. Biol. Chem., 248(7), 2627-2629 (1973).

Shuler, M. L., H. M. Tsuchiya, and R. Aris, "Diffusive and Electrostatic Effects with Insolubilized Enzymes Subject to Substrate Inhibition," J. Theor. Biol., 41(2), 347-356 (1973).

Srere, P. A., B. Mattiasson, and K. Mosbach, "Immobilized Three-Enzyme System. Model for Microenvironmental Compartmentation in Mitochondria," Proc. Natl. Acad. Sic. U. S. A., 70(9), 2534-2538 (1973).

ANALYTICAL, DIAGNOSTIC, AND ENGINEERING APPLICATIONS

Gellf, G., and J. Boudrant, "Enzymes Immobilized on a Magnetic Support. Preliminary Study of a Fluidized Bed Enzyme Reaction," Biochim. Biophys. Acta, 334(2), 467-470 (1974).

Gholson, R. K., and P. E. Guire, "Enzymic Removal of Oil Slicks," U.S. Natl. Tech. Inform. Serv., AD Rep. 757071 (1973), 14 pp.

Hinberg, I., A. Kapoulas, R. Korus, and K. O'Driscoll, "Gel Entrapment of Enzymes. Kinetic Studies of Immobilized Glucose Oxidase," Biotechnol. Bioeng., 16(2), 159-168 (1974).

Inman, D. J., and W. E. Hornby, "Immobilization of Enzymes on Nylon Structures and Their Use in Automated Analysis," Biochem. J., 129(2), 255-262 (1972).

Inman, D. J., and W. E. Hornby, Preparation of Immobilized Linked Enzyme Systems and Their Use in the Automated Determination of Disaccharides," Ibid., 137(1), 25-32 (1973).

Lin, S. H., "Nonisothermal Immobilized Enzyme Reaction in a Packed-Bed Reactor," Biophysik, 8(4), 302-309 (1972).

Nachtigal, J., "Stabilized, Enzyme-Containing Dentifrice Paste," Fr. Demande 2,156,149 (June 29, 1973).

Rosalki, S. B., and D. Tarlow, "Amylase Determination Using Insoluble Substrates," Ann. Clin. Biochem., 10(2), 47-52 (1973).

Veith, W. R., S. S. Wang, and R. Saini, "Immobilization of Whole Cells in a Membraneous Form," Biotechnol. Bioeng., 15(3), 565-569 (1973).

Wang, S. S., and W. R. Vieth, "Collagen-Enzyme Complex Membranes and Their Performance in Biocatalytic Modules," Ibid., 15(1), 93-115 (1973).

Weetall, H. H., N. B. Havewala, H. M. Garfinkel, W. M. Buehl, and G. Baum, "Covalent Bond between the Enzyme Amyloglucosidase and a Porous Glass Carrier. Effect of Shearing," Ibid., 16(2), 169-179 (1974).

Weibel, M. K., W. Dritschilo, H. J. Bright, and A. E. Humphrey, "Immobilized Enzymes. Prototype Apparatus for Oxidase Enzymes in Chemical Analysis Utilizing Covalently Boun d Glucose Oxidase," Anal. Biochem., 52(2), 402-414 (1973).

THE CONTROLLED RELEASE OF BIOACTIVE COMPOUNDS FROM POLYMERIC SYSTEMS

G. G. Allan, J. F. Friedhoff, W. J. McConnell, and J. C. Powell

College of Forest Resources
University of Washington
Seattle, Washington 98195

The ability to effect profound changes in the growth and behavior of living systems through the application of minute quantities of biologically active compounds has lead to development of the multibillion dollar pest control industry. Indeed, the introduction of modern chemical pesticides was originally thought to represent the "ultimate weapon" in man's age-old battle with pests [1]. Not until the late fifties and early sixties was it recognized that the continued use of this weapon, in its present form, might well result in a Pyrrhic victory for man. As a consequence of this realization, the concept of efficient pest control has undergone considerable modification in recent years. The undesirable secondary and tertiary effects of the highly successful long-lived pesticides has caused their use to be severely restricted [2]. An additional general concern is the often gross excess of applied pesticide in relation to that necessary to achieve control of the pest in the target area. For example, depending on the method of application and the climatic conditions, as much as 60 to 90% of the applied pesticide never reaches its objective [3]. In addition, the effective concentration of the pesticide is further reduced by leaching and soil erosion after application [4,5].

Unfortunately, when one examines the other side of the coin, represented by short-lived easily biodegraded pesticides, the results, in terms of the duration of pest control, tend to be economically discouraging, precisely because of the rapidity of degradation after application. Consideration of Fig. 1 provides a visual estimation of this deficiency. Here the calculated duration of a typical nonpersistent pesticide with a half-life of 15 days is represented by curve A. Assuming the loss of pesticide is the result of a unimolecular reaction, then the application of a level just above the minimum necessary for pest control (say 1 mg) would provide protection for only a day or so. To achieve practical periods of protection of 50, 100, or 150 days, the level of application would have to be increased 10-, 100-, or 1000- fold, respectively. In contrast, the theoretical level required to achieve these periods of protection is only that amount needed to continuously maintain the minimum effective level of 1 mg. Ideally then, the pesticide should be continuously provided from some efficient reservoir at a rate exactly equivalent to the rate of loss that occurs after the initial minimum treatment. If a pesticide were applied using just such an efficient reservoir, then the amounts needed to provide 50, 100, and 150 days of protection would be represented by curve B in Fig. 1. The area between curves A and B represents logarithmically the amount of pesticide wasted in conventional treatments and indicates how much room exists for improvements in application techniques.

A viable solution to the problems outlined above utilizes the potential of pesticide-polymer combinations as a means of securing the localized, continuous release of an easily biodegraded pesticide in the correct amount over an appropriate time period. In recent years a remarkable number of such pesticide-polymer formulations have been proposed [6-79]. In the more primitive forms, a pesticide Ⓟ is adsorbed on an inert substrate Ⓢ. Although this is often described as a physical operation, it is more helpful from a fundamental point of view, in our opinion, to regard it as a chemical process, albeit one of low energy. Thus we now have a quasi-

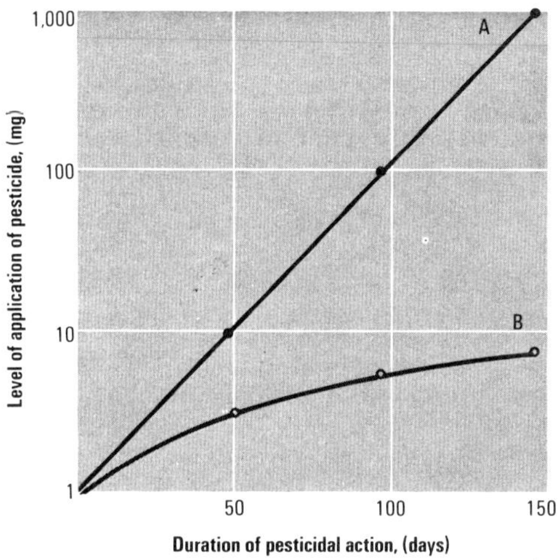

Fig. 1. Duration of pesticidal action for practical (A) and ideal (B) application techniques.

chemical structure $\textcircled{S}-\textcircled{P}$, which by definition is biologically inactive, and insulated from the deteriorating effects of the macroenvironment. If the combination is to provide control, then the following sequence of steps must take place:

$$\textcircled{S}-\textcircled{P} \xrightarrow{K_r} \textcircled{S} + \textcircled{P} \xrightarrow{K_\rho} \text{consumption}$$

However, the cleavage of $\textcircled{S}-\textcircled{P}$ requires an energy input and is governed by a rate constant K_r, which is theoretically calculable from the theory of absolute reaction rates [80] where:

$$K_r = (RT/N_o h)e^{-F/RT}$$

and the symbols have their usual meanings. Thus there is an inverse dependence of K_r on the free energy of formation of $\textcircled{S}-\textcircled{P}$. This in turn implies that, for a given quantity of combination $\textcircled{P}$, the higher the energy of the linkage, the longer will be the period of release. Since all forms of controlled release pesticides can be regarded as being a spectrum of bonded pesticides (some bonded with more energy and some bonded with less), the level and duration of release of a toxicant can be varied with the type of bonding employed in the formulation. Of course, boundary conditions on the energy of bonding do exist, and the bond type should be chosen so that K_r has a reasonable value. Otherwise, for formulations with low K_r values it is difficult to maintain Ⓟ at a suitable level without changing the capacity factor tremendously, i.e., without adding huge quantities of the combination initially. Conversely, a significant reduction in the amount of pesticide applied and a longer term controlled release cannot be anticipated from formulations with high K_r values. Thus combinations relying on low energy bonds such as Van der Waals attractions, as is the case for pesticides adsorbed on materials like walnut shells, simple clays, or silica gels [6,8-10], show limited utility in extending the duration of control or in reducing the amount of pesticide applied. However, a range of bond energies which do provide reasonable values of K_r can be achieved through the chemical bonding of pesticides to polymer backbones.

CHEMICAL ATTACHMENT

Chemically bonded pesticide-polymer combinations can readily be synthesized by conventional, organic chemical procedures wherein the attachment can take any one of the forms illustrated in Fig. 2. The simplest of these has the pesticide attached as a pendent substituent on a natural or synthetic water-soluble or -insoluble polymer having a replaceable hydrogen as formulated in

$$\text{Polymer—X—H} + \text{RCOOH} \underset{\text{environment}}{\overset{\text{synthesis}}{\rightleftarrows}} \text{Polymer—X—COR} \qquad (1)$$

where X = N, O, or S. Obviously, the pesticide must contain a structural moiety suitable for use as a link to the macromolecule, and the carboxyl group is only one of the many alternatives possible. In the mid-sixties several patents were issued which embodied this concept. The earliest claim [31] utilized chlorinated phenol esters of tri- and tetracarboxylic benzene acids to achieve fungicidal properties in alkyd resin coatings. A recent patent [17,18] of a

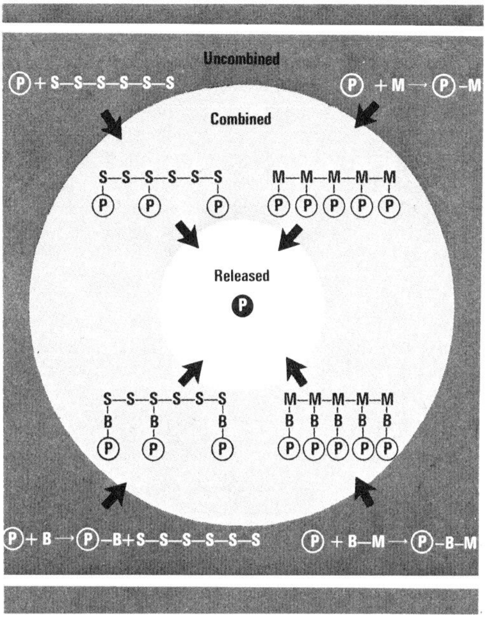

Fig. 2. Synthetic routes to controlled release of chemical combinations.

more general nature reports the utilization of forest wastes and other low cost materials for the production of sustained release herbicides. An example of this type of combination is the direct bonding of herbicide acids to the hydroxyl groups of natural waste products such as bark and lignin.

For pesticides which cannot be directly attached on a substrate to form a bond of suitable stability, a bridging entity may be interposed (Fig. 2). An example of this would be the linking of pesticide alcohols to a polysaccharide substrate by means of a phosphate or diurethane bridge [81]. Alternatively, the pesticide may be initially converted to a polymerizable derivative, e.g., vinyl 2,4-dichlorophenoxyacetate, which is then homo- or copolymerized to give a wholly synthetic pesticide-polymer [82] as illustrated in Fig. 2.

For all of these variants, side-chain degradation occurs in the biological environment. The chemical bonds holding the pesticide within its polymeric prison are sequentially broken to provide a sustained release of the biocide over an extended period of time. The rate of release will be determined not only by the nature of the pesticide-polymer bond, but also by the chemical characteristics of the pesticide and the polymer, the dimensions and structure of the resultant macromolecular combination, and the hydrophilicity of neighboring groups. For example, carboxylic acid groups introduced (via acrylic acid) in copolymers of 2,3,5-trichloro-4-pyridyl methacrylate and acrylic acid produce dramatic changes in the rate of release of 2,3,5-trichloro-4-pyridinol. Thus, because the carboxylic acid groups increase the hydrophilicity and catalyze the hydrolysis of the ester linkage between the herbicide and the polymer, the higher the percentage of acrylic acid the shorter the period of protection [83]. The dimensions of the pesticide-polymer combination as well as the solubility of the polymer backbone also play important roles in determining the release rate of a biocide [29].

Where the polymeric backbone is water soluble, the rate of hydrolytic degradation (R_h) for n sperical particles (radius r and density ρ) in a water-saturated heterogeneous surface reaction is given by

$$R_h = n\rho 4\pi r^2 \, dr/dt \tag{2}$$

and by

$$R_h = nK_h 4\pi r^2 C_o \tag{3}$$

where C_o is the initial per unit area concentration of pesticide-polymer linkages on the surface of a particle of radius r_o, and K_h is the hydrolysis rate constant. Since each pesticide molecule released by hydrolysis exposes another linkage beneath, C_o is a constant and the duration of pesticide release (D_{pr}), by combination of Eqs. (2) and (3), can be predicted from the relationship

$$D_{pr} = \rho r_o / K_h C_o - (R_h \rho^2 / nK_h^3 C_o^3) \tag{4}$$

or by

$$D_{pr} = A - BW^{-1/2} \tag{5}$$

where A and B are constants and W is the amount of pesticide-polymer combination used. The validity of the above equations as a predictive tool was demonstrated by measuring the persistence of herbicidal activity exhibited by a combination synthesized by the partial acylation of polyvinyl alcohol with 2-methyl-4-chlorophenoxyacetyl chloride [29]. The validity of this analysis is demonstrated by a plot of A as the reciprocal of the level of application of active ingredient versus the duration of herbicidal action. The curve obtained is clearly linear (Fig. 3).

Conversely, when the polymer backbone selected is water insoluble, the rate of degradation (R_d) can be expressed as

$$R_d = Wdc/dt \tag{6}$$

or by

$$R_d = K_d WC \tag{7}$$

where W is again the amount of pesticide-polymer combination used, C is the concentration of pesticide per unit weight contained therein at time t, and K_d is the degradation rate constant.

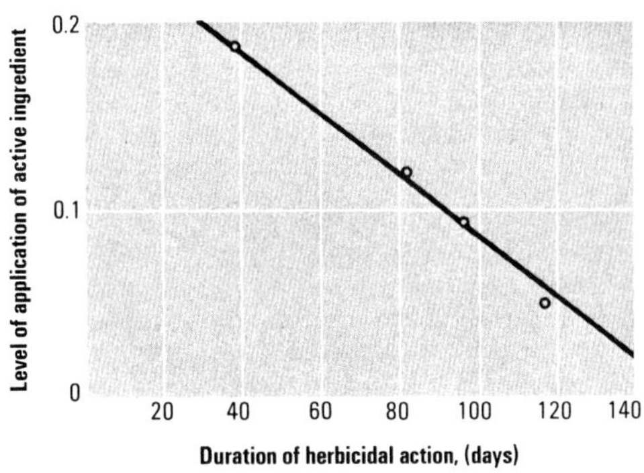

Fig. 3. Dependence of the duration of herbicidal action on the level of application for a chemical combination of 2-methyl-4-chlorophenoxyacetic acid with a water-soluble polymer. (Level of application = liter/mg applied.)

The duration of effectiveness (D_{pr}) for this type of controlled release herbicide-polymer combination, by combination of Eqs. (6) and (7), is therefore given as

$$D_{pr} = (1/K_d)\ (\ell n\ W - \ell n\ R_d/K_d) \tag{8}$$

or

$$D_{pr} = M \log_{10} W - N \tag{9}$$

where M and N are constants. Experimental confirmation of the usefulness of Eqs. (8) and (9) in the design of controlled release pesticide-polymer combinations is demonstrated in the release rates of herbicide acids, via ester cleavage, from cellulose and lignin (Fig. 4) [29]. Clearly the constants M and N change for each combination and are determined by the nature of the pesticide-polymer bond and by the chemical and physical characteristics of the resultant macromolecule. Therefore, by proper selection of bond energies coupled with variations in the chemical and physical microenvironment of the bonding site, a wide range of different release rates is obtainable. Practical application of these findings is now being utilized for the selective suppression of unwanted weeds and deciduous brush.

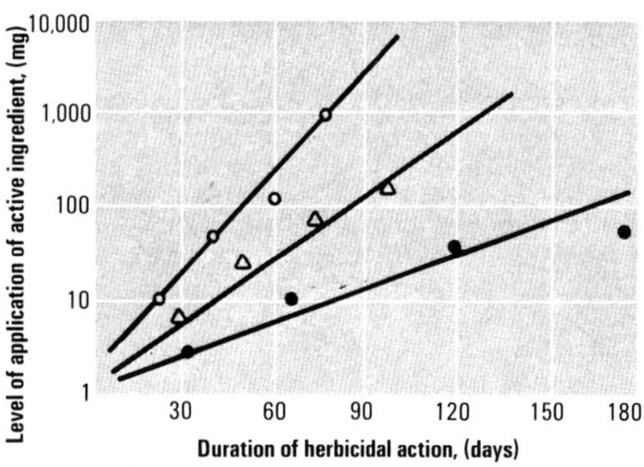

Fig. 4. Dependence of the duration of herbicidal action on the level of application for 2,4-dichlorophenoxyacetic acid (2,4-D) in chemical combination with water-insoluble polymers. (O) 2,4-D Alone; (Δ) 19.7% 2.4-D/Douglas fir bark; (●) 39% 2,4-D/kraft lignin.

As an example of this utilization, the reforestation of commercial forest lands with coniferous species is often severely retarded by the rapid invasion of unwanted vegetation in these plantations [84]. In spite of the recognition of this problem, its control by herbicide application is far from straightforward. The characteristics of the problem require the application of a herbicide with selective toxicity to the unwanted vegetation only, and with a period of duration on the order of several years. The most promising materials, in terms of their selective toxicity, are the chlorophenoxybutyric acids [85]. Unfortunately, these butyric acid herbicides are readily enzymically oxidized in the soil to the corresponding chlorophenoxyacetic acids, which have considerable toxicity to conifers [84]. In this case the controlled release concept is being utilized to perform a dual function: to greatly increase the duration of effectiveness achieved through a single application of the herbicide and to protect the butyric acid herbicide from soil oxidation to the corresponding acetic acid analog.

To achieve this goal, ester combinations of 2,4-dichlorophenoxybutyric acid with bark were synthesized, and their selective toxicity to western red alder (Alnus rubra Bong.) in the presence of Douglas fir (Pseudotsuga menziesii (Mirb. Franco) was evaluated. The results clearly indicated a favorable effect from this type of treatment, and the lower levels of application showed no damage to the conifer while effectively eliminating the red alder [84]. A field evaluation of the practicality of this approach was therfore initiated [86]. The field site near Sedro Woolley, Washington covers an area of 10,000 yd^2, experiences an annual rainfall of 30 to 40 in., and contains approximately 800 Douglas fir seedlings. The seedlings were treated with bar combinations containing 37% esterified herbicide at four levels of application. After one growing season the height increase for the treated seedlings was more than double that of the controls. At the same time the vegetation level, on a scale of 1 to 10, was an average of 2 for the area around treated seedlings in contrast to a reading of 8 for the area around untreated seedlings. To assess the possible effect of varying climatic conditions on these results, field trials of a similar nature have been initiated near Glenwood (annual rain fall 25 to 30 in.) and Gray's Harbor (annual rainfall 90 to 100 in.), Washington.

The compete suppression of vegetation is also being investigated in field trials utilizing controlled release formulations of the herbicides 2,4,5-trichlorophenoxyacetic acid and 2-chloropropionic acid as esters of bark and crab shells. These materials, when applied on railroads, firebreaks, and power-line right-of-ways, provided complete suppression of all vegetation for a period of 2 years at the highest level.

Although these two vegetation control examples demonstrate the potential of chemical attachment as a viable method of improving the efficiency of pest control, this approach is precluded for a large number of biologically active compounds due to their lack of functional moieties amenable to polymeric attachment. Fortunately, this restraint can be overcome by the incorporation of such compounds into polymeric matrices.

CONTROLLED RELEASE FROM POLYMERIC MATRICES

Efforts to secure improvements in the volatility, solubility, toxicity, and period of duration for a wide variety of biologically active materials, through polymer incorporation, have been reported in the literature [11-28]. While at first glance they appear to be a hodgepodge of different concepts and inventions, they all can be separated into two major classes. The simplest classification involves the pesticide interacting with the polymer to afford a solid solution wherein the low energy of individual interactions is supplemented by the additional constraint of diffusion through appreciable masses of plastic. Notable examples in this class include Shell's No Pest Strip [88-90] and the increasingly popular flea collars for cats and dogs. Physical combinations in which the active ingredient is dispersed or encapsulated in a polymeric matrix in which they have low solubility comprise a second classification. This concept has been discussed in the pharmaceutical literature for the prolonged dissemination of birth control hormones [91], and it forms the basis of Pennwalt's new microencapsulated methyl parathion [92].

SOLID SOLUTION FORMULATIONS

The equations which describe diffusion from solid polymeric solutions are well-known and characterized [93-95]. In particular, descriptions and mathematical analyses for diffusion-controlled release systems, especially in the pharmaceutical field, are numerous [95-101]. Although each analysis is predicated on a unique set of predetermined boundary conditions, in all case release is moderated by the rate of diffusion of the active component to the polymer surface as well as the rate of its subsequent detachment therefrom. Fundamentally, these two constraints on release are governed by the concentration and diffusivity of the active compound in the polymer, and by the geometry, surface area, and surface resistance of the pesticide-polymer combination.

The functional relationship between concentration and diffusivity, common in the description of diffusion within polymers, results from the additional mobility imparted to the polymer segments by the plasticizing diffusant [102]. Thus, as the plasticizer concentration decreases with time, the free volume within the polymer becomes exponentially harder to generate and the diffusivity decreases, often by several orders of magnitude. Quantitatively, this concentration dependence can be expressed in equation form as

$$D_c = e^{ac}$$

where D_c is the diffusivity at concentration c , and a is a positive real constant. In addition, even if the diffusivity is assumed to remain constant throughout the period of release, the flux or driving force of diffusion has been shown by Fick's law [103] to decrease with decreasing concentration of the diffusant. Thus the mathematical description of diffusion in solid solutions, as a result of the above considerations, is represented by the following nonlinear, parabolic partial differential equation [104]:

$$\frac{\partial c}{\partial t} = D\nabla^2 c + \frac{dD}{dc}\left(\frac{\partial c}{\partial x}\right)^2$$

where the change in concentration of the active component with time ($\partial c/\partial t$) varies with the change in concentration (c) and diffusivity (D) throughout the polymeric matrix. Numerical solutions of this equation predict release rates which decrease exponentially with time [104]. However, variations in the geometry of the pesticide-polymer combination show limited ability in overcoming this proclivity. For example, normalized computer models of the release rate from cylindrical and planar geometries show relative release rates, after 50% of the biocide has been dissipated, which are only 30 and 100% greater than the rate of release from a spherical conformation with an equal surface to volume ratio [104]. Moreover, a substantial improvement in the efficiency of solid solutions can be realized by coating the formulation with an enveloping film through which diffusion of the active component is slow [89]. In this instance the rate of release is no longer controlled by diffusion of the biocide to the matrix surface, but instead it is controlled by the subsequent step of detachment into the environment. Thus the rate-controlling resistance is concentrated at the surface of the polymer matrix, and the solid solution functions as a reservoir maintaining a constant flux on the outer envelope of diffusion-resistant polymer. As a result, the concentration and diffusivity of the active component remain constant in the surface coating, thereby linearizing the rate of loss of the pesticide from the polymer matrix. The magnitude of this effect is amply demonstrated in Fig. 5, where the proportion of biocide released from plane geometry solid solutions is plotted versus time for various dimensionless surface transport coefficients (k) [104]. Clearly, for high surface resistances (k = 1, 0.5, or 0.1) the release rate is effectively constant. Therefore, formulations in which the surface resistance is increased by surface coatings represent an efficient method of applying a constant dosage of active ingredient over an extended period of time.

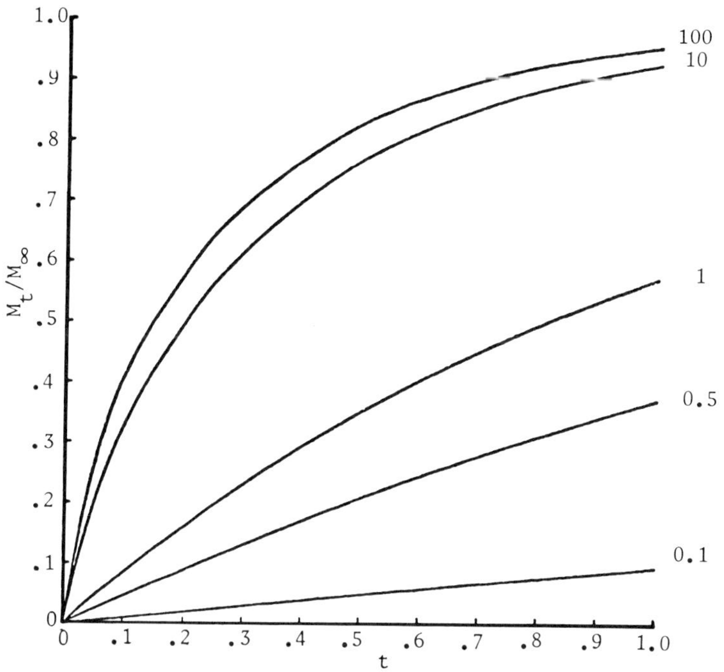

Fig. 5. The proportion of pesticide released as a function of time for the indicated values of k (the dimensionless surface transport coefficient).

The basic factors influencing the design of this type of controlled-release system are the chemical and thermodynamic compatibility of the resultant pesticide-polymer matrix. While chemical reactions between the pesticide and polymer are to be avoided, thermodynamic compatibility is a prerequisite to the formation of stable solid solutions of pesticides within polymer matrices. Through modification of the solubility parameter theory originally developed by Hildebrand and Scatchard [105-107], relatively simple guidelines are available for assessing the thermodynamic free energy of mixing. Thus pesticides which are probably compatible with a polymer may be easily selected by comparing their respective solubility parameters [104]. This ability is demonstrated in a recent patent which claims a new controlled-release method based on novel polymer mixtures in which the applicable polymers and pesticides are delineated by solubility parameter ranges [108]. It should be recognized, however, that every pesticide will not be amenable to solid solution formulation. In fact, the formation of stable solid solutions is largely limited in scope to relatively nonpolar compounds. The primary difficulty in extrapolating the solid solution concept to more polar compounds lies in their inherent incompatibility with the common commercial flexible polymers [104]. However, this disability forms the basis of an alternative general formulation concept which utilizes solid suspensions of the active component in a polymer matrix.

SOLID SUSPENSION FORMULATIONS

A solid suspension or dispersion refers to a mixture of one or more active ingredients in a polymeric carrier or matrix. The attainment of a finely divided state of biocide particles within the polymer is critical to the success of this type of sustained release formulation since a pseudo-steady state between the rate of solute release from the matrix and the rate of particle fragment dissolution within the pellet must be established. In this manner the rate of release is still governed by diffusion of the biocide through the polymer. However, the suspended particles now maintain a constant concentration of the biocide within the matrix phase.

Predictive equations for the release rates from such solid suspension formulations have been developed by Higuchi and others [98-101] using the following set of restrictions: (1) the rate-limiting step is diffusion of the pesticide within the matrix; (2) a pseudosteady state exists between the release rate and the rate of particle dissolution; (3) the initial concentration of the active component in the matrix greatly exceeds its solubility in the polymer; (4) the initial matrix porosity is negligible; and (5) the receiving medium is a perfect sink. Under these assumptions [99] the total release of biocide (Q) after some time (t), for the case of a planar geometry, is given by

$$Q = [(D_m t(2A - C_s)C_s)]^{1/2}$$

where D_m is the diffusivity of the active component in the polymer phase, A is the concentration per unit volume of the active component in the polymer matrix, and C_s is the solubility of the active compound in the polymer phase. This class of formulations offers an improvement over simple solid solutions in terms of the constancy of release. However, it still suffers an everdecreasing rate of release. Consideration of the diffusion pathlength, as a function of time, offers a clear explanation of the phenomena. During the period of release, the suspended biocide is depleted from the surface inward. Thus the thickness of the depleted zone, through which diffusion must take place, steadily increases with time. Hence the change in the amount of active ingredient released per unit area corresponds to this change in the thickness of the depleted zone and so the decrease in the release rate varies with the square root of time. Nonetheless, the efficiency and lifetime of the active material are substantially improved, and numerous pharmaceutical investigators have considered variations of this release mechanism [109]. When the active component is only sparingly soluble in the environmental eluent, one such variation is particularly useful.

Thus, when the eluent solution bathing the formation is capable of penetrating the matrix through pores or capillaries, then the rate of release becomes dependent on the dissolution, leaching, and subsequent diffusion of the biocide out of the pellet along a tortuous path. Therefore, the equation describing the release rate of an active component from a porous, solid suspension takes the following form [99]:

$$Q = [(D_f e/T(2A - eC_s)D_s t)]^{1/2}$$

where Q is the mass per unit surface area of biocide leached from the matrix at time t, D_f is the diffusivity of the biocide in the permeating fluid, e and T are the porosity and tortuosity factors for the matrix, respectively, A is the concentration per unit volume of the bioactive component in the matrix, and C_s is the solubility of the diffusant in the permeating fluid. The validity of this equation for the release of biologically active compounds from a tablet matrix of polyethylene was demonstrated by Desai et al. [110]. As predicted, a linear relationship for the release of sodium salicylate was found experimentally when Q was plotted against the square root of time.

It is clear from the above discussions that no one type of formulation can be the ideal candidate for all applications. Rather, the delivery system employed will largely be determined by the physicochemical characteristics of the active component and the duration and rate of release desired. In addition, the character of the environment in which the formulation is placed cannot be overlooked.

The above derivations, while applicable for aqueous media, may require modifications to compensate for the soil environment. The pertinent differences between aqueous and soil conditions are numerous and include the presence of soil particles, variances in water content, cyclical temperature changes, biological activity, erosion forces, and possibly even chemical reactivites [104,111-114]. The integration of all of these factors to provide a suitable formulation-ecosystem balance is the necessary prerequisite to achieving efficient pest control. Clearly, such pesticide-polymer combinations are likely to be of special value for the economic control of pest infestations where systemic chemicals are effective but do not possess a sufficient period of activity. This sitution abounds in forest management, particularly in the tropics where increasing efforts are being made to specifically cultivate commercially valuable tree species.

A typical acre of tropical forest contains perhaps 150 trees species of which only one or two are commercially useful. However, planting efforts to extend the numbers of valuable trees such as Spanish cedar or mahogany in tropical forests have been frustrated in the Americas by devastating attacks of a lepidopteran shootborer,

Hypsipyla grandella Zeller [104, 115-117]. The larvae of this moth tunnel into the stems of developing seedlings causing crippling distortion or death. Conventional protection by pesticides is economically impossible because applications of systemic insecticides last only 2 to 3 weeks under the climatic conditions of frequent, heavy rainfall and high biological activity [118]. However, these same readily biodegraded organophosphate and carbamate insecticides, when formulated as solid suspensions in a fatty polyamide, give much longer periods of protection - over 13 months in the best case [113]. These trees have thus been protected from the massive attack potential of eight pest generations. Modified designs which will extend and linearize this period of protection for a period of 3 years are now being evaluated in Costa Rica, Puerto Rico, and the Virgin Islands.

CONCLUSIONS

Although not all-inclusive, the release mechanisms and examples reviewed in this paper reveal the very extensive breadth and depth of the controlled release technology which has been researched over the last 10 years and which is now more than adequate to deal with the vast majority of existing pesticides.

The current combined and mounting pressures of pesticide shortages and concern for the quality of the environment will inevitably increase the rate at which the obsolete and wasteful conventional application techniques are discarded.

However, it is the belief of the authors that the pace of introduction of controlled-release technology has been funereal from a societal point of view. To remedy this situation in the pesticide area, research should now emphasize development of the existing reservoir of technology so that practical and sensible new delivery systems can be expeditiously introduced.

REFERENCES

[1] M. Grosser, Zoecon Corp. 1970 Annual Report, 19 (1970).

[2] Chemical Week, 110(25), 34 (1974).

[3] Canadian Wildlife Service, Ministry of Fisheries and Forestry, Ottawa, Canada, 4, 1971.

[4] Chemical and Engineering News, 42(46), 26 (1964).

[5] K. A. Hassall, World Corp Protection, Vol. 2, Chemical Rubber Co., Cleveland, 1969.

[6] H. A. Benesi, V. P. Sun, and E. S. Loeffler, U.S. Patent 2,868,688 (January 13, 1959).

[7] H. L. Morrill, U.S. Patent 2,739,052 (March 20, 1956).

[8] J. W. Nemec and E. A. Nolan, U.S. Patent 3,194,730 (July 13, 1965).

[9] W. Benzel, U.S. Patent 3,127,235 (March 31, 1964).

[10] W. P. Reeves, U.S. Patent 3,376,191 (April 2, 1968).

[11] C. G. L. Furmidge and J. M. Osgerby, J. Sci. Food Agric., 18, 269 (1967).

[12] F. Marson, J. Appl. Chem., 19, 93 (1969).

[13] R. J. Geary, U.S. Patents 3,074,845 (January 22, 1963), 3,162,573 (December 22, 1964).

[14] R. T. McFadden, R. R. Langher, and L. L. Wade, U.S. Patents 3,212,967 (October 19, 1965), 3,223,513 (December 19, 1965), 3,228,830 (January 11, 1966).

[15] N. F. Cardarelli, U.S. Patents 3,417,181 (December 17, 1968), 3,074,848 (January 22, 1963), 3,223,510 (December 14, 1968).

[16] G. E. Barnsley, J. K. Eaton, and J. Malone, British Patent 898,915 (June 14, 1962).

[17] G. G. Allan, Belgian Patent 706,509 (December 15, 1967).

[18] G. G. Allan, Canadian Patents 855,181 (November 3, 1970), 863,310 (February 9, 1971).

[19] H. D. Grown, A. B. Patel, S. V. Chattopudhyay, and S. N. Pennington, Enzymologia, 35, 215 (1968).

[20] G. H. McKibben, D. D. Hardee, T. B. Davich, R. C. Gueldner, and P. A. Hedin, J. Econ. Entomol., 64(1), 317 (1971).

[21] Chemical and Engineering News, 49(37), 20 (1971).

[22] S. C. Khanna, T. Jecklin, and P. Speiser, J. Pharm. Sci., 59, 614 (1970).

[23] S. C. Khanna and P. Speiser, Ibid., 58, 1114 (1969).

[24] S. J. Desai, P. Singh, A. P. Simonelli, and W. J. Higuchi, Ibid., 55, 1230 (1966).

[25] A. F. Caprio and W. Horback, U.S. Patent 2,460,377 (February 1, 1949).

[26] F. B. Folckemer, R. E. Hanson, and A. Miller, Belgian Patent 648,606 (November 30, 1964), U.S. Patent 3,318,769 (May 9, 1964).

[27] J. W. Miles and J. E. Woehst, Adv. Chem. Ser., 86, 183 (1969).

[28] T. J. Roseman, Chem. Eng. News, 49(14), 13 (1971).

[29] G. G. Allan, C. S. Chopra, A. N. Neogi, and R. M. Wilkins, Nature (London), 234, 349 (1971).

[30] E. Baltazzi, U.S. Patent 3,343,941 (September 26, 1967).

[31] R. L. Broadhead, U.S. Patent 3,179,608 (April 20, 1965).

[32] E. Kolb, U.S. Patent 2,387,591 (October 23, 1945).

[33] J. E. Lloyd and J. G. Matthysse, J. Econ. Entomol., 59(2), 363 (1966).

[34] C. W. Murray, U.S. Patent 2,286,636 (June 16, 1942).

[35] G. Faerber, British Patent 826,831 (January 20, 1960).

[36] D. Werner, U.S. Patent 3,607,863 (September 21, 1971).

[37] A. Williams, Sustained Release Pharmaceuticals, Noyes Development Corp., Park Ridge, New Jersey, 1969, p. 272.

[38] I. Fox, G. A. Rivera, and I. G. Bayona, J. Econ. Entomol., 62, 1246 (1969).

[39] C. G. Wright, Ibid., 64, 278 (1971).

[40] T. L. Harvey and J. R. Brethour, Ibid., 63(5), 1688 (1970).

[41] C. H. Stockman, E. H. Bollinger, and R. N/ Naranch, Mosq. News, 30, 584 (1970).

[42] R. N. Wilkinson, W. W. Barnes, A. R. Gillogly, and C. D. Minnemeyer, J. Econ. Entomol., 64(1), 1 (1971).

[43] A. R. Clark and M. M. Clark, U.S. Patent 3,592,910 (July 13, 1971).

[44] U. R. Hattangadi, V. R. Shah, and D. S. Hattangadi, Indian Patent 116,080 (June 20, 1970).

[45] M. V. Fogle and G. E. McGowan, South African Patent 7,002,725 (December 2, 1970).

[46] S. S. Wasti and F. R. Shaw, J. Econ. Entomol., 64(1), 224 (1971).

[47] M. Vlig and N. V. VanBreen, German Patent 1,802,684 (May 13, 1971).

[48] R. H. Czurak and R. M. Thompson, German Offen. 2,040,000 (April 22, 1971).

[49] J. C. McCool, German Patent 1,010,314 (June 13, 1957).

[50] M. Dima, A. Carpov, and S. Maxim, Rev. Roum. Chim., 13, 485 (1968).

[51] O. B. Wurzburg, W. Herbst, and H. M. Cole, U.S. Patent 3,091,567 (May 28, 1963).

[52] M. L. Beasley and R. L. Collins, Science, 169, 769 (1970).

[53] CIBA Ltd., German Patent 1,936,748 (January 29, 1970).

[54] J. S. Pierce, U.S. Patent 2,503,297 (April 11, 1950).

[55] R. W. Simmons, A. E. Markham, and E. G. King, U.S. Patent 3,287,209 (November 22, 1966).

[56] C. H. Boehringer Sohn., French Patent 1,447,479 (July 29, 1966).

[57] Upjohn Co., British Patent 929,407 (June 19, 1963).

[58] J. E. Vandegaer and F. G. Meier, U.S. Patent 3,515,882 (April 20, 1971).

[59] J. E. Vandegaer, U.S. Patent 3,577,515 (May 4, 1971).

[60] J. G. Wagner, French Patent 1,468,601 (February 10, 1967).

[61] W. B. Hemson, Abstracts of 162nd ACS Meeting (September 13, 1971).

[62] A. T. Jancosek and J . S. Brown, U.S. Patent 2,906,744 (September 29, 1959).

[63] C. W. MacMullen, U.S. Patent 3,299,566 (January 24, 1967).

[64] T. H. Shepherd, U.S. Patent 3,299,566 (January 24, 1967).

[65] PPG Industries, German Patent 1,945,873 (January 22, 1970).

[66] J. F. Vartiak, U.S. Patents 3,360,356 (December 26, 1967), 3,413,109 (April 19, 1967).

[67] Diamond Shamrock, Farm Chem., 52 (1967).

[68] CIBA Ltd., Netherland Patent 6,401,097 (August 12, 1964).

[69] L. L. Danielson, Weeds, 15, 60 (1967).

[70] G. G. Allan, C. S. Chopra, A. N. Neogi, and R. M. Wilkins, Tappi, 54, 1293 (1971).

[71] C. Haremcki, J. R. Nuss, and P. H. Williams, Proc. Northeast Weed Control Conf., 1969, 76.

[72] K. S. Knyazeua, A. M. Klechetova, L. N. Pdogina, and I. V. Khorkina, Inst. Byt. Obsluzh., 3, 115 (1963).

[73] S. V. Knylov and K. A. Timisyazeva, Nauch. Konf., 29, 69 (1957).

[74] K. Nakayama, L. Koiwa, and K. Noguchi, Nippon Sakumotsu Gakkai Kiji, 37, 645 (1968).

[75] H. L. Pierce, U.S. Patent 3,172,754 (March 9, 1965).

[76] M. Rubin, U.S. Patent 3,269,900 (August 30, 1966).

[77] J. D. Stone and L. M. Stahler, U.S. Patent 2,904,423 (September 15, 1959).

[78] M. L. Dunham and F. M. O'Connor, Union Carbide, U.S. Patent 3,359,092 (December 19, 1967).

[79] J. E. Yaffe and R. E. Sellers, U.S. Patent 3,274,052 (September 20, 1966).

[80] F. Daniels and R. A. Alberty, Physical Chemistry, 3rd ed., Wiley, New York, 1966.

[81] G. G. Allan and J. C. Powell, Sven. Papperstidn., In Preparation.

[82] R. M. Wilkins, "The Design of Polymers for the Sustained Release of Selected Herbicides," M.S. Thesis, University of Washington, Seattle, 1969.

[83] A. N. Neogi, "Polymer Selection for Controlled Release Pesticides," Ph.D. Thesis, University of Washington, Seattle, 1970.

[84] G. G. Allan, C. S. Chopra, and R. M. Russell, Int. Pest Control, 14(2), 15 (1972).

[85] J. H. Rediske and G. R. Staebler, For. Sci., 8, 353 (1962).

[86] G. G. Allan, C. S. Chopra, M. W. Maggi, A. N. Neogi, and R. M. Wilkins, Int. Pest Control. 15(3), 24 (1973).

[87] G. G. Allen, J. F. Friedhoff, and J. C. Powell, In press.

[88] J. W. Gillette, J. R. Harr, F. T. Lindstrom, D. A. Mount, A. D. St. Clair and L. J. Weber, Residue Rev., 44, 161 (1972).

[89] F. T. Lindstrom, A. D. St. Clair, and C. E. Wicks, Special Report 399, Oregon State University, Corvallis, Oregon, 1973.

[90] I. Fox, G. A. Rivera, and I. G. Bayona, J. Econ. Ent., 62, 1246 (1909).

[91] D. R. Mischell and M. E. Simkin, Fertil. Steril., 21, 99 (1970).

[92] Pencap M, Technical Bulletin, Pennwalt Corp., Morovia, California, 1973.

[93] J. Crank and G. S. Park, eds., Diffusion in Polymers, Academic, Oxford, 1956.

[94] J. Crank, The Mathematics of Diffusion, Clarendon Press, Oxford, 1956.

[95] G. L. Flynn, S. H. Yalkowsky, and T. J. Roseman, J. Pharm. Sci., 63, 479 (1974).

[96] Y. W. Chien, H. J. Lambert, and D. E. Grant, Ibid., 63, 365 (1974).

[97] J. W. Ayres and P. A. Laskar, Ibid., 63, 351 (1974).

[98] P. Dothisiri, and J. T. Carstensen, Ibid., 62, 1468 (1973).

[99] T. Higuchi, Ibid., 52, 1145 (1963).

[100] T. Higuchi, Ibid., 50, 874 (1961).

[101] T. Higuchi, Ibid., 56, 315 (1967).

[102] A. Aitken and R. M. Barrer, Trans. Faraday Soc., 51, 116 (1955).

[103] A. Fick, Poggendorffs Ann., 94, 59 (1855).

[104] S. C. Roberts, "Principles and Practice of Controlled Release Insecticide-Polymer Blends," Ph.D. Thesis, University of Washington, Seattle, 1974.

[105] J. H. Hildebrand and R. L. Scott, The Solubility of Nonelectrolytes, 3rd ed., Reinhold, New York, 1950.

[106] J. H. Hildebrand and R. L. Scott, Regular Solutions, Prentice-Hall, Englewood Cliffs, New Jersey, 1962.

[107] G. Scatchard, Chem. Rev., 8, 321 (1931).

[108] A. W. VanBreen, U.K. Patent 1,321,044 (June 20, 1973).

[109] "Symposium on Controlled Release of Biologically Active Agents," Southern Research Institute, April 19-20, 1973.

[110] S. J. Desai, P. Singh, A. P. Simonelli and W. J. Higuchi, J. Pharm. Sci., 55, 1230 (1966).

[111] C. G. L. Furmidge, Pestic. Sci., 3, 745 (1972).

[112] C. G. L. Furmidge, A. C. Hill and J. M. Osgerby, J. Sci. Food Agric, 17, 518 (1966).

[113] J. M. Osgerby, Pestic. Sci., 3, 753 (1972).

[114] J. R. Coppedge, R. A. Stokes, R. E. Kinzer, and R. L. Ridgway, Econ. Ent., 67(2), 292 (1974).

[115] R. M. Wilkins, "Suppression of the Shootborer, Hypipyla grandella (Zeller) with Controlled Release Pesticides," Ph.D. Thesis, University of Washington, Seattle, 1972.

[116] G. G. Allan, R. I. Gara and R. M. Wilkins, Turrialba, 20, 478 (1970).

[117] G. G. Allan, R. I. Gara and R. M. Wilkins, Int. Pest Control, 16(4) (1974).

[118] D. DeBach, Biological Control of Insect Pests and Weeds, Reinhold, New York, 1964.

ENZYMES IMMOBILIZED BY MICROENCAPSULATION WITHIN SPHERICAL ULTRATHIN POLYMERIC MEMBRANES

T. M. S. Chang

Faculty of Medicine
McGill University
Montreal, Quebec, Canada

INTRODUCTION

Biological cells in nature have diameters of about 7 to 50 μm, with a cell membrane of about 0.01 μm in thickness. This means that the total membrane area available for diffusion can be extremely large in comparison to the volume and that the cell membrane can also be very thin without the loss of mechanical strength. In addition, rapid mixing is possible within the individual microscopic compartments. Biological cells have very complex and selective enzyme systems to act on substrates crossing the cell membrane. Artificial cells in the form of spherical ultrathin polymeric membrane envelopes containing enzymes have been prepared in this laboratory with some of these properties [1-3]. Detailed papers and upadated information published since then on this subject have been described in a recent monograph [4].

PREPARATION

The cell model system first prepared in this laboratory in 1956 consisted of red blood cell hemolysate enclosed within spherical ultrathin collodion membranes of cellular dimensions. Each "artificial cell" is of cellular dimension containing all the protein and enzyme systems of red blood cells, except that the red blood cell membrane has been replaced with an ultrathin collodion membrane. In this initial preparative procedure there are three main steps: (I) The aqueous protein and enzyme solution is emulsified in a water-immiscible solution to form aqueous microdroplets. (II) With the addition of a polymer solution to the stirred emulsion, a membrane is formed on the surface of each microdroplet and then allowed to set in an appropriate solution. (III) Finally, the microencapsulated red blood cell hemolysate formed is dispersed in an aqueous phase. With modifications, this general procedure has been varied for the preparation of artificial cells with a great variety of membrane material, content, size, and configuration. The preparative procedures have been updated and described in detail [4].

Briefly, the first approaches [1,2,4] involved organic phase separation. Polymers dissolved in a water-immiscible fluid are deposited around each aqueous microdroplet dispersed in the water-immiscible fluid. This is one of the most convenient ways of preparing microencapsulated enzymes. The details have been updated and simplified [4]. The best approach is to use the 10 g % hemoglobin solution for microencapsulation, and any enzymes to be enclosed could be added to the hemoglobin solution, since without the 10 g % hemoglobin solution, which is a basic component of the cell model, a great deal of variation and modification of the preparative procedure would be needed. Enzymes, proteins, or other material for microencapsulation are added to the 10 g % hemoglobin solution before carrying out the microencapsulation procedure.

Another approach [2-4] is to use interfacial polymerization. A diamine dissolved in the aqueous phase of the 10 g % hemoglobin reacts at the interphase of each aqueous microdroplet with sebacoyl chloride in the organic phase to form a membrane. Nylon membrane is used most commonly for this, but other membranes could also be formed, for instance, polyamides, polyureas, polyurethanes, polysulfonamides, and polyphenylesters [4]. Again, in this typical example, it is most convenient to use a 10 g % hemoglobin solution and to add to this the enzymes or other proteins that one wants to immobilize within the microcapsules. This chemical procedure, unlike the first procedure of organic phase separation, may inactivate some enzyme systems.

A third approach [4,5] would be by secondary emulsion. Here a fine emulsion of enzyme solution is dispersed in an organic phase containing a polymer. The organic phase containing the fine aqueous microdroplets is then dispersed in the form of large droplets in an aqueous phase. The organic phase containing the polymer solutions is then solidified to form solid microspheres, each containing a number of aqueous microdroplets. Unlike the two examples described earlier, this procedure results in solid polymer microspheres, each of which contains several aqueous microdroplets. A typical example is the use of silicone rubber to microencapsulate red blood cell hemolysate [5]. Other polymers include the use of cellulose membrane and other membranes [4,5]. Another procedure [7] is the emulsification of enzyme solution in a polymer organic solution, but instead of forming microspheres the polymer solution containing the enzyme microdroplets is extruded in the form of fibers. Further extensions include the microencapsulation of enzymes within liposomes [8]. However, in this case the enzyme is entrapped between onion-skin-like multiple laminar lipid. Hemolysate has also been microencapsulated within spherical ultrathin lipid membrane [9]. Enzymes have also been microencapsulated within liquid membranes [10].

Since the "artificial cell" is a general concept [4], the artificial cells prepared by the procedures that have been described are only some examples for demonstrating this concept. It is expected that modifications or extensions of these procedures can be made available to further demonstrate or improve the preparation of different types of artificial cells.

PROPERTIES

The ultrathin membranes (0.02 μm) and the large surface to volume relationship of artificial cells (2 m^2 area in 10 ml of 20 μm diameter artificial cells) results in extremely high transport rates across the artificial cell membranes. Thus 10 ml of these 10 μm diameter spherical ultrathin polymer membranes has a theoretical transport capacity 200 times higher than that of a whole standard 1 m^2 dialysis machine [5,11]. The enclosing polymeric membrane of artificial cells is prepared in such a way that while impereable to macromolecules or suspensions, it is extremely permeable to most of the solutes normally present in the biological fluid. Enveloped by the spherical ultrathin polymer membrane is either an aqueous solution or a suspension of biologically active material like enzyme systems. While remaining at all times enveloped by the enclosing polymeric membrane and prevented from coming into direct content with the external environment, the enclosed enzyme systems act on external permeant substrates diffusing into the artificial cells. This means that enzymes cannot leak out of the spherical ultrathin polymeric membrane envelopes or come in contact with external macromolecules like antibodies. However, permeant substrates can equilibrate rapidly across the enclosing membranes to be acted on by the enclosed enzymes, and the products can diffuse out of the microcapsules. The permeability characteristics of the polymeric membranes can be varied at will as to porosity, charge, composition, and lipid contents. In addition, the polymeric membranes can also be made biologically compatible [4].

MICROENCAPSULATED ENZYMES

Table 1 lists enzymes and proteins which have been immobilized by microencapsulation. Additional pertinent references are also to be noted [16,18,32,36-38,41,49-51,53-55]. Thus hemoglobin, albumin, uricase, urease, catalase, asparaginase, carbonic anhydrase, trypsin, lactase, and many other enzymes have been immobilized by microencapsulation. Even complex enzyme systems like the contents of red blood cells and muscle cells have been microencapsulated (Figs. 1-3). The microencapsulated enzymes do not leak out but act efficiently on permeant substrates. Products formed can leave the microcapsules. Most enzyme activities are retained after microencapsulation. The kinetics of reaction are governed by a large number of factors: concentration of enzymes

TABLE 1

Enzymes and Proteins Immobilized within Semipermeable Microcapsules

	References
Enzymes and hemoglobin of red blood cell hemolysate	1-3, 6, 9, 12, 13, 28
Urease	1, 3, 5, 6, 10, 14, 15, 17, 28-31, 34, 35, 44
Carbonic anhydrase	2, 19
Uricase	2, 3
Trypsin	2
Catalase	6, 20, 42, 45, 46
Asparaginase	4, 11, 21, 22, 30, 33-35 39, 43, 47, 48, 52
Albumin	2, 3, 23
Lipase	6
Zymase complex (yeast)	24
Muscle extract	24
Phenolase	25
Nitrate reductase	26
β-Fructofuranosidase	8
β-Galactosidase	27, 39, 40, 52

in the microcapsules, permeability of membrane to the substrates and products, charge of membrane, composition of membranes, and other factors.

Enzymes can be stabilized by microencapsulation with a high concentration of protein in the form of the 10 g % hemoglobin solution [3,30,48]. Proteins cannot leak out of semipermeable microcapsules even when the microcapsules themselves are dispersed in a very dilute suspension. Thus, unlike native enzymes in a dilute solution, the microencapsulated enzyme is at all times retained in an intracellular environment with a high concentration of protein. In this way the stability of the enzyme is much greater than the enzyme in the dilute solution. For example, the half-life of

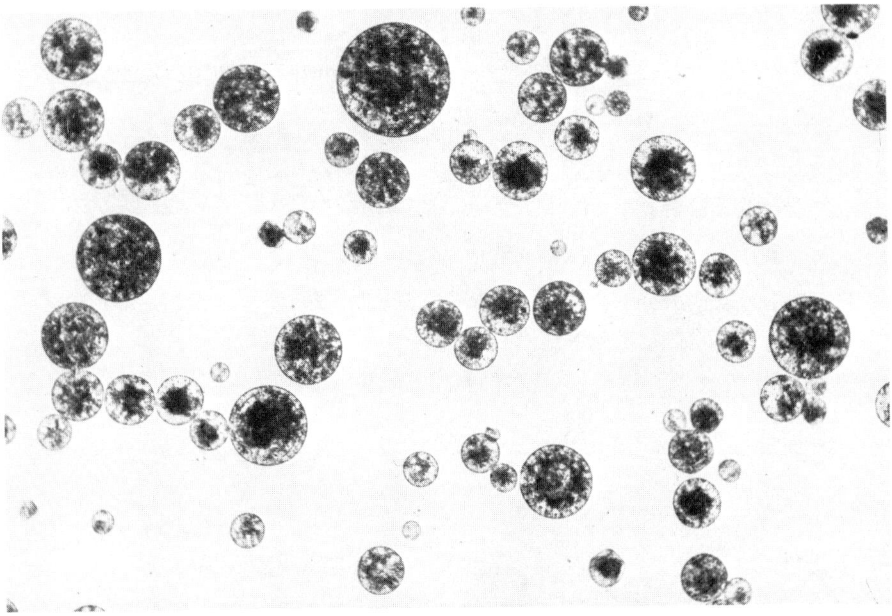

Fig. 1. Enzyme immobilized by microencapsulation within spherical ultrathin polymeric membranes. Mean diamter: 100 μm. spherical cellulose nitrate membrane of 0.02 μm thickness enveloping a complex enzyme system.

catalase in free solution is 15 days when stored at 4°C. When microencapsulated with a high concentration of protein and stored at 4°C, it retained more than 90% of its initial activity for more than 100 days. The half-life of asparaginase in free solution was 20 days when stored at 4°C, but when microencapsulated with a high concentration of protein the enzyme retained 100% of its initial activity for more than 100 days [4,39,48]. Urease and trypsin also become more stable after microencapsulation with a high concentration of protein. Enzymes can also be stabilized by directly microencapsulating insolubilized enzymes [4,34]. An easier way to insolubilize microencapsulated enzymes is to microencapsulate the enzyme at a very high concentration. Following this, it is treated

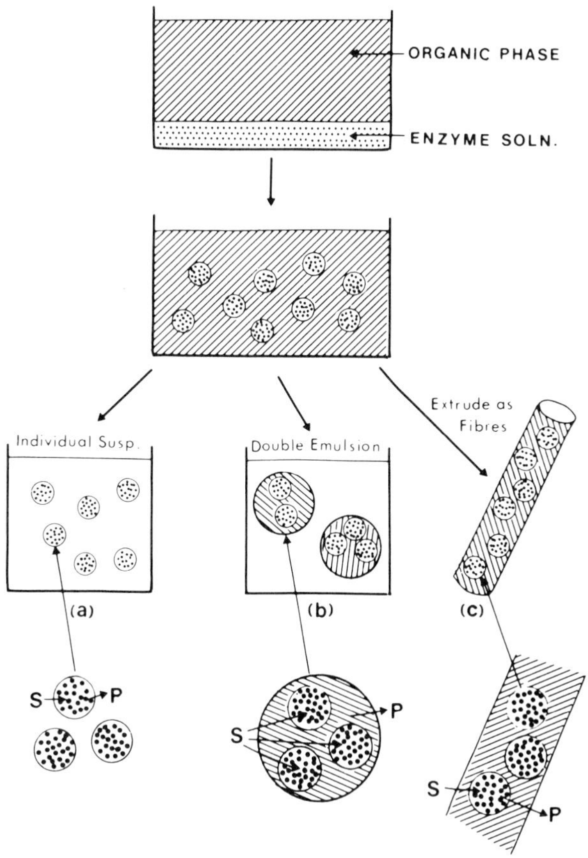

Fig. 2. Schematic representation of the preparation of spherical ultrathin polymeric membranes containing (a). (b) and (c) are modifications and extensions to form solid spheres (b) or fibers (c).

with glutaraldehyde, thereby cross-linking the enzymes within the microcapsule [48]. In this way, microencapsulated catalase and asparaginase retained more than 90% of their original activity after storage for more than 100 days at 4°C [48]. At body temperature of 37°C, the microencapsulated enzymes are also more stable.

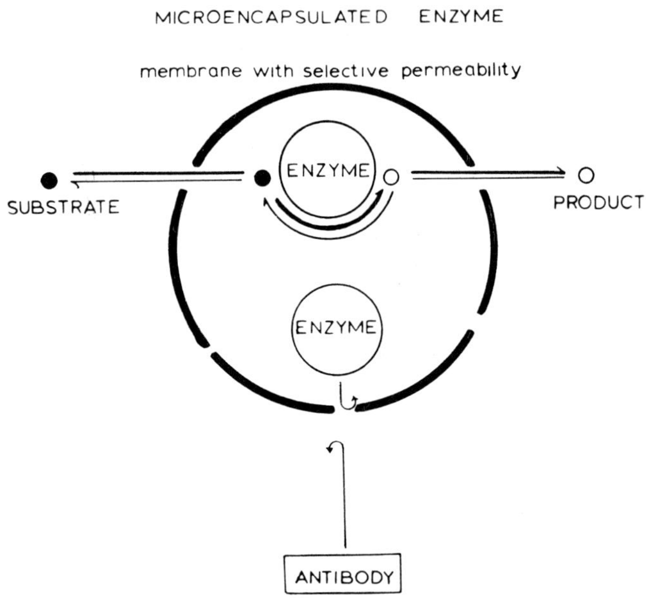

Fig. 3. Schematic representation of the action of microencapsulated enzymes. Enzyme does not leak out, but acts on permeant substrates. Antibody cannot enter to adversely affect the enzyme.

Microencapsulated enzymes have been injected intraperitoneally or used in extracorporeal shunts [4]. These microencapsulated enzymes did not leak out of the microcapsules but continued to act on substrates diffusing into the microcapsules. Urease acted on body urea [4]; microencapsulated catalase acted efficiently in replacing the deficient enzyme in an inborn error of metabolism; acatalasemia [4] and microencapsulated asparaginase acted efficiently in suppressing the growth of an asparagine-dependent lymphosarcoma [4]. Microencapsulated enzymes remain inside the artificial cells after injection. They are not inactivated by cells or enzymes in the body and are not removed rapidly. As a result, injected microencpasulated enzymes, such as catalase and asparaginase, acted significantly longer than when the same enzymes were injected in

TABLE 2

Stages of Enzyme Research

I.	Research on enzymes in dilute solution.
II.	Research on enzymes immobilized by covalent linkage, adsorption, or gel-entrapment as models of membrane-bound enzymes.
III.	Research on enzymes immobilized by microencapsulation within spherical ultrathin polymer membranes as models of intracellular enzymes (artificial cells).

the free form [4]. Furthermore, microencapsulated enzymes, unlike free enzymes, are prevented from taking part in immunological reactions [4,42].

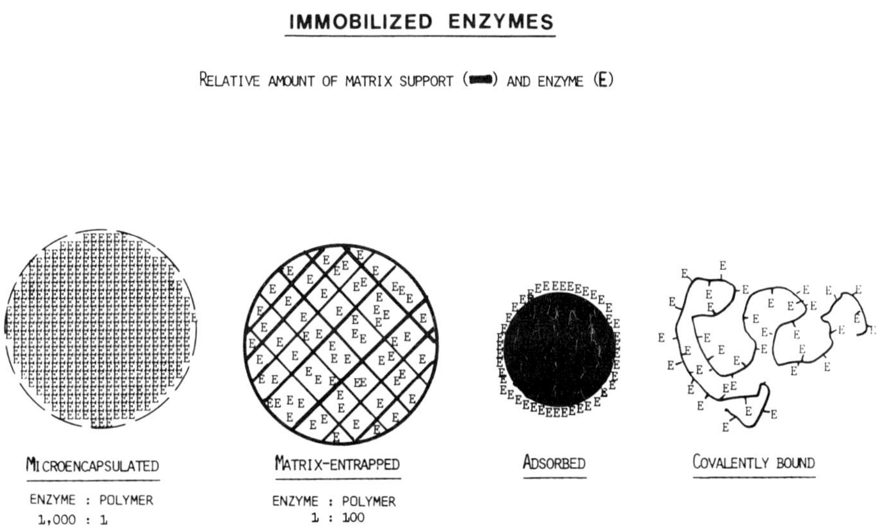

Fig. 4. Stages of enzyme research, illustrating the relative amounts of matrix support (■) and enzyme (E). Stage II: matrix entrapped, adsorbed, and covalently bound. Stage III: microencapsulated.

GENERAL

Enzyme research has proceeded from dilute enzyme solutions to the second level of gel-entrapped, adsorbed, and covalently bound enzymes as a model for membrane-bound enzymes, and finally, with the application of advances in polymer chemistry, we have demonstrated the feasibility of proceeding to the level of artificial cells containing enzymes as models of biological cells (Table 2, Fig. 4). With the experimental demonstration of the possible application of this spherical ultrathin polymeric system for basic research, applied research, and clinical use, further efforts are indicated, especially to apply the enormous amount of knowledge in polymer chemistry to this system.

ACKNOWLEDGMENT

This work was supported by the Medical Research Council of Canada (MRC-MT-2100).

REFERENCES

[1] T. M. S. Chang, "Hemoglobin Corpuscles," Report of research project for B.Sc. Honors Physiology, McGill University, Montreal (1957).

[2] T. M. S. Chang, "Semipermeable Microcapsules," Science, 146, 524 (1964).

[3] T. M. S. Chang, F. C. MacIntosh, and S. G. Mason, "Semipermeable Aqueous Microcapsules: Preparation and Properties," Can. J. Physiol. Pharmacol., 44, 115 (1966).

[4] T. M. S. Chang, Artificial Cells, monograph, Thomas, Springfield, Illinois, 1972, p. 207.

[5] T. M. S. Chang, "Semipermeable Aqueous Microcapsules ("Artificial Cells"): with Emphasis on Experiments in an Extracorporeal Shunt System," Trans. Am. Soc. Artif. Intern. Organs, 12, 13 (1966).

[6] M. Kitajima, S. Miyano, and A. Kondo, "Studies on Enzyme-Containing Microcapsules," J. Chem. Soc. Japan, Ind. Chem. Soc., 72, 493 (1969).

[7] W. Marconi, C. Saronio, M. Salmona, E. Mussini, and S. Garattini, Insolubilised Enzymes, Raven, New York, 1974.

[8] G. Gregoriadis and B. E. Ryman, "Fate of Protein-Containing Liposomes Injected into Rats. An Approach to the Treatment of Storate Diseases," Eur. J. Biochem., 24, 485 (1972).

[9] P. Mueller and D. O. Rudin, "Resting and Action Potentials in Experimental Bimolecular Lipid Membranes," J. Theor. Biol., 18, 222 (1968).

[10] S. W. May and N. N. Li, "The Immobilization of Urease Using Liquid-Surfactant Membranes," Biochem. Biophys. Res. Commun., 47, 5 (1972).

[11] T. M. S. Chang, in Recent Developments in Separation Science (N. Li, ed.), CRC Press, Cleveland, 1972.

[12] A. W. L. Jay and M. A. Edwards, "Mechanical Properties of Semipermeable Microcapsules," Can. J. Physiol. Pharmacol., 46, 731 (1968).

[13] A. W. L. Jay and K. S. Sivertz, "Membrane Resistance of Semipermeable Microcapsules," J. Biomed. Mat. Res., 3, 577 (1969).

[14] R. E. Sparks, R. M. Salemme, R. M. Meier, M. H. Litt and O. Lindan, "Removal of Waste Metabolites in Uremia by Microencapsulation Reactants," Trans. Am. Soc. Artif. Intern. Organs, 15, 353 (1969).

[15] S. N. Levine and W. C. LaCourse, "Materials and Design Consideration for a Compact Artificial Kidney," J. Biomed. Mat. Res., 1, 275 (1967).

[16] R. E. Sparks, O. Lindan, N. S. Mason, M. H. Litt, and P. M. Meier, "Removal of Uremic Waste Metabolites from the Gastrointestinal Tract by Encapsulated Carbon and Oxidized Starch," Trans. Am. Soc. Artif. Intern. Organs, 17, 229 (1971).

[17] P. V. Sundaram, "The Kinetic Properties of Microencapsulated Urease," Biochim. Biophys. Acta, 293, 1 (1973).

[18] D. L. Gardner, D. R. Falb, B. C. Kim, and D. C. Emmerling, "Possible Uremic Detoxification via Oral Ingested Microcapsules," Trans. Am. Soc. Artif. Intern. Organs, 17, 239 (1971).

[19] R. C. Boguslaski and A. M. Janik, "A Kinetic Study of Microencapsulated Bovine Carbonic Anhydrase," Biochim. Biophys. Acta, 250, 260 (1971).

[20] A. O. Mogensen and W. R. Vieth, "Mass Transfer and Biochemical Reaction with Semipermeable Microcapsules Containing Asparaginase," Ibid., 14, 663 (1972).

[21] T. Mori, T. Sato, Y. Matui, T. Tosa, and I. Chibata, "Preparation and Characteristics of Microcapsules Containing Asparaginase," Ibid., 14, 663 (1972).

[22] T. Mori, T. Tosa, and I. Chibata, "Enzymatic Properties of Microcapsules Containing Asparaginase," Biochim. Biophys. Acta, 321, 653 (1973).

[23] M. Shiba, S. Tomiika, M. Koishi, and T. Kondo, "Studies on Microcapsules. V. Preparation of Polyamide Microcapsules Containing Aqueous Protein Solution," Chem. Pharm. Bull., 18, 803 (1970).

[24] M. Kitajima and A. Kondo, "Fermentation without Multiplication of Cells Using Microcapsules That Contain Zymase Complex Muscle Enzyme Extract," Bull. Chem. Soc. Japan, 44, 3201 (1971).

[25] O. R. Zaborsky, Immobilized Enzymes, CRC Press, Cleveland, 1973.

[26] R. R. Mohan and N. N. Li, in Immobilized Enzymes (O. R. Zaborsky, ed.), CRC Press, Cleveland, 1973.

[27] J. C. W. Ostergaard and S. C. Martiny, "The Immobilization of β-Galactosidase through Encapsulation in Water-Insoluble Microcapsules," Biotech. Bioeng., 15, 561 (1973).

[28] T. M. S. Chang, F. C. MacIntosh, and S. G. Mason, "Semipermeable Aqueous Microcapsules," Proc. Can. Fed. Biol. Sci., 6, 16 (1963).

[29] T. M. S. Chang and F. C. MacIntosh, "Effects on Injected Semipermeable Aqueous Microcapsules," Ibid., 7, 58 (1964).

[30] T. M. S. Chang, "Asparaginase-Loaded Semipermeable Microcapsules for Mouse Lymphomas," Ibid., 12, 62 (1969).

[31] T. M. S. Chang, L. J. Johnson, and O. Ransome, "Semipermeable Aqueous Microcapsules: Nonthrombogenic Microcapsules with Herparin-Complexed Membranes," Can. J. Physiol. Pharmacol., 45, 70 (1967).

[32] T. M. S. Chang and M. J. Poznansky, "Semipermeable Aqueous Microcapsules (Artificial Cells): Permeability Characteristics," J. Biomed. Mat. Res., 2, 187 (1968).

[33] T. M. S. Chang, A. Pont, L. J. Johnson and N. Malave, "Response to Intermittent Extracorporeal Perfusion through Shunts Containing Semipermeable Microcapsules," Trans. Am. Soc. Artif. Intern. Organs, 15, 163 (1968).

[34] T. M. S. Chang, "Clinical Potential of Enzyme Technology," Sci. Tools, 16, 34 (1969).

[35] T. M. S. Chang, "Removal of Endogenous and Exogenous Toxins by a Microencapsulated Adsorbent," Can. J. Physiol. Pharmacol., 47, 1043 (1969).

[36] T. M. S. Chang and N. Malave, "The Development and First Clinical Use of Semipermeable Microcapsules (Artificial Cells) as a Compact Artificial Kidney," Trans. Soc. Artif. Intern. Organs, 16, 141 (1970).

[37] T. M. S. Chang, A. Gonda, J. H. Dirks and N. Malave, "Clinical Evaluation of Chronic Intermittent or Short Term Hemoperfusions in Patients with Chronic Renal Failure Using Semipermeable Microcapsule (Artificial Cells) Formed from Membrane-Coated Activated Charcoal," Ibid, 18, 465 (1972).

[38] T. M. S. Chang, "Hemoperfusions over Microencapsulated Adsorbent in a Patient with Hepatic Coma," Lancet, ii, 137 (1972).

[39] T. M. S. Chang, "Semipermeable Microcapsules Containing Asparaginase: In-vitro and In-vivo Stability," Enzym J., 14, 95 (1973).

[40] T. M. S. Chang, "Biomedical Application of Artificial Cells: Routes of Administration," Biomed. Bioeng. J., 8, 334 (1973).

[41] T. M. S. Chang, J. F. Coffey, P. Barre, A. Gonda, J. H. Dirks, M. Levy, and C. Lister, "Microcapsule Artificial Kidney: Treatment of Patients with Acute Drug Intoxication," Can. Med. Assoc. J., 108, 429 (1973).

[42] M. J. Poznansky and T. M. S. Chang, "Comparison of Enzyme Kinetics and Immunologied Properties of Catalase Immobilized by Microencapsulation and Catalase in Free Solution for Enzyme Replacement," Biochem. Biophys. Acta, 334, 103 (1974).

[43] E. Siu Chong and T. M. S. Chang, "In-vivo Effects of Intraperitoneally Injected L-Asparaginase Solution and L-Asparaginase Immobilized within Semipermeable Nylong Plasma L-Asparagine Levels," Enzyme J., 18, 218 (1974).

[44] T. M. S. Chang and S. K. Loa, "Urea Removal of Urease and Ammonia Absorbents in the Intestine," Physiologist, 13, 70 (1970).

[45] T. M. S. Chang, "Semipermeable Microcapsules as Artificial Cells," Sci. J., 3, 62 (1967).

[46] T. M. S. Chang and M. J. Poznansky, "Semipermeable Microcapsules Containing Catalase for Enzyme Replacement in Acatalasaemic Mice," Nature (London), 218, 243 (1968).

[47] T. M. S. Chang, "The In-vivo Effects of Semipermeable Microcapsules Containing L-Asparaginase on 6C3HED Lymphosarcoma," Ibid., 229, 117 (1971).

[48] T. M. S. Chang, "Stabilization of Enzyme by Microencapsulation with a Concentrated Protein Solution or by Microencapsulation Followed by Cross-Linking with Glutaraldehyde," Biochim. Biophys. Res. Commun., 44, 1531 (1971).

[49] T. M. S. Chang, "In-vitro Kinetics of Enzymes Immobilized by Microencapsulation," Biotech. Bioeng., 14, 520 (1972).

[50] T. M. S. Chang, A. Gonda, J. H. Dirks, J. F. Coffey, and T. Burns, "ACAC Microcapsule Artificial Kidney for the Long Term and Short Term Management of Eleven Patients with Chronic Renal Failure," Trans. Am. Soc. Artif. Intern. Organs, 18, 465 (1972).

[51] P. V. Sundarum, E. K. Pye, T. M. S. Chang, V. H. Edwards, A. E. Humphrey, N. O. Kaplan, E. Katchalski, Y. Levin, M. D. Lilly, G. Manecke, K. Mosbach, J. Porath, H. Weetall, and L. B. Wingrad, Jr., "Recommendations for Standardization of Nomenclature in Enzyme Technology," Biotech. Bioeng., 14, 15 (1972).

[52] T. M. S. Chang, "Immobilization of Enzymes, Adsorbents, or Both within Semipermeable Microcapsules (Artificial Cells) for Clinical and Experimental Treatments of Metabolite-Related Disorders," in Enzyme Therapy for Congenital Diseases (D. Bergsma, ed.), Williams and Wilkins, Baltimore, 1973, p. 66.

[53] T. M. S. Chang, J. F. Coffey, C. Lister, A. Stark, and E. Taroy, "Glutethimede, Methyprylon, and Methaqualone Clearance by the ACAC Microcapsule Artificial Kidney: In-vitro and in Patients with Acute Intoxication," Trans. Am. Soc. Artif. Intern. Organs, 19, 87 (1973).

[54] T. M. S. Chang and M. Migchelsen, "Characterization of Possible 'Toxic' Metabolites in Uremia and Hepatic Coma Based on the Clearance Spectrum of the ACAC Microcapsule Artificial Kideny," Ibid., 19, 314 (1973).

[55] T. M. S. Chang, "Enzymes Immobilized by Microencapsulation: Preparation and Biomedical Applications," in Insolubilized Enzymes (M. Salmona, C. Saronia, and S. Garattini, eds.), Raven, New York, 1973.

SOLID PHASE PEPTIDE SYNTHESIS

John Morrow Stewart

Department of Biochemistry
University of Colorado School of Medicine
Denver, Colorado 80220

The last two decades have been an era of rapid progress in peptide research. This era was begun by the work of Sanger on the amino acid sequence determination of insulin and by du Vigneaud on the structure determination and synthesis of oxytocin. This period has seen impressive progress in the structure elucidation and synthesis of many peptides of natural origin and of great biological significance, as well as in methods for sequence determination and chemical synthesis of peptides [1-4]. Perfection of techniques and instruments for automatic determination of the amino acid sequence of peptides and proteins has made possible a greatly broadened understanding of genetics and evolution as well as the more chemical areas of mechanism of action of enzymes and hormones, and physical chemistry of peptides and proteins. Effective methods of peptide synthesis are crucial to progress in this area, because only by synthesis can adequate amounts of important peptides be made available for chemical, biological, and physical studies, as well as for exploration of the structure-function aspects of biological molecules. In general, progress in peptide synthesis has lagged far behind that in amino acid sequence determination. This is not surprising since effective peptide synthesis requires a very sophisticated system of selectively removable protecting groups for functions of the amino acids involved, and the synthesis of a large heteropolymer of defined sequence requires near perfection of each one of the many steps of the assembly. The classical approach to peptide synthesis, using standard organic chemical methods of synthesis and purification of intermediates, has yielded impressive results during these two decades. However, the special problems associated with the assembly of large molecules make staggering investments in time and materials necessary for the synthesis of large peptides or proteins by classical methods.

Fourteen years ago, in an attempt to alleviate some of these problems, Merrifield introduced solid phase peptide synthesis [5]. In this method the amino acid which will form the carboxyl-terminal residue in the desired peptide sequence is attached by a covalent ester bond to an insoluble support which serves as a carrier for the peptide as well as a blocking group for the carboxyl function throughout the synthesis. During this attachment the amino group of the C-terminal amino acid is protected by a selectively removable blocking group, and all amino acid side-chain functions are blocked by protecting groups which will be stable throughout the synthesis but can be removed at the end without damage to the peptide chain. From this point the operations involved in peptide synthesis by the solid phase method are application of a reagent to remove the amine blocking group and coupling of a new N-protected amino acid. Reaction conditions are chosen so that both of these reactions will be forced, as nearly as possible, quantitatively to completion. All excess reagents and by-products should be removable from the resin-supported peptide by washing with appropriate solvents. If this goal of complete reactions can be achieved, a homogeneous macromolecule of defined amino acid sequence can be synthesized, and the complex operations of purification at the intermediate stages of classical solution peptide synthesis can be completely avoided. Moreover, all of the operations of synthesis after attachment of the first amino acid to the resin involve merely selection and addition of a solvent or solution, agitation with the resin for a suitable period of time, and removal of the liquid by filtration. Suitable equipment has been designed and constructed to carry out all of these operations automatically [6], and at least two commercial versions of such automatic instruments are available [7]. After the desired amino acid sequence has been assembled on the resin, a reagent is applied which will cleave the peptide from the resin and remove the protecting groups from the peptide. The rationale of solid phase peptide synthesis (SPPS) is given in Fig. 1. Using currently available resin supports, amino acid derivatives, equipment, and technology [8], several very impressive syntheses have been achieved [9]. Perhaps the most spectacular of these was the synthesis by Gutte and Merrifield of a protein with the properties of ribonuclease A [10]. Significant milestones in the progress of SPPS are listed in Table 1. Solid phase peptide synthesis is clearly one of the most important and imaginative applications of grafts on polymers ever to be described. It is also safe to say that its ultimate scope and potential can only be imagined at the present time.

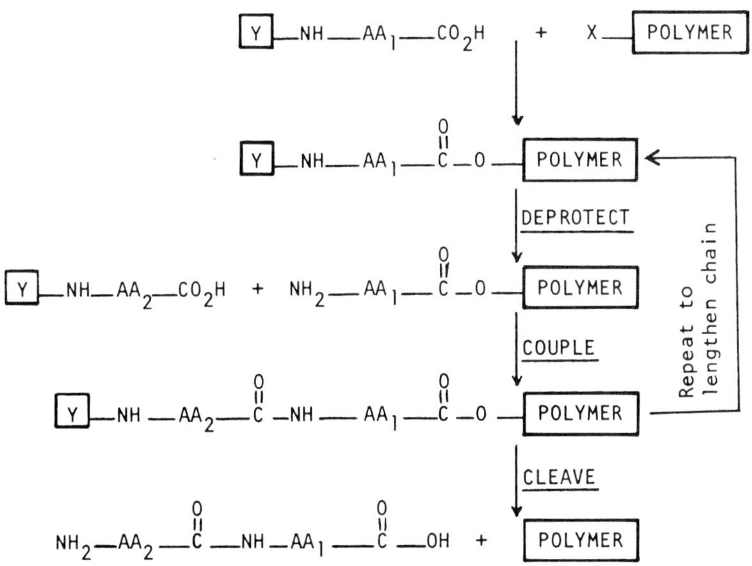

Fig. 1. The principle of solid phase peptide synthesis.

TABLE 1

Progress of Solid Phase Synthesis

Date	Peptide	Amino Acids
1963	A tetrapeptide	4
1964	Bradykinin	9
1965	Angiotensin II	8
1966	Insulin	21 + 30
1968	Clostridium ferredoxin	55
1969	Ribonuclease A	124
1969	Phe59 cytochrome c	104
1970	Pancreatic trypsin inhibitor	58
1971	"Human growth hormone"	188
1971	Parathormone 1-34	34
1971	Lysozyme	129
1971	Acyl carrier protein	74
1971	Salmon calcitonin	32-NH_2
1973	Human ACTH	39

In this review I shall not attempt to cover in detail the chemistry, methodology, and achievements of SPPS, as this has already been done several times, most recently by Meinhofer [9]. His review serves as an excellent summary of the method and the literature, including earlier reviews. On the contrary, after a brief introduction to the chemistry of SPPS, I shall concentrate on several problem areas which need application of the talents of organic chemists, and in particular on those problems associated with the resin support. I shall also give my personal opinions and suggestions for future developments.

THE CHEMISTRY OF PEPTIDE SYNTHESIS

The basic chemical operation in peptide synthesis is the formation of amide bonds. This very simple operation is complicated by several factors. Outstanding among these are the polyfunctional nature of amino acids, the problems associated with repetitive operations to synthesize macromolcules, the insolubility of the intermediates obtained, and the difficulty of purification of these intermediates.

All amino acids contain at least two functional groups, the amine and carboxyl functions, and for selective peptide synthesis one of these groups in each component of a reaction mixture must be blocked by a selectively removable group. In addition, many amino acids contain reactive side-chain functional groups which must be blocked during the synthesis. The blocking groups for these side-chain functions must be stable enough to remain intact throughout all synthetic operations, but must be fully labile to some reagent which can remove them quantitatively at the end of the synthesis without damage to the peptide chain. The proper design of a system of blocking groups is one of the major considerations in peptide synthesis, since several categories of groups must be available with different orders of stability. This is true both of classical and solid phase methods of synthesis. Closely associated with this problem is that of strategy of synthesis. There are several ways in which a large peptide chain may be assembled. For example, amino acids may be added one at a time, from one end or the other of the chain. This is the stepwise approach. Alternatively, small peptides may be synthesized and then these small peptide fragments may be combined into a large final product; this is the fragment condensation approach. Each of these methods has its advantages and disadvantages, which will not be discussed here, but have been adequately treated in reviews. Most large peptides synthesized by classical solution methods have been synthesized by the fragment condensation approach, while solid phase synthesis is a special case of the stepwise method in which the polymeric resin support serves also as blocking group for the carboxyl of the C-terminal amino acid, and amino acids are added one at a time to this residue bound to the polymer.

As amino acids with blocked side-chain functional groups are assembled into a peptide, the molecular weight increases rapidly and the solubility of the product in organic solvents usually shows a concomitant decrease. Once the peptide has reached any significant size, only powerful solvents such as dimethylformamide can be used as reaction media. In many cases even this is not adequate, and insolubility of intermediate products frequently causes very serious problems in solution methods of peptide synthesis. Failure to obtain complete reaction in coupling steps is a particularly onerous problem in fragment condensation methods in solution after the fragments have reached large size. The molar concentrations of the reactive groups are very low, and these groups are prevented from effectively finding each other by the large size and complex folding of the attached peptide chains. The resulting low yields cause serious losses of very expensive large-molecule intermediates. On the other hand, if the stepwise approach is used, physical and chemical properties of the large-molecule starting material and the product of a given coupling reaction may not be sufficiently different to allow effective purification, although coupling reactions can usually be more effectively forced to completion by use of a large excess of the small-molecule reactant.

All of these problems combine to make the prospect of synthesis of a large (50 amino acids or more) peptide or of a protein (100 amino acids or more) by solution methods a very forbidding one. Solid phase synthesis offers an approach to a solution of some of these problems, and is at the present time clearly the only practical method for synthesis of molecules in these size ranges.

THE CHEMISTRY OF SOLID PHASE PEPTIDE SYNTHESIS

The chemistry of the most commonly used system of SPPS, that introduced by Merrifield [11], is shown in Fig. 2. The support most widely used in SPPS so far has been a copolymer of styrene and divinylbenzene (DVB). Early work was done with a 2% DVB polymer, while recently a 1% cross-linked resin has been found to be more satisfactory. These resins swell very extensively in the solvents used for the synthesis. This is a crucially important point, since the synthesis takes place throughout the resin beads. There is hope

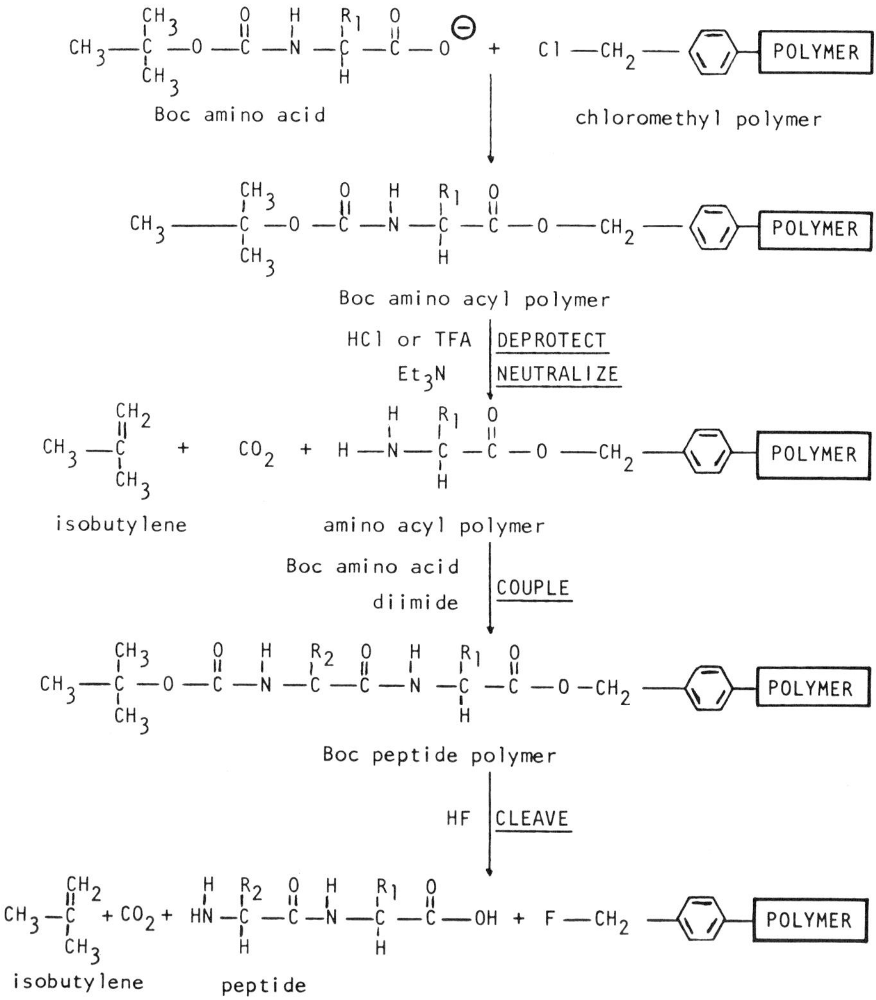

Fig. 2. Chemistry of the standard system of solid phase peptide synthesis.

of obtaining a homogeneous product from a synthesis only if all the reactions go fully to completion. In order to achieve this goal, reagents must be able to penetrate freely into the swollen resin matrix. The polystyrene is functionalized by chloromethylation, and the chloromethyl polymer is caused to react with a salt of the N-protected amino acid. The most common protecting group for the amino function is the tertiary butyloxycarbonyl (Boc) group. Reaction of the salt (usually with triethylamine) of the Boc amino acid with the chloromethyl polystyrene attaches the amino acid to the resin as an ester. This resin ester, in effect a substituted benzyl ester, is fairly stable to all of the reactions used in the synthesis. The Boc protecting group can be completely removed in most cases without loss of an excessive amount of the first amino acid (and resulting peptide) from the resin. However, in the synthesis of large peptides, or proteins, loss of the peptide from the resin by acidolysis can be a significant problem. Anhydrous strong acid in a suitable medium (such as HCl in dioxane or trifluoroacetic acid in chloroform or dichloromethane) will remove the Boc group. The protonated form of the amino group on the aminoacyl polymer is converted to the free base by treatment with a tertiary amine, usually triethylamine, and the next Boc amino acid is coupled to the amino group. The most commonly used coupling reagent is dicyclohexylcarbodiimide. This is a satisfactory procedure for coupling all amino acids except glutamine and asparagine. In these cases active esters are usually used to avoid the side reaction of dehydration of the omega amide to a nitrile (and incorporation of this derivative into the peptide). Following the coupling reaction, excess Boc amino acid and by-products can be washed out of the polymer suspension, leaving the pure peptide attached to the polymer. An excess (usually 2.5 to 5 equivalents) of each new Boc amino acid and coupling agent must be used to help assure complete coupling. The cycle of deprotection, neutralization, and coupling of the new amino acid is repeated as necessary to assemble the desired peptide chain on the polymer. Finally, the peptide is removed from the polymer by treatment with a stronger acid, usually anhydrous HF, or HBr in trifluoroacetic acid. The advantage of HF is that it also simultaneously removes all of the commonly used blocking groups from side-chain functions of the peptide, yielding the free peptide in one step. When HBr is used for cleavage of the peptide from the resin, some further step must often be used to remove certain blocking groups. It is noteworthy that all of the reactions of the synthesis and cleavage take place under anhydrous conditions, thus assuring that no hydrolysis of the peptide chain can occur. Further details of the chemistry of solid phase synthesis can be obtained from any of the reviews on the subject [8,9].

Several different kinds of peptide derivatives can be obtained from the synthetic peptide-resin. The different kinds of cleavage commonly used and the derivatives formed are shown in Fig. 3. Other types of linkage of peptide to the resin support have been developed and used for certain special purposes. One area that has received special emphasis is the synthesis of peptide amides. This is important since so many peptide hormones are amides. This fact, coupled with the problems which arise in ammonolysis from the resin of peptides containing glutamic and aspartic acids, has led to the development of resins which yield peptide amides directly by treatment with HF. The most widely used of these is the benzhydrylamine (BHA) resin [12], in which the additional activation of the added phenyl group on the carbon bearing the amine function is adequate to enable cleavage of the C-N bond by HF (Fig. 4). Although this resin has been successfully used for the synthesis of a number of peptide amides, certain peptides, notably those with carboxyl-terminal phenylalanine, are not completely removed from the resin by the normal HF treatment. Clearly a more labile type of peptide-resin link is needed for synthesis of some peptide amides. We investigated this problem and found [13] that the ease of removal of peptides from the BHA resin depends on the nature of the C-terminal amino acid of the peptide. Histidine peptides appear to be removed most easily, with glycine, glutamic acid, threonine, proline, methionine, valine, leucine, and phenylalanine peptides being increasingly more stable. Hardly any of the amide of angiotensin II (terminating in phenylalanine) could be removed from the BHA resin by standard HF treatment (0°, 30 min).

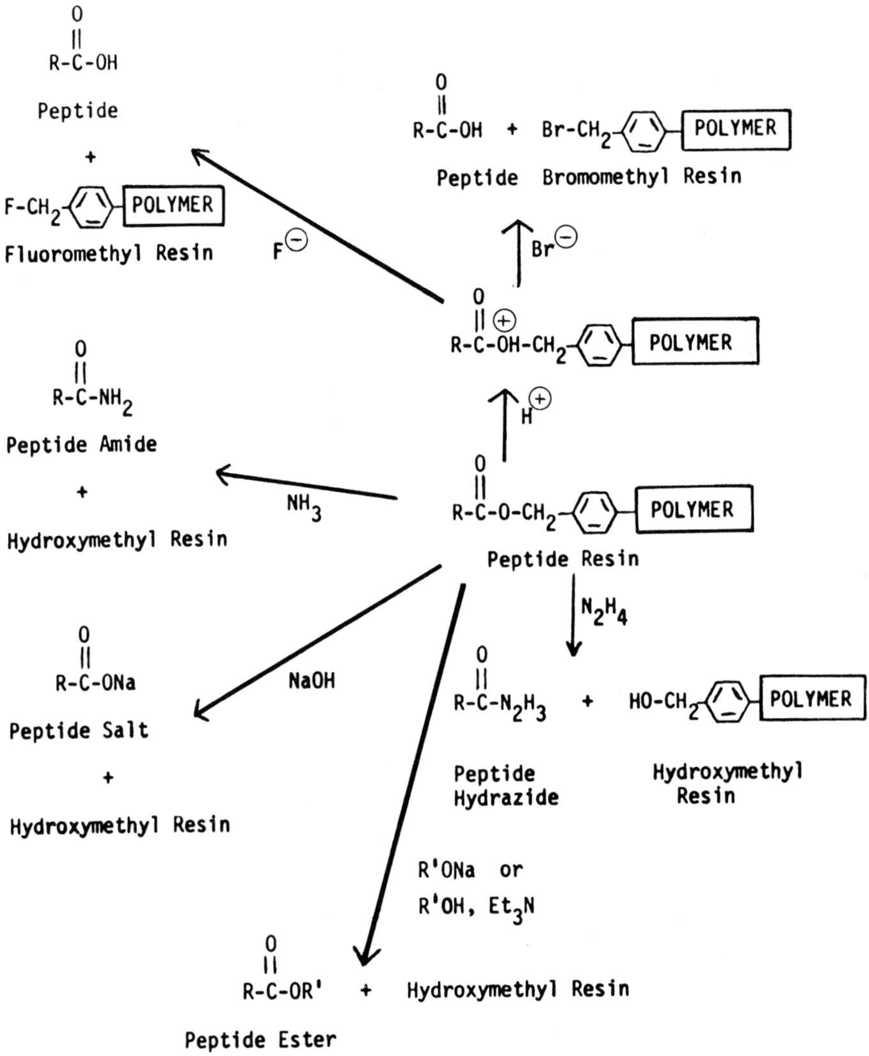

Fig. 3. Cleavage of peptides from the resin.

Use of higher temperature and longer time for HF cleavage is not satisfactory because of harmful side reactions. Of the several resin types examined as potential solutions to this problem, the

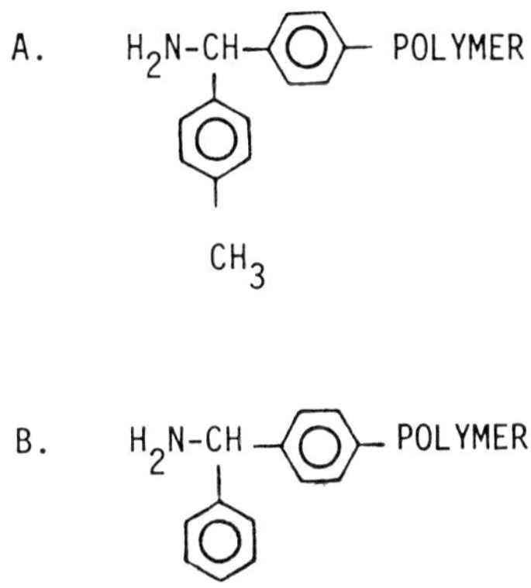

Fig. 4. The benzhydrylamine (A) and p-methylbenzhydrylamine (B) resins for synthesis of peptide maides.

p-methylbenzhydrylamine (MBHA) resin (Fig. 4) appears to be a useful complement to the BHA resin. It does not seem that a single amine resin will offer the requisite stability to Boc deprotection reagents while maintaining sufficient HF lability. When used with Boc alpha protection, MBHA is suitable for all short peptides and long peptides having labile C-terminal residues. MBHA should be satisfactory for synthesis of all peptides if an alpha blocking group more liable than Boc is used.

Some investigators have worked on the solid phase synthesis of small peptides which would subsequently be coupled together in solution (or even on solid phase resins) to synthesize larger peptides. This was the goal of the hydrazine cleavage of the peptide-resin indicated in Fig. 3. A better approach to this problem is probably that of Wang and Merrifield [14], who synthesized a resin carrying the equivalent of Boc hydrazide groups on the polymer (Fig. 5), which after peptide synthesis would yield protected peptide hydrazides directly by acidolysis. These could be converted to azides for direct coupling to other peptides.

PROBLEMS IN SOLID PHASE SYNTHESIS

In spite of the impressive progress made in solid phase synthesis in the decade since its introduction, many serious

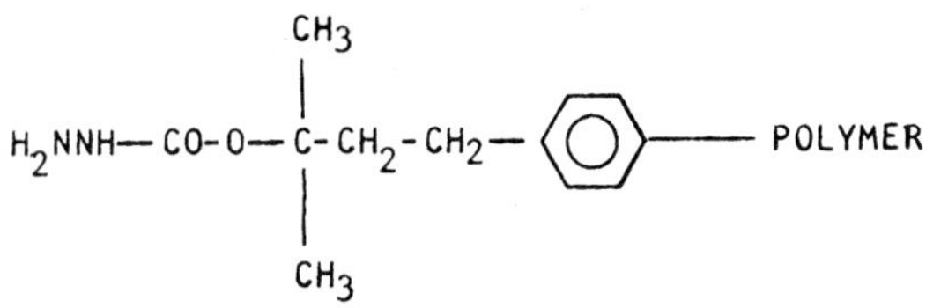

Fig. 5. The resin for synthesis of protected peptide hydrazides.

problems still remain to be overcome. The desirable goal of 100% reaction at each stage has not been generally attainable, although in certain cases (e.g., 98% yield of a decapeptide [15]) the mark has been missed by only a very narrow margin. In the case cited, the average coupling yield must have been at least 99.93%. Many of the peptides listed in Table 1, particularly the large ones, were obtained only as very impure materials. At the present time reliance must be placed on purification of the final product if one hopes to obtain homogeneous material. While this is feasible for small peptides, it is not realistic if one hopes to synthesize pure proteins (chains of more than 100 amino acids) or large peptides. Currently available purification techniques are clearly not selective enough to separate, for example, a synthetic chain of the size of ribonuclease (124 amino acids) from a 123 amino acid chain in which the deletion was a simple aliphatic amino acid such as glycine or alanine. While some peptide chemists believe it will be possible to synthesize essentially homogeneous large peptides by SPPS when sufficient improvement has been made in blocking groups and resin supports, others feel that this is not a realistic goal, and that pure products can be obtained only by combination of small peptides into larger chains. In this latter approach, the small fragments can be purified exhaustively, and then when they are coupled together to provide the larger peptides the differences between the desired product and the incorrect products will be sufficiently large that available purification techniques can offer some hope for achieving homogeneity.

Other difficulties have arisen in the synthesis of some peptides because of unexpected idiosyncrasies in the behavior of the peptide. These difficulties appear to be sequence dependent and are probably due to interactions of one part of the peptide chain with another or with the resin matrix, and cause certain sequences to show unusual difficulty of removal of blocking groups or coupling of protected amino acids. Some of these problems may well be improved by new types of blocking groups for the various functional groups of amino acids, or by new types of coupling reactions. Other problems are clearly due to the nature of the polymeric support used and to interactions between the peptide chain and the support, and demonstrate the need for improved types of supports.

PROBLEMS ASSOCIATED WITH THE PEPTIDE CHAIN

A suitable system of stable and labile groups for the various functions of the amino acids is an essential requirement for any kind of peptide synthesis. The "stable" groups should be completely stable throughout the synthetic operations and should be removable at the end of the synthesis by means of a reagent which will cause no harm to the peptide itself. In SPPS these stable groups are the blocking groups for the side-chain functions and the bond attaching the carboxyl-terminal amino acid to the resin. The "labile" blocking group is that which protects the α-amino function of each amino acid as it is coupled onto the peptide and is removed before attachment of the next amino acid. This labile blocking group must be completely removable by a reagent which will neither remove any of the other "stable" blocking groups nor cleave peptide from the resin. These requirements are not fulfilled completely by the system of SPPS chemistry shown in Fig. 2. An illustration will point up these problems.

Perhaps the most outstanding achievement of the solid phase method was the synthesis by Gutte and Merrifield [10] of a protein (124 amino acid residues) with ribonuclease activity (Table 1).

During the synthesis of this protein it became apparent that losses of the peptide from the resin at each step were significant, so that at the end of the synthesis only 16% of the original amino acid substituted on the resin still remained. Furthermore, the crude product contained a very large proportion of shorter peptides. These can arise either when peptide chains fail to continue to grow ("truncation sequences") because of failure of complete removal of the labile blocking (Boc) group, or of failure to achieve further coupling of amino acids to a chain, or when chains do not grow for a time and then resume growth ("deletion sequences"). Deletion sequences can arise at places in the peptide chain where either deprotection or coupling reactions are very slow if insufficient time is allowed for their completion, or if truncated sequences become again reactive at a later stage of the synthesis. Non-reactivity of a peptide chain on the resin is apparently due to peptide-peptide or peptide-resin steric interactions. As the chain grows, the microenvironment of the peptides changes, and truncated ends previously hindered may again become available due to changes in the secondary or tertiary structure of neighboring peptide chains on the resin. Another problem encountered in the synthesis of ribonuclease and other large molecules was slow removal of side-chain "stable" blocking groups during the synthesis, exposing the reactive groups with consequence branching of the peptide chain and contamination of the product with molecules having molecular weights larger than that of the desired product.

With respect to loss of peptide from the resin, evidently one needs for this kind of synthesis a greater difference in the stability of the α-blocking group and the link attaching the peptide to the resin. A solution to this problem might be found either in the use of more labile α-blocking groups or in the use of a more stable bond linking the peptide to the resin. The latter approach is not practical since the currently used HF treatment for removal of peptides from the resin is apparently too harsh for some peptides, and will probably be undesirable for most proteins. Several α-blocking groups are available which are more labile than the Boc group, but there are also problems associated with their use. The so-called Bpoc group (see Ref. 9, p. 111) is extremely labile and has been used in a number of syntheses, but this greater lability makes the derivatives very unstable and inconvenient to use. Recently Birr [16] has introduced the Ddz group, of a reactivity intermediate between Boc and Bpoc, which he feels is labile enough to have most of the advantages of Bpoc protection without the extreme instability. The very mild acid needed for complete removal of either of these groups should not cause significant loss of peptide from the resin. Several research groups, including my own, are currently working on other labile blocking groups and believe that the α,2,4,5-tetramethylbenzyloxycarbonyl (TmZ) group offers an ideal combination of properties [13]. The TmZ group is cleaved acidolytically 4200 times as fast as the Boc group, 1/5 as fast as the DdZ group, and 1/30 as fast as the Bpoc group. Removal of the TmZ group by 0.5% trifluoroacetic acid can be monitored photometrically, and the optical activity of the group should help assure optical purity of the amino acids used in synthesis.

The use of a modifiable link for attachment of the peptide to the resin has been proposed by several investigators, and has been called the "Safety-Catch Principle" by Sheppard [17]. In this approach the first amino acid is attached to the resin by a bond which is quite stable. At the end of the synthesis, the bond is modified selectively by a chemical reaction to another form which can be very easily cleaved. One approach to such a method was described by Marshall and Liener, and by Flanigan and Marshall, who used a thioether bond in the peptide-resin link. After the synthesis the thioether was oxidized to a sulfone which activated the ester bond linking the peptide to the support and made it quite labile. This example suffers from the need to use an oxidative reagent, and is not generally applicable, but it illustrates the principle. Different types of "safety-catch" links have been proposed by others; the most promising may be the use of acid-catalyzed elimination of a tertiary alcohol to labilize the attaching link before cleavage [18].

The problem of loss of side-chain blocking groups during synthesis appears to be much improved at the present time over the situation prevailing in the "standard" SPPS system [8] due to the introduction of more stable blocking groups to replace those that had been found to be too labile [19]. The introduction of blocking groups for the amides of asparagine and glutamine is a significant improvement because it allows diimide-mediated coupling reactions to be used for these amino acids. In one case this led to a 20% increase in the yield of a simple peptide. It is not certain yet what the ideal blocking group for the amides will be. Allowing for the fact that blocking groups can be abnormally stable in certain sequences, the proposed benzhydryl group mav be too stable for glutamine. A more labile group, such as xanthydryl, may be more useful at the present time. It does allow the use of diimide coupling reactions with attendant yield increases, but is so labile that it is removed at the first acidolytic step for deblocking Boc groups. Thus it may not effectively protect against glutamine cyclization or acidolytic deamidation, two other problems of these amino acids. A blocking group of intermediate stability is needed. Another amino acid which can still give problems is histidine. The best derivative at the present time is toluenesulfonyl (Tosyl) histidine, but the Tosyl group appears to be marginally stable for some purposes. For example, Boc Tosyl histidine must be kept as the salt for storage at room temperature; after liberation of the acid it should be stored in the freezer. Synthesis of C-terminal histidine peptides using this derivative should be done on hydroxymethyl resin, due to base-catalyzed detosylation when coupling is attempted in the usual way with chloromethyl resin. The Tosyl group is currently the best blocking group for the guanidino group of arginine, but the requirement for use of HF for cleavage is a difficulty for very large-scale synthesis. The new boron tris-trifluoroacetate cleavage reagent [20] may offer a solution to this problem. The nitro blocking group is not acceptable due to degradation of nitroarginine to ornithine. A new approach is needed for methionine, which can cause problems in some syntheses. The sulfoxide has seen some application as a protecting derivative, but cannot always be reduced readily. In all of these cases, work is needed on development of new blocking groups.

The basic cause of the problems in currently used blocking groups is that the system depends on quantitative (rate) differences in ease of cleavage of the different bonds in the same reaction, namely acidolysis. What is needed is for all the side-chain blocking groups and the peptide-resin link to be susceptible to a type of cleavage reagent that is qualitatively different from that used for removal of the labile (α-amino blocking) group. The specifically labilized "safety-catch" peptide-resin link mentioned above would be a first approximation to this principle. This area and the ones discussed in the preceding paragraph urgently need the attention of skilled organic chemists who also understand the peculiar problems of peptide synthesis.

There is need for general improvement of coupling reactions in SPPS. Dicyclohexylcarbodiimide (DCC) has been by far the most widely used condensing agent in SPPS, largely because of the speed of DCC-mediated coupling reactions and the convenience of its use. With the advent of blocking groups for asparagine and glutamine, it can be used with all amino acids. Racemization has not generally been a problem except with histidine or im-benzyl histidine (Tosyl histidine is satisfactory). Particularly when used with an additive such as hydroxybenzotriazole [21], it may be the condensing agent of choice for SPPS. Some investigators have claimed improved yields in SPPS through the use of Boc amino acid symmetrical anhydrides, even though it has also been claimed that the symmetrical anhydride is the actual acylating agent in solid phase coupling reactions mediated by DCC. At the moment there is no clear explanation for this, although one might be found in possible harmful effects of the dicyclohexyl urea which precipitates in the resin and has been thought by some persons to cause occlusion of otherwise reactive peptide amino groups. This has been offered as the explanation for the observed greater efficacy of double coupling over single coupling with prolonged times for obtaining complete SPPS coupling reactions. In the double coupling procedure the two coupling reactions were separated by washing with a solvent which will dissolve and remove the urea. Repetition of the triethylamine neutralization step is usually effective for this purpose, and should always be done before repetition of the coupling reaction. Whether the actual acylating agent in DCC-mediated couplings is the symmetrical anhydride or the O-acyl isourea, either of these is a very bulky reagent, and some new type of coupling agent of very small size might lead to significant improvement of SPPS coupling reactions.

In order to evaluate difficulties such as those described above, one must have methods for accurate monitoring of the deprotection and coupling reactions [22]. At the present time, no fully satisfactory solution to this problem is available. Monitoring of deprotection is a very difficult problem, as the last fraction of a percent of undeprotected amino group must be determined in the presence of a large amount of free amino groups. Neither methods based on estimation of amount of blocking group removed or amount of free amino group produced on the resin appear to have the requisite sensitivity, although they may be useful to warn the operator of a gross failure of the deblocking reaction. The use of radioactive Boc groups has been suggested, and may have the requisite sensitivity, but this appears to be a very expensive and impractical approach. Titration of amino groups formed in the deprotection reaction has been used by some to monitor deprotection, but some peptide chemists feel that operations such as this may have harmful effects on the peptide and cause the monitoring process to be partially self-defeating. Better success has been obtained in monitoring of coupling reactions, for in this case one goes to an end point of zero and small residual amounts of uncoupled amino groups are theoretically easier to determine. The widely used ninhydrin method lacks sensitivity, the naphthaldehyde method is slow, and the very sensitive fluorescamine method [23] does not detect proline amino groups and appears to suffer from spurious fluorescence, especially when used on the standard Merrifield resin. The use of xanthydryl asparagine and glutamine also causes problems for fluorescamine monitoring, since the xanthydryl group is fluorescent. Improvements are needed in both these types of monitoring.

The method of attachment of the first amino acid to the conventional Merrifield chloromethyl polystyrene resin has recently been shown to be a source of problems in SPPS. The chloromethyl groups are usually present on the resin at a much higher concentration than that desired for the first amino acid. Therefore, these groups are not usually completely replaced during the amino acid attachment step. Originally, because he was worried about alkylation of peptide chains by chloromethyl groups on the resin, Merrifield converted all of the remaining chloromethyl groups to acetoxy groups. Subsequently it was found that good yields of peptides could be obtained without this added step, and it was usually omitted. However, in some cases where investigators have looked carefully at the amount of amino groups on the resin throughout a synthesis, loss of amine not attributable to loss of peptide from the resin could be observed. There have also been cases of sudden catastrophic chain termination, notably when neutralized peptide-resins (after the triethylamine step) were left standing for extended periods of time. A logical explanation for these observations is alkylation of the peptide amino group by residual chloromethyl groups. Another difficulty with residual chloromethyl groups is that they are slowly converted to quaternary groups by the neutralizing amine at each step of the synthesis, particularly if triethylamine is used. These quaternary groups, along with those already present from coupling the initial amino acid to the chloromethyl resin, confer undesirable ion-exchange properties upon the polymer. Rather than try to avoid this side reaction by use of the expensive diisopropylethylamine (on which people may skimp due to the expense and risk incomplete neutralization--a situation made even worse by the hindered structure of the amine), it would appear to be more logical to avoid both these problems by elimination of residual chloromethyl groups. This can be done either by Merrifield's original method of conversion to acetoxy groups after attachment of the first amino acid or preferably by using a different method of attachment of the first amino acid to the resin. During the conventional method of attachment of the first amino acid by reaction of the triethylammonium salt of the Boc amino acid with the chloromethyl resin, some triethylamine reacts with chloromethyl groups to form quaternary ammonium groups on the resin. These, as well as the additional quaternary ammonium groups formed gradually during the synthesis, can cause serious problems. If either acetic or trifluoroacetic acids is used in the deprotection step, some amount of these acids may be strongly bound to the quaternary ammonium groups by ion exchange and resist complete removal by the amine neutralization step. Subsequent use of diimide coupling agents may activate these acids and cause termination of the peptide chain by acetylation or trifluoracetylation. Both of these side reactions have been observed. It has been suggested by some investigators that the superior results they have obtained by the use of preformed Boc amino acid symmetrical anhydrides in the coupling reaction are due to lack of activation of residual trifluoroacetic acid groups on the resin by DCC. Quaternary ammonium groups on the resin also interfere with some kinds of monitoring during the synthesis. Such monitoring is essential for effective synthesis of long peptide chains. A logical solution to all these problems would appear to be avoidance of direct use of chloromethyl resin for attachment of the first amino acid. Two alternate attachment techniques that have been suggested (see Ref. 8) are the use of hydroxymethyl resins or sulfonium resins. One of these methods should probably be chosen for the synthesis of long peptides.

Alternatively the first amino acid could be attached to the chloromethyl resin by the Loffet procedure [24], followed by conversion of all remaining chloromethyl groups to acetoxy or other innocuous groups before proceeding with the synthesis. Recent observations in several laboratories that synthesis of a particular peptide sequence on a benzhydrylamine resin may give a far better overall result than synthesis of the same sequence on the classical Merrifield resin may be due to the fact that no quaternary groups exist or can be formed on the former.

PROBLEMS ASSOCIATED WITH THE POLYMERIC SUPPORT

In many cases of failure to achieve quantitative reactions in SPPS it is difficult to differentiate clearly between those problems caused primarily by the peptide sequence and those caused primarily by the nature of the resin used as the support for the peptide. In certain cases, however, the latter cause seems to be clearly responsible. An early example of this type of problem was encountered by Merrifield in the synthesis of angiotensinyl bradykinin. At that time both angiotensin and bradykinin had been satisfactorily synthesized on 2% cross-linked resins using HCl in glacial acetic acid as the reagent for the deprotection. However, when the synthesis of the longer chain was attempted, in which the sequences of angiotensin and bradykinin were added together, a difficulty was encountered during the second half of the synthesis. This difficulty was failure to remove the Boc group completely at one step of the synthesis. The problem was overcome by making two changes. First, the cross-linking of the resin was reduced to 1%, giving a product which swells much more extensively and which apparently allows better penetration of reagents into the resin. Second, the reagent for deprotection was changed from HCl-HOAc to HCl in dioxane. Whereas the previous reagent does not swell the resin extensively, the dioxane reagent swells the resin to the maximum attainable with any solvent. Four years later a claim was made in the literature that a group of peptides could not be synthesized by solid phase because of failure to deprotect. It was later shown that this difficulty could be overcome when reagents were used which swelled the resin maximally and when the resin cross-linkage was reduced to 1%. Very recently a 0.5% cross-linked resin has been described by Birr [18], who has said that it improves the synthesis of certain peptides. It is not yet possible to give a complete evaluation of this resin. This new resin swells twice as much as the 1% cross-linked resin in dichloromethane, but is very gelatinous and appears to require special equipment for handling.

The logical ultimate development in this direction would be the use of a noncross-linked resin. However, linear plystyrene is soluble in the solvents used for synthesis, and serious chemical and mechanical problems arise in handling such material. Some work has been done in this area, the most recent being a proposal for use of polyvinyl alcohol as the polymeric support [25], with ultrafiltration as the means for separating the peptide-polymer from low molecular weight materials. The use of such soluble polymeric carriers introduces several new problems and greatly complicates the mechanical operations of synthesis. These materials have not as yet received extensive application or thorough evaluation.

One probable cause for discrepancies in attempts to compare SPPS results from different laboratories or even from different batches of resins is the lack of precise knowledge by the chemist of the exact degree of cross-linking of the resin he is using at any time. Due to the different reactivities of styrene and divinylbenzene, the cross-links are probably concentrated near the core of the resin beads; this may actually be a good thing for SPPS. What is more important and probably even less appreciated is that the chloromethylation reaction is an effective means for cross-linking styrene-DVB polymers. Patterson has stated [26] that the common Dowex 2-X8 (nominally 8% cross-linked) is synthesized by chloromethylation and subsequent aminolysis of a styrene-2% DVB polymer; the additional cross-linking is introduced in the chloromethylation step. Other chemists [27], attempting to use soluble linear polystyrene as a carrier for peptide synthesis, found that it was very difficult to chloromethylate this polymer without cross-linking and consequent insolubilization. They also found that cross-links were introduced during the neutralization steps of the peptide synthesis (chain termination?!). During the chloromethylation reaction, the chlorine content of the resin first rose and then fell as the chloromethyl groups already introduced reacted with other aromatic nuclei to establish cross-links.

The significance of this for standard SPPS is that chemists should be very careful in conducting the chloromethylation reaction to see that it is done under the most gentle conditions possible for introduction of the requisite number of chloromethyl groups. It is obviously not possible to know about the conditions of chloromethylation used in the production of commercial resins, but a high chlorine content is unnecessary and may mean that the reaction was pushed vigorously, with concomitant cross-linking. Moreover, a low chlorine content may mean that the reaction was done under such vigorous conditions that most of the chloromethyl groups introduced were used up in cross-linking reactions.

The best measure of cross-linking of chloromethylated resins is the degree to which they swell in a solvent such as dichloromethane. Some guidelines for such estimation have been published by Fankhauser and Brenner [28]. The chemist would be wise to check each new batch of commercial resin in this way before using it in synthesis.

Another probable cause of confusion in interpretation of results in SPPS is the fact that the usual polystyrene-DVB beads contain a significant amount of linear polystyrene trapped in the cross-linked matrix. These linear chains may be chloromethylated and participate in peptide synthesis, only to dissolve gradually in the reaction solvents and be lost, along with their load of peptide. To avoid the loss, the beads can be extracted with hot trifluoroacetic acid before use. This treatment may remove as much as 20% of the weight of some batches of beads.

Since many of the difficulties encountered in SPPS appear to be related to penetration of reagents into the polymer matrix, several investigators have tried supports which are nonporous and in which the reactive sites are limited to the surface. A pellicular layer of polystyrene coated on glass beads has been suggested, but appears to lack the requisite mechanical stability, and additionally did not appear to offer significant chemical advantages. Glass beads have also served as carrier for peptide synthesis in systems involving covalent linkage of the functional group to the silicon of the glass particles. At the present time there has been little evaluation of such systems. Another system which has received some attention consists of "linear" polystyrene grafted onto Kel-F particles. This type of carrier has been used for several syntheses but does not appear to offer the needed improvements in the system. One of the problems with this resin may be that the added polystyrene is not truly linear, or that the added layer is too thick. Other investigators have reported on work with macroporous, macroreticular, and "popcorn" types of polymers. All of these surface-limited supports suffer from the disadvantage of having very low capacity.

SOME SUGGESTIONS FOR NEW POLYMERIC SUPPORTS

An ideal insoluble support for solid phase peptide synthesis would have several characteristics.

1. The support should have good mechanical properties. It should be possible to conduct manipulations such as agitation and filtration without mechanical degradation of the polymer or excessive filtration times.

2. The support should be chemically compatible with a good system of peptide synthesis. The chemist should be able to use a variety of kinds of links for attaching the peptide to the resin and of blocking groups for the amine function as well as side-chain functions. For example, a glass support is not compatible with the use of HF unless one visualizes complete dissolution of the glass as a means of liberating the peptides.

3. The support should not cause problems due to steric interactions with the peptide chain. This probably means that a very

low degree of cross-linkage, or no crosslinkage at all, will be necessary.

4. A high level of substitution should be possible. This is desirable so that huge amounts of support need not be handled in order to synthesize a reasonable amount of peptide. This is difficult to attain in systems having the reaction restricted to the surface of the material, and division of the support into particles small enough to have a very large surface will probably cause serious mechanical problems in handling.

5. The matrix and the peptide chain should probably be solvated to a similar degree by solvents suitable for reaction media. This requirement may mean that polystyrene-based resins are not ultimately desirable, and that new types of polymer need to be examined. A suitable reaction medium should probably be protic, and should be helix-breaking for protected peptides.

It is difficult to visualize any conventional polymer which would meet all of these criteria, and new polymeric supports will probably need to be designed and synthesized. One novel suggestion was that solid phase synthesis might be carried out in continuous belts, films, or fibers of support material, using automatic processing equipment.

It has been suggested by Sheppard [17] that a major difficulty in solid phase synthesis may be the different nature of the polymeric support and the peptide chain which has grown on the support. For successful synthesis, presumably both the polymer carrier and the peptide chain should be fully extended and solvated (Fig. 6A). The polystyrene matrix usually used is nonpolar and swells in nonpolar solvents. On the contrary, the peptide chain is polar and should be expanded by polar solvents which may collapse the polystyrene matrix (Fig. 6B). The nonpolar solvents which expand the matrix may collapse the peptide chain (Fig. 6C), thus causing many of the serious problems, apparently due to steric interaction, which have been observed. Following his own suggestion, Sheppard [29] has examined cross-linked polyacrylamide as a carrier for SPPS and has reported that the synthesis of one peptide which had been

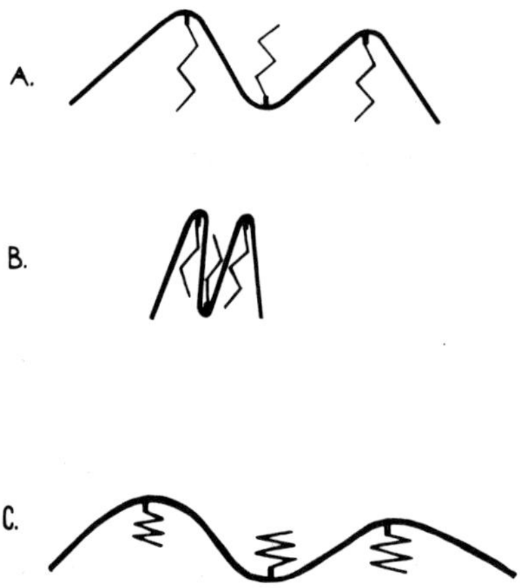

Fig. 6. Peptide-resin interactions in solid phase peptide synthesis (A) The ideal case, with both polymer matrix and peptide chain extended. (B) In polar solvents, with the hydrophilic peptide chain extended but the matrix collapsed. (C) In nonpolar solvents, with the matrix expanded but the peptide collapsed.

very difficult in the conventional SPPS system was much improved. Polar solvents were used throughout the synthesis. It has been observed that the nature of the resin beads changes during a synthesis. After attachment of several amino acids the conventional resin no longer swells as extensively in nonpolar solvents such as dichloromethane and chloroform. It has been found that with certain peptide resins, coupling reactions which could not be forced to completion in dichloromethane as solvent did go to completion when dimethylformamide or urea was added to the reaction mixture. Recently we found do significant change in swelling of a benzhydrylamine resin after assembly of a long peptide chain.

Perhaps a useful approach to a solution of some of the resin-associated problems of SPPS would be to synthesize a new type of carrier polymer such as that depicted schematically in Fig. 7.

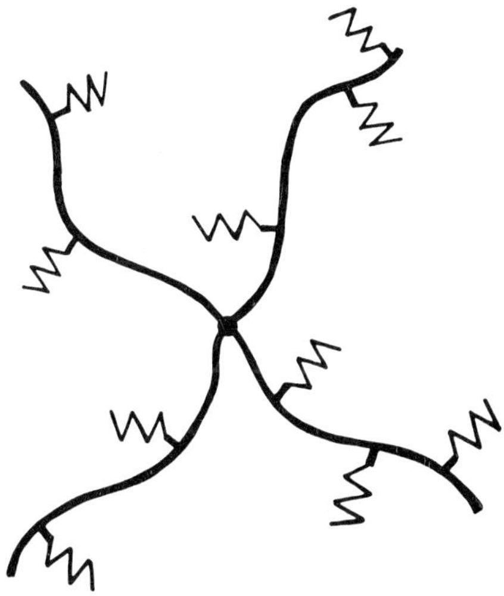

Fig. 7. A proposed ideal support resin for solid phase peptide synthesis.

This type of material, called a "Hippie" resin by Rudinger, would have a dense, impermeable core, with long "hairs," or fibers, extending from it. These "hairs" should have reactive sites at suitable intervals for attachment of the first amino acid of the peptide chain. In this way the peptide chains would be able to grow freely outward from the center without encountering serious steric hindrance from a polymer matrix. The "whiskers" extending from the core might be composed of linear polystyrene chains, with the peptide attached in the usual way (Fig. 8), but more realistically should probably be relatively hydrophilic in nature so that they would be solvated by the same types of solvents which solvate the peptide chain. This might be achieved in a vinyl-type linear polymer by incorporating acrylamide groups (Fig. 9), or it might be necessary to go to a polyamide type of chain (Fig. 10). The reactive groups for attachment of the first amino acid should probably be incorporated into these whiskers during polymerization, rather than introduced later, as is the case with the present chloromethyl

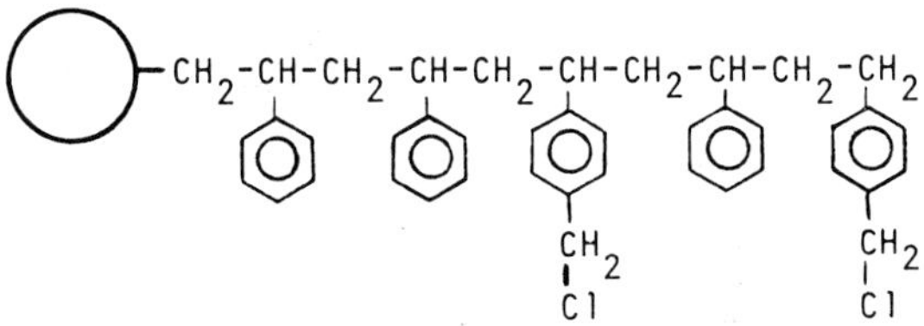

Fig. 8. Suggested polymer chains based on polystyrene.

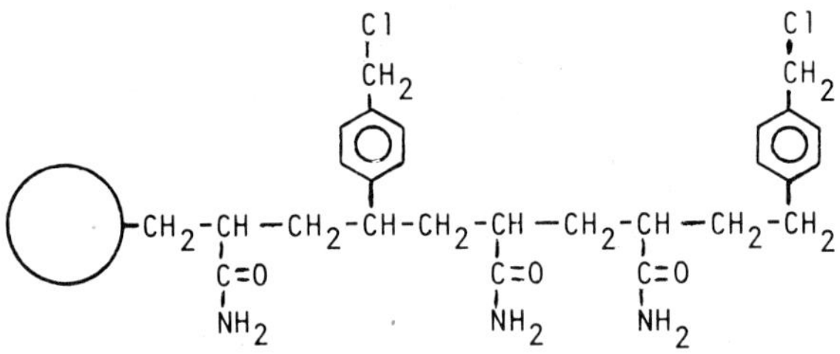

Fig. 9. Suggested polymer chains based on polyacrylamide.

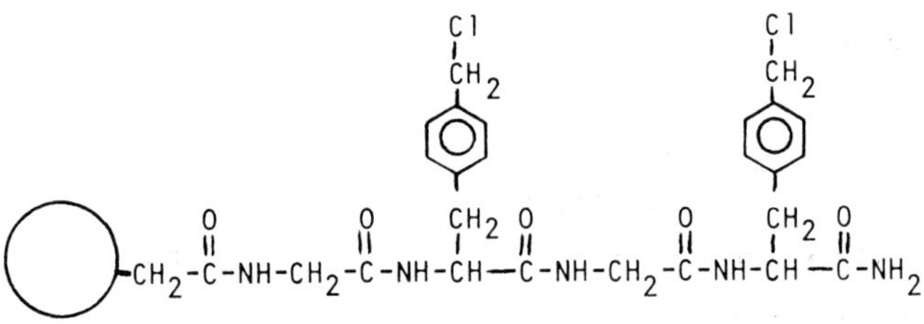

Fig. 10. Suggested polar polyamide resin chains.

resin. Perhaps some polymer chemist, reading this account of our problems and needs, will be inspired to design and synthesize such a polymer.

CONCLUSION

This account, which has concentrated on the present difficulties of SPPS, may give the reader a feeling of pessimism about the future of this method. It is by no means the intention of the author to cause such a reaction. On the contrary, I am very enthusiastic about SPPS, and hasten to express the debt of gratitude I and all peptide chemists owe to Bruce Merrifield for the invention of this ingenious and important technique. I feel sure that eventually the wise men in Stockholm will see fit to honor him for this achievement, as others have already done. My real purpose has been to help the inexperienced and perhaps overenthusiastic follower of SPPS to avoid pitfalls that may lead to discouragement or ever bitter disillusionment, and to encourage organic and polymer chemists to work on the present problems of SPPS so that its full potential may be more rapidly realized. The future importance of peptide synthesis will surely exceed all our present expectations, possibly even our wildest pipe dreams. The areas of biological control, such as endocrinology and neurochemistry, offer many exciting prospects. Let us all work together, freely exchanging ideas and accomplishments, that we may rapidly solve the problems and meet the needs that only we can fulfill.

REFERENCES

[1] E. Wunch, ed., Syntheses von Peptiden, Theimie, Stuttgart, 1974, 2 vols.; E. Schröder and K. Lübke, The Peptides, Academic, New York, Vol. 1, 1965; Vol. 2, 1966.

[2] R. Walter and J. Meienhofer, eds., Peptides: Chemistry, Structure and Biology, Ann Arbor Science Publishers, Ann Arbor, Michigan, 1975, and previous volumes in this series.

[3] Y. Wolman, ed., Peptides 1974, Wiley, New York, 1975, and previous volumes in this series.

[4] G. T. Young, ed., Amino Acids, Peptides, and Proteins, Specialist Periodical Reports, The Chemical Society, London, Vol. 6, 1975 and earlier annual volumes.

[5] R. B. Merrifield, J. Am. Chem. Soc., 85, 2149 (1963).

[6] R. M. Merrifield, J. M. Stewart, and N. Jernberg, Anal. Chem., 38, 1905 (1966).

[7] Beckman Instruments, Palo Alto, California; Vega Engineering, Tucson, Arizona.

[8] J. M. Stewart and J. D. Young, Solid Phase Peptide Synthesis, Freeman, San Francisco, 1969.

[9] J. Meienhofer, in Hormonal Proteins and Peptides (C. H. Li, ed.), Academic, New York, 1973, p. 46.

[10] B. Gutte and R. B. Merrifield, J. Biol. Chem., 246, 1922 (1971).

[11] R. B. Merrifield, Biochemistry, 3, 1385 (1964).

[12] P. G. Pietta, P. F. Cavallo, G. R. Marshall, and M. I. Pace, Gazz. Chim.Ital., 103, 483 (1973).

[13] G. R. Matsueda and J. M. Steward, Unpublished Data.

[14] S. S. Wang and R. B. Merrifield, J. Am. Chem. Soc., 91, 6488 (1969).

[15] T. Wieland, C. Birr, and F. Flor, Justus Liebigs Ann. Chem., 727, 130 (1969).

[16] C. Birr, Ibid., 763, 162 (1972).

[17] R. C. Sheppard, in Peptides 1971 (H. Nesvadba, ed.), North Holland, Amsterdam, 1973, p. 111.

[18] C. Birr, F. Flor, P. Fleckenstein, and T. Wieland, Ibid., p. 175.

[19] See papers by Erickson and Yamashiro in Chemistry and Biology of Peptides, (J. Meinhofer, ed.), Ann Arbor Science Publishers, Ann Arbor, Michigan, 1972.

[20] W. Bauer and J. Pless, in Ref. 2, p. 341.

[21] W. Konig and R. Geiger, Chem. Ber., 103, 2034 (1970).

[22] See Ref. 9, p. 169.

[23] S. Udenfriend, S. Stein, P. Bohlen, W. Dairman, W. Leimgruber, and M. Weigle, Science, 178, 871 (1972).

[24] A. Loffet, Int. J. Pept. Protein Res., 3, 297 (1971).

[25] E. Bayer and M. Mutter, Nature (London), 237, 512 (1972).

[26] J. H. Patterson, in Biochemical Aspects of Reactions on Solid Supports (G. Stark, ed.), Academic, New York, 1971, p. 189.

[27] B. Green and L. R. Garson, J. Chem. Soc., C, 1969, 401.

[28] P. Fankhauser and M. Brenner, in The Chemistry of Polypeptides (P. G. Katsoyannis, ed.), Plenum, New York, 1973, p. 389.

[29] E. Atherton, D. L. J. Clive, and R. C. Sheppard, J. Am. Chem. Soc., 97, 6584 (1975).

RIGID SUPPORT MATERIALS FOR THE IMMOBILIZATION OF ENZYMES

Garfield P. Royer, Gail M. Green, and Biranda K. Sinha

Department of Biochemistry
The Ohio State University
Columbus, Ohio 43210

ABSTRACT

The purpose of the work presented here was to prepare a support material for enzymes and "affinity ligands" with the following characteristics: low cost, durability, rigidity, and high capacity. Our study encompassed conjugates of porous and nonporous silicas with organic polymers and macroporous ion-exchange resins. Polyethyleneimine (PEI), polyacrylic acid (PAA), poly(methyl vinyl ether/maleic anhydride) were attached to porous glass and silica in various combinations. The composite of silica beads with PEI and PAA is a good support for the enzyme trypsin as judged by the activity against N-α-benzoyl-L-arginine ethyl ester.

Amberlyst (macroporous, sulfonated polystyrene) was activated by treatment with thionyl chloride; the resulting resin was either used directly or reacted with a diamine. The diamine derivative was used for enzyme coupling or transformed further to the succinyl or p-aminobenzoyl derivative. None of these derivatives were particularly good as supports for the enzyme trypsin. Duolite converted to a PAA, succinyl, or succinimide derivative was a good support. The enzyme-resin adduct has good activity and stability.

The resin is quite durable and of low cost. The Duolite-trypsin has good activity against protein. In addition, this derivative was active in 7 *M* urea. The proteolytic activity was nearly doubled by urea, presumably as a result of substrate (casein) denaturation. The michaelis constants and pH dependences are compared for trypsin conjugates with Duolite A-7, Silica-PEI-PAA, agarose, and porous glass. A cost comparison reveals that the Duolite and silica derivatives are much less expensive than agarose and glass.

INTRODUCTION

Enzyme immobilization and affinity chromatography have had a dramatic impact on the application and isolation of enzymes [1-3]. Of obvious importance to both areas is the support material to which enzymes or affinity ligands are bound (for reviews, see Refs. 4 and 5). The ideal support would have all the characteristics listed in Table 1. The degree to which a carrier must meet these requirements is, of course, a function of the application for which it is intended. For a small-scale application which yields a valuable product, cost and durability are not important. For a large-scale process such as enzymatic treatment of industrial waste, the cost and life of the carrier might well represent the limiting factors. Porous glass and agarose are widely used supports. Both materials are expensive. Glass is friable unless coated with zirconium oxide which raises the cost still further. Agarose gel is compressible and susceptible to microbial attack.* Both glass and agarose are porous. It is our goal to develop an inexpensive, rigid support which is porous or of high surface area. Our approach was to attach various polymers to porous glass in the hope of retaining rigidity and amplifying binding capacity. Thus reducing the amount of carrier needed. Second, we bound organic polymers to colloidal silica (Figs. 1 and 2). This support is inexpensive, and

---

*Although the manufacturers of agarose recommend protection against microbial contamination with a disinfectant, Prof. Jerker Porath has suggested that the problem is not as severe as many believe.

TABLE 1

Criteria for the Ideal Support Material for Enzyme Immobilization and Affinity Chromatography. Some Materials in Use Have Many of These Characteristics. For Some Applications a Carrier Which Meets Only a Few of these Requirements may be Acceptable.

---

1. Chemical functionality (or route thereto) sufficiently general to permit reaction with a variety of groups on proteins and ligands for affinity chromatography

2. Low cost

3. Resistance to microbial attack

4. Dimensional stability - the retention of physical shape following changes in pressure temperature, solvent composition, etc.

5. Durability - the support material should not erode, react chemically with the solvent, or break when used in suspension

6. Hydrophilic - many proteins denature at hydrocarbon-water interfaces

7. Regenerable - the feature of carrier reuse could be advantageous. However, if the support is of low initial cost, this may not be critical

8. High capacity for enzyme or ligand

9. Accessibility to solvent or high porosity

---

suitable for use in suspension-type reactors. Finally, we studied rigid, macroporous ion-exchange resins which are of low cost (Figs. 3 and 4).

EXPERIMENTAL

Materials

Trypsin (lot TRL-20A) with specific activity of 40 units/mg against N-α-benzoylarginine ethyl ester at pH 8.0 was obtained from Worthington Biochemical. Agarose (Sepharose-4B) was a product of

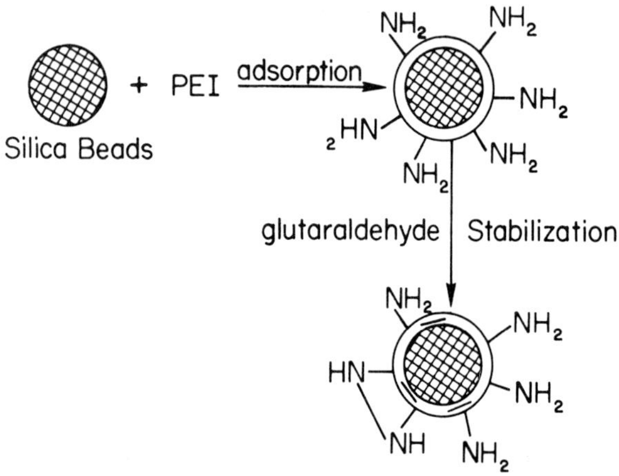

Fig. 1. Preparation of silica-PEI. An adsorption step is followed by a cross-linking step with a bifunctional reagent such as a glutaraldehyde.

Pharmacia. The alkylamine porous glass (550 Å) was obtained from Pierce Chemical Co. Duolite A-7 was a generous gift of Dr. Richard Harris of the Diamond Shamrock Chemical Co. Amberlyst (16-50 mesh, 4.9 meq-SO_3H/g) was supplied by Mallinkrodt. PEI-600 and PAA (50,000 MW) were from Polyscience; GAN from GAF Corp. DMF (MCB, spectrograde) was treated with anhydrous $MgSO_4$ and molecular sieves. N,N-Dimethyl casein was prepared by the method of Lin et al. [6]. All other chemicals were reagent grade and used without further purification.*

Methods

Preparation of Supports

PEI was bound to glass through glutaraldehyde. The alkylamine porous glass (1 g) was added in increments to 20 ml of 2.5%

---

*Abbreviations used: PEI, polyethyleneimine; PAA, polyacrylic acid; GAN, poly(methyl vinyl ether/maleic anhydride); CMC, 1-cyclohexyl-3-(2-morpholinoethyl)-carbodiimide metho-p-toluenesulfonate; DMF, dimethylformamide; TNBS, trinitrobenzenesulfonate.

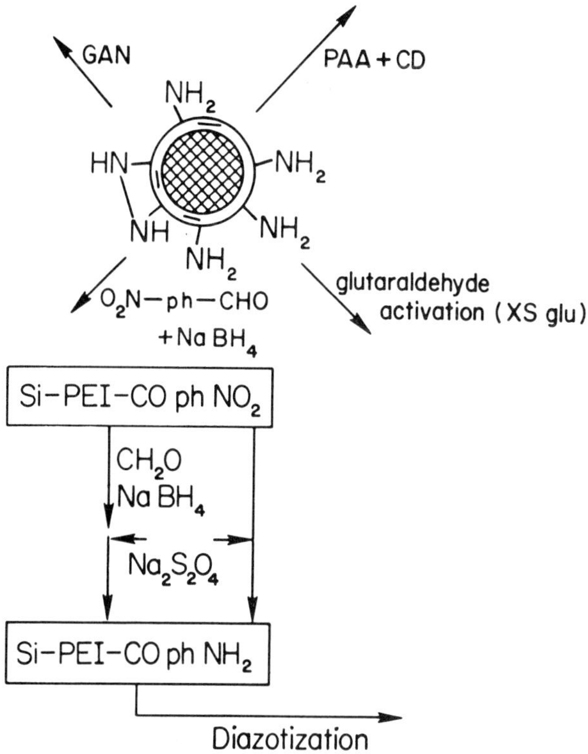

Fig. 2. Reaction pathways to various derviatives of silica-PEI.

glutaraldehyde at 0°C. The suspension was stirred with an overhead mechanical stirrer and the pH was maintained at 7.0. Addition of glass was over a 40-min period. After an additional 30 min the glass was filtered and washed with 500 ml of cold distilled water. The derivitized glass was then added to 100 ml of PEI solution (1 mg/ml) adjusted to pH 8.0. This suspension was stirred for 2 hr. The glass-PEI (5% w/w) was acylated by refluxing in $CHCl_3$ with p-nitrobenzoyl chloride and triethylamine, both in 10-fold excess of the primary amine on the support. The nitro group was reduced by refluxing for 1 hr in 10% sodium dithionite. In an effort to improve the yield of enzyme coupling, we reductively methylated [7] the remaining amines prior to reduction of the nitro group.

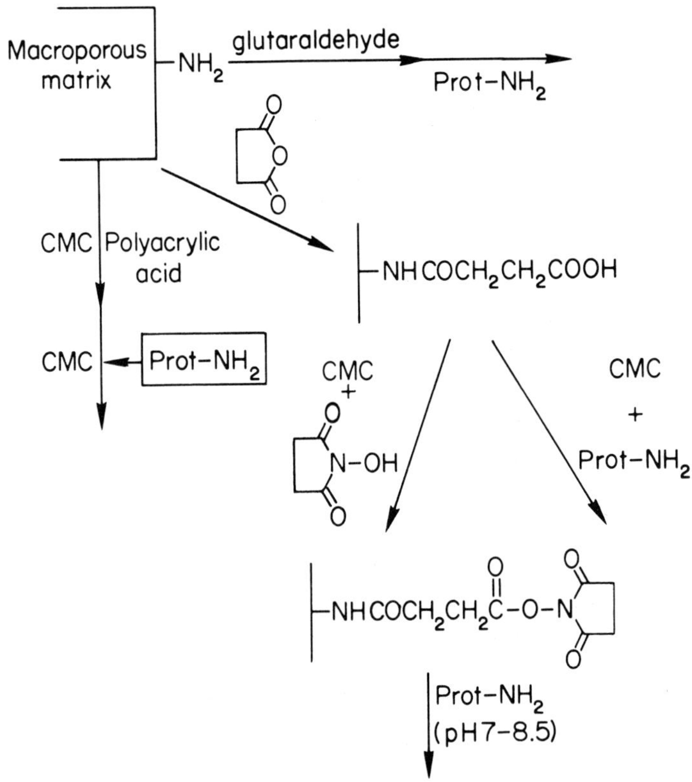

Fig. 3. Activation of Amberlyst. The initial step is the production of the sulfonyl chloride. This derivative may then be reacted with proteins or further derivitized by reaction with HDA (hexamethylenediamine). The resulting derivative may be treated still further with conventional techniques.

Glass-PAA was prepared with a water-soluble carbodiimide. Alkylamine glass (1 g) and 125 mg of PAA was stired with 100 mg CMC in 10 ml H_2O. The pH was maintained at 4.0 for 30 min at room temp-

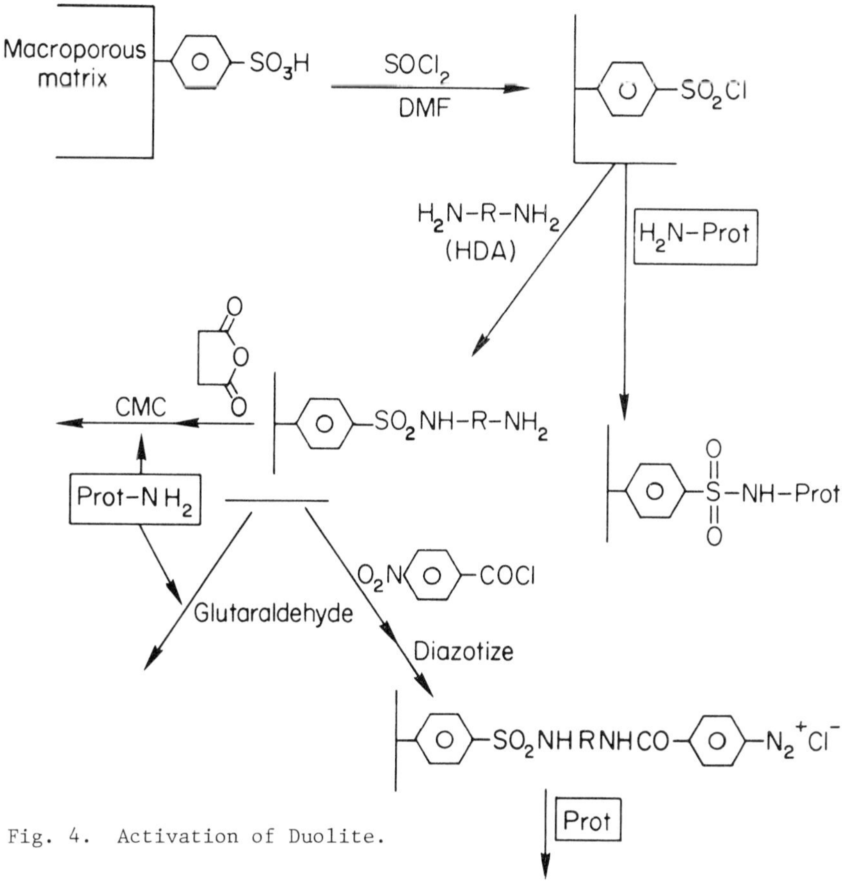

Fig. 4. Activation of Duolite.

erature. The suspension was then stirred overnight at 4°C. The glass-PAA was filtered and washed repeatedly with 4 M KCl and distilled water. This derivative bound 0.15 mmoles of glycine ethylester/g by the method of Line et al. [8].

Silica-PEI was prepared by adsorption of PEI on silica followed by cross-linking with glutaraldehyde. To 2 liters of 10 mM sodium borate were added 20 g of silica beads (Sigma) and 8.75 g PEI. This suspension was stirred in a blender for 15 min. Glutaraldehyde (410 ml, 2.5%) was then added followed by 10 g $NaBH_4$. This mixture was blended for 10 min. The silica-PEI was collected by centrifugation and washed repeatedly with distilled water. The resulting particles were 5 to 50 μm in diameter. Silica-PEI was reacted with GAN by rotation of 75 mg solid with 30 mg GAN and 5 ml dry pyridine in a closed tube for 2 hr. at room temperature. The solid was washed with methanol and dried. The existence of anhydride groups on the solid support was demonstrated by the hydroxamate test. Silica-PEI-PAA was prepared in a manner similar to Glass-PAA. Silica-PEI (250 mg), 125 mg PAA, and 50 mg CMC were reacted in 10 ml volume.

Derivatization of Amberlyst-15 was preceded by extensive washing with DMF under vacuum. One gram of resin was suspended in dry DMF, and 0.72 ml of freshly distilled thionyl chloride was added dropwise at 0°C with rapid stirring. The suspension was allowed to return to room temperature and stirred overnight. The beads were then washed several times with dry DMF and cooled to 0°C. Hexamethylenediamine (1.16 g in 10 ml of DMF, precooled) was added dropwise over a period of 1 hr. The suspension was then stirred overnight at room temperature. Succinylation of hexamethylenediamine-Amberlyst and Duolite A-7 was carried out with 5-fold excesses of succinic anhydride and triethylamine in DMF. The disappearance of amine may be followed with the ninhydrin procedure of Moore [6] using 5 mg of resin. The succinylated beads were washed with water and stirred for 45 min at room temperature in 0.1 N NaOH. The derivatives showed no base-sensitive groups remaining, and the ninhydrin reading remained negligible. The derivatives were stored under distilled water at 4°C. No evidence of microbial growth was apparent over a period of 6 months.

The N-hydroxysuccinimide ester of succinyl-Duolite A-7 was prepared by reacting 2 g of damp resin with 2.3 g of N-hydroxysuccinimide and 8.4 g of CMC in 40 ml of dry DMF. After 4 hr. considerable insoluble urea appears. The resin is washed with cold, "super dry" ethanol and dry DMF. The product is stored under dry DMF in a dessicator.

Polyacrylic acid was bound to Duolite A-7 through CMC. Duolite A-7 was stirred with distilled water under reduced pressure for several hours. The hydrated resin (1 g) was mixed with 100 mg of

CMC and 200 mg of a 25% solution of polyacrylic acid. The pH was adjusted to 4.5 and maintained there for 3 hr. with 0.1 N NaOH. The derivative was washed with 2 N KCl and distilled water.

Enzyme Coupling

Arylamine derivatives were diazotized as previously described [9]. Coupling with trypsin was performed at pH 8.0 in 2 mM Tris HCl with 25 mM $CaCl_2$ at 0°C. Trypsin was bound to the carboxyl derivatives by the method of Line et al. [8]. Trypsin coupling via glutaraldehyde was performed by the same procedure used for attaching PEI to arylamine glass presented earlier. Reaction of trypsin with the N-hydroxysuccinimide ester of succinylated Duolite A-7 was done in 0.2 N Hepes HCl buffer, pH 8.0, containing 25 mM $CaCl_2$.

Enzyme Assays

Benzoyl-L-arginine ethyl ester (Fig. 3) is a synthetic substrate of low molecular which is hydrolyzed rapidly by the enzyme trypsin. The reaction can be conveniently followed titrimetrically. We employed the Radiometer SBR3 titrigraph, ABU12 autoburette with the TT11 titrator. The reaction vessels were stirred mechanically at speeds sufficient to minimize external diffusion. A nitrogen stream was blown over the surface of the reaction mixture. Conditions were as follows: 1 mM substrate, 25°C, 10 mM standard base, 10 ml reaction volume, and 5×10^{-4} M Tris HCℓ.

Casein was reductively methylated to reduce the background absorbance in the trinitrobenzenesulfonate (TNBS) assay [6] (Fig. 5). We found this assay for proteolytic activity superior to titrimetric or ninhydrin procedures. In the proteolytic assays the amount of bound enzyme used was based on a standard activity against the low molecular weight substrate, benzoylarginine ethyl ester. The activities reported in Table 4 are therefore relative activities which could reflect the accessibility of the enzyme bound to the solvent. A representative plot of caseinolytic activity versus enzyme concentration is shown in Fig. 6. The shape of this curve is quite similar to that of the free enzyme.

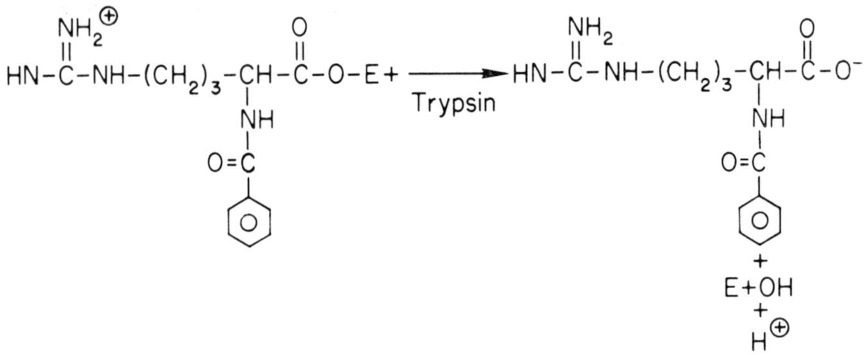

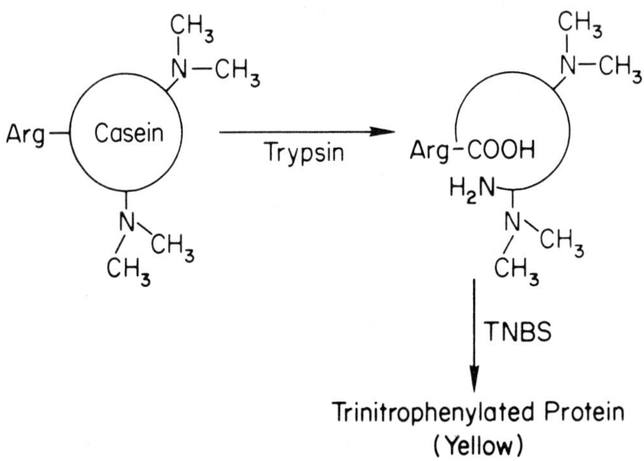

Fig. 5. Assays of trypsin. In the top reaction the trypsin-catalyzed hydrolysis of benzoylarginine ethyl ester is shown. One equivalent of acid is produced. In the bottom reaction the hydrolysis of N,N-dimethyl Casein is shown. Peptide bond hydrolysis produces a free amino group which may be derivatized by reaction with trinitrobenzene sulfonate.

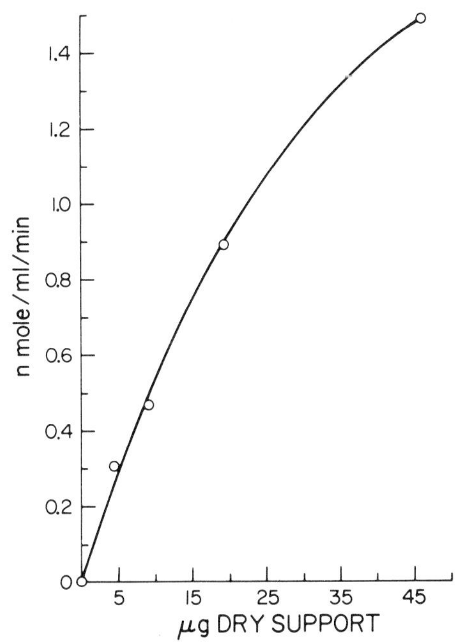

Fig. 6. A representative plot showing the dependence of the activity of agarose-trypsin on concentration of enzyme. The substrate is N,N-dimethyl casein. Use of the TNBS assay permits the spectrophotometer determination of the nmoles of amides broken/mg/min [6].

RESULTS

One goal of this study was the amplification of the enzyme-binding capacity of porous glass. We reasoned that a polymer coating should provide more functional groups through which to bind enzyme with no significant loss in the rigidity of the support. The arylamine derivative of PEI-glass proved inferior to unmodified porous glass as an enzyme support (Table 2). The possibility of reaction within the particle during the activation and coupling steps was addressed by methylation of the remaining primary amines

TABLE 2

Summary of Activities of Trypsin Conjugates with Polymer-Coated Inorganics

Support	Activity (units/g derivative)
Porous glass	110
Glass-PEI[1]-Ph-NH_2	69
Glass-PEI(CH_3)-Ph-NH_2	55
Glass-PAA[2]	59
Silica-PEI-GAN[3]	82
Silica-PEI-PAA	300

1. PEI; polyethyleneimine.
2. PAA; polyacrylic acid.
3. GAN; poly(methyl vinyl ether/maleic anhydride), GANTREZ-AN 119, GAF Corp.

prior to reduction and diazotization of the arylamine groups. The methylation provided a derivative less effective than the unmethylated one.

PEI is a globular polymer; we decided to investigate a linear polymer, PAA. The resulting glass-PAA was also inferior to porous glass as a support for trypsin. Our attention then turned to Silica-PEI. We coated silica beads with a mantle of PEI which was subsequently stabilized by cross-linking with glutaraldehyde (Fig. 1). Linear polymers were then added in hopes of producing a rigid bead with projecting linear polymers as sites for enzyme binding. The PAA derivative was superior to GAN derivative (Table 2). The activity of the trypsin complex of Silica-PEI-PAA was initially about 3 times greater than porous glass. After extensive washing and repeated use, the activities became about equal (Table 4). The Michaelis constant for the enzyme on silica-PEI-PAA, however, is more than double that for enzyme on glass.

Amberlyst is a cross-linked polystyrene with a macroporous, rigid structure. Our primary concern at the outset was the apolar character of polystyrene supports. As shown in Table 3, Amberlyst is not a particularly good support for trypsin. Reactions of the Amberlyst in DMF went smoothly. However, the use of water as a solvent was not successful for enzyme binding. This result was not surprising since it is well known that many proteins denature at hydrocarbon-water interfaces [10]. In an investigation of an insoluble Edman reagent, Dowling and Stark used glucosaminol to reduce the hydrophobicity of an isocyanatopolystyrene [11]. This approach applied to our system failed to promote protein binding or enzyme

TABLE 3

Activity of Trypsin Bound to Derivatives of Macroporous Ion Exchange Resins

Support	Derivative	Coupling Reaction	Specific Activity[1]
Amberlyst-15	$-SO_2Cl$	Direct-sulfonamide	1.7
	$-SO_2-NH(CH_2)_6NH_2$	Glutaraldehyde	4.9
	$-SO_2NH(CH_2)_6NHCO-CH_2CH_2COOH$	CMC[2]	1.7
	$-SO_2NH(CH_2)_6NH-COphNH_2$	Diazonimum salt	0.6
Duolite A-7	Unmodified	Glutaraldehyde	0.9
	$-NH(CH_2CH_2)_nCOOH$	CMC	11.4
	$-NHCOCH_2CH_2COOH$	CMC	21.7
	$-NHCOH_2CH_2COOH$	N-Hydroxysuccinimide ester	10.3 (23)[3]
Porous glass	$-Si(OR)_2R'NHCO-phNH_2$	Diazonium salt	100 (110)[3]

1. μ mole/min/g carrier, substrate is 1 m$\underline{M}$ benzoylarginine ethyl ester, at pH 8.0, 25 m$\underline{M}$ $CaCl_2$ and 25°C.

2. 1-Cyclohexyl-3-(2-morphollinoethyl)-carbodiimide metho-p-toluene sulfonate.

3. Values in parentheses were determined at pH 9.5.

stability. Although these derivatives of Amberlyst are not satisfactory for the enzyme trypsin, they should be good supports for enzymes which naturally occur in a hydrophobic environment.

The results shown in Table 3 indicate that Duolite-A7 derivatives of trypsin have activities of the same order of magnitude as porous glass. These rates were determined at saturating levels of benzoylarginine ethyl ester with new derivatives washed with dilute buffer only. The succinylated Duolite activated by a water-soluble carbodiimide or by an N-hydroxysuccinimide ester is the most promising form of enzyme carrier, at least for trypsin.

A typical loss of activity of a bound enzyme as a function of washing with concentrated salt solutions is shown in Fig. 7. After extensive washing with concentrated salt solutions and numerous assays, "plateau" values of specific activity (act/g resin) were determined. The results are shown in Table 4. All four derivatives are in the same range of activity. The comparison is admittedly only an approximation since the ages of the preparations were not exactly identical. It may be noted that Duolite A7, activated by N-hydroxysuccinimide ester, is better than porous glass as a support for trypsin.

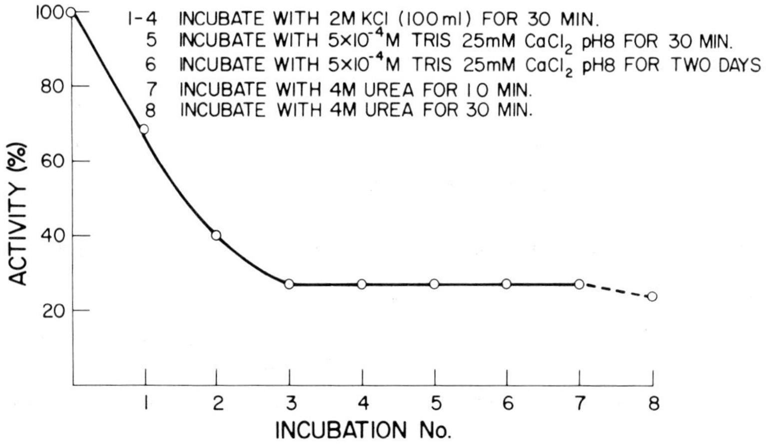

Fig. 7. Washing of agarose-trypsin with concentrated salt solutions.

TABLE 4

A Comparison of Duolite and Silica-PEI-PAA with Agarose and Glass as Supports for the Enzyme Trypsin. All Derivatives Were Extensively Washed with Concentrated Salt Solution and Assayed Many Times until a Constant Level of Activity Was Reached

Support	BAEE[1] Activity	Apparent K_m	Apparent pH opt.	Casein[2] Activity
Duolite A-7 (SIE)	1.01	0.8×10^{-4}	9.5	0.5 (0.9)[3]
Silica-PEI-PAA	0.79	1.4×10^{-3}	10	0 (0.7)[4]
Agarose	1.59	0.4×10^{-3}	10	0.9
Glass (AZO)	0.88	0.6×10^{-3}	9.5	0.5

1. Activity against N-benzoyl-L-arginine ethylester at pH 9.0, μ mole/min/g support.
2. Activity against N,N-dimethyl casein normalized to a constant BAEE activity, pH 9.0 (nmole amine released/ml/min).
3. Determined with reaction mixture which contains 7 M urea.
4. Determined at pH 8.0.

The apparent Michaelis constants which appear in Table 4 were determined under conditions where external diffusion was not limiting. The pH of the bulk solution for these determinations was 9. It is clear, however, from the data in Fig. 8 that "local" pH effects are operating. In the hydrolysis of benzoylarginine ethyl ester, a proton is produced (Fig. 5). Accumulation of protons within the porous matrix would tend to raise the apparent pH optimum. Electrostatic effects would also contribute to the apparent K_m (see next section).

The proteolytic activities of the various bound enzyme derivatives are shown in Table 4. The assay employing TNBS and N,N-dimethylcasein is very sensitive and more acceptable than the assays based on TCA-soluble peptides of ninhydrin. Figure 6 is a representative plot showing the dependence of activity on enzyme concentration. The soluble and insoluble trypsins show very similar concentration

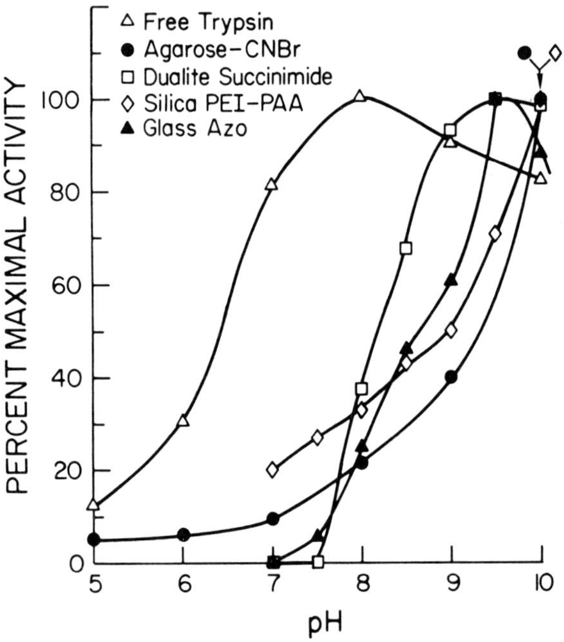

Fig. 8. The pH-dependence of the esterolytic activity of various insoluble derivatives of trypsin. Notice that the pH optima of all the derivatives, regardless of charge, are increased from that of free trypsin.

dependences. The proteolysis rates shown in Fig. 4 are from the linear parts of the plots of activity vs enzyme concentration.

It is significant that the Duolite-trypsin is activated by the presence of 7 M urea. The activity increases from 0.5 nmole amine released/ml/min in dilute buffer to 0.9 nmole amine released/ml/min in 7 M urea. The most reasonable explanation for this observation is that the casein is unfolded by urea and made more susceptible to proteolysis. Since casein is relatively accessible in dilute buffer, one would expect even larger effects with more compact proteins. Prolonged exposure of the Duolite-trypsin, unfortunately, leads to inactivation of the enzyme.

DISCUSSION

The pH profiles shown in Fig. 8 cannot be examined by the conventional formulations of electrostatic effects based on equilibrium thermodynamics (Ref. 12 and references therein). In the simplest case the pH optimum of an enzyme will be unchanged when the enzyme is on a neutral support, raised when on a negatively charged support, and lowered when on a positively charged support. The rationale for these observations may be expressed both in qualitative and quantitative terms. When the enzyme is in the midst of negative charges, the "local" pH is lower near the enzyme than in the bulk solution. For the enzyme to reach maximal activity, therefore, the pH of the bulk solution must be higher than in the case of an enzyme free in solution or on a neutral carrier. For an enzyme bound to a positively charged carrier, the "local pH" is above that of the bulk solution (H^+ is repelled). This condition of course results in a decrease in the apparent pH optimum based on the pH of the bulk solution.

Using the concept of electrochemical potential, it can be shown that

$$\frac{a'_{H^+}}{a_{H^+}} = e^{-\varepsilon\psi/kT} \qquad (1)$$

in which a'_{H^+} is the activity of the proton within the matrix and a_{H^+} is the activity of the proton in the bulk solution. The charge on the electron is represented by ε, ψ is the electrostatic potential, k is Boltzmann's constant, and T is the absolute temperature. Taking logs and rearranging, Eq. (1) becomes

$$pH' - pH = \Delta pH = 0.43\varepsilon\psi/kT \qquad (2)$$

The ΔpH is reflected by the change in pH optimum of the bound enzyme. At ψ = 150 mV, ΔpH would be about 2.5 units.

It is clear that the pH profiles shown in Fig. 8 cannot be explained by the forgoing treatment. Agarose, for instance, is neutral or slightly positively charged after CNBr treatment. However, the ΔpH shown in Fig. 8 would suggest a very high degree of negative charge. In stronger buffer at about the same ionic strength, we see a much smaller ΔpH. The large effect we see in Fig. 8 may be explained by proton accumulation within the matrix. Knights and Light have recently reported the same observation [13]. In addition, for the trypsin-catalyzed hydrolysis of substrates such as amides where protons are not produced, very little change is pH optimum is observed for glass and agarose derivatives of trypsin.

Using glass and agarose as standards, it appears that the silica-PEI-PAA and the succinyl derivative of Duolite-A7 have considerable promise as support materials. In small quantities, glass and agarose are orders of magnitude more expensive than the silica-PEI-PAA or the Duolites. The Duolite has the advantages that it is rigid, durable, and macroporous. It can be stirred with any type of stirring device including a magnetic stirrer bar.

We have recently tested carbohydrate adducts of porous glass as enzyme supports. Glycophase-G and CPG/Dextran are produced for gel permeation and affinity chromatography by Corning Glass and Distributed by Pierce Chemical. The Glycophase-G is prepared by coating porous glass with a glyceryl silane. We have activated this material with CNBr and immobilized trypsin. The resulting derivative had 75% of the activity of agarose-trypsin. We found this encouraging since the Glycophase-G could be used in columns at high flow rate. The lack of nonspecific adsorption could be an advantage in many cases. In addition, the hydrophylic nature of the surface should make it compatible with a wide range of enzymes. With the exception of cost, this material fulfills every criterion listed in Table 1.

ACKNOWLEDGMENT

This work was supported by grants from the National Institutes of Health (GM 19507-01) and the National Science Foundation (GI-32141).

REFERENCES

[1] R. Goldman, L. Goldstein, and E. Katchalski, in Biochemical Aspects of Reactions in Solid Supports (G. R. Stark, ed.), Academic, New York, 1971, p. 1.

[2] P. Cuatracasas, Ibid., pp. 79-107.

[3] H. Guilford, Chem. Soc. Rev., 3, 249 (1973).

[4] J. P. Andrews, R. Uy, and G. P. Royer, Enzyme Technol. Dig. 1, 99 (1973).

[5] G. P. Royer, in Bonded Stationary Phases in Chromatography (E. Grushka, Ed.) Ann Arbor Science Publishers, Ann Arbor, Michigan, In Press.

[6] Y. Lin, G. E. Means, and R. E. Feeney, J. Biol. Chem., 244, 789 (1969).

[7] G. E. Means and R. E. Feeney, Biochemistry, 7, 2192 (1968).

[8] W. F. Line et al., Biochim. Biophys. Acta, 242, 194 (1971).

[9] G. P. Royer and J. P. Andrews, J. Biol. Chem., 248, 1807 (1973).

[10] S. Ghosh and H. B. Bull, Arch. Biochem. Biophys., 99, 121 (1962).

[11] L. M. Dowling and G. R. Stark, Biochemistry, 8, 4728 (1969).

[12] G. P. Royer, in Immobilized Enzymes, Antibodies, Antigens and Peptides, (H. H. Weetall, ed.), New York, In Press.

[13] R. L. Knights and A. Light, Arch. Biochem. Biophys., 160, 377 (1974).

POLYMER-DRUG GRAFTS FOR IRON CHELATION

Rene S. Ramirez and Joseph D. Andrade

Department of Materials Science and Engineering
University of Utah
Salt Lake City, Utah 84112

INTRODUCTION

Accidental poisonings account for a large proportion of children's accidents. There is a need to improve the emergency treatment of poisonings, particularly to decrease the period of coma and to minimize mortality.

Iron poisoning is a common cause of poisoning among young children. Use of chelating agents, especially Deferoxamine B, to remove excess iron from the body is one of the most effective treatments. However, in very severe cases of acute iron overloads, the amount of deferoxamine required to remove the iron is so high that the drug and its iron chelate would be toxic.

Our objective is to develop a chelating polymer which could be used in extracorporeal detoxification of severe acute iron poisoning cases.

POLYMER CHELATES

Ion-Specific Resins

In the search for specific ion-exchange resins there has been a great deal of effort toward the synthesis of specific chelating or complexing resins. The commercial manufacture of these resins has not yet been successful[1] with the exception of Dowex A-1[2].

Polymer chelates consist of a polymer backbone and a grafted chelating group (Fig. 1a) although resins containing the chelating group as part of the repeat unit have been reported[3-11] (Fig. 16). Resins of the first type may be considered as random graft copolymers and those of the second type as homopolymers.

Chelating resins differ from ion exchange resins in three properties[9]: greater selectivity, higher metal-polymer interaction energy, and slower kinetics.

Applications of chelating resins have been mainly in water purification (i.e., removal of heavy metals from water), analytical determination of metal ions, chromatographic separation of ions,

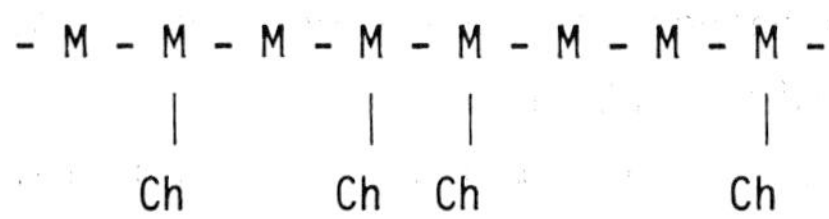

(a)

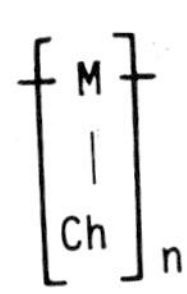

(b)

Fig. 1. General structure of polymer chelates. M = repeat unit; Ch = chelating group. (a) The chelating group is grafted on the polymer chain. (b) The chelating group is part of the monomer unit.

and recovery of metals from solutions. A number of chelating groups have been grafted onto polymers (Table 1).

Synthesis of Polymer Chelates

There exist two general approaches for the synthesis of chelating polymers. One is the incorporation of the active (chelating) group in the polymer matrix by making use of reactive functional groups in the polymer. The second method is the transformation of the active group into a monomer with subsequent polymerization. In the first method the hydrophilicity of the chelating groups is sometimes incompatible with the polymer matrix (usually hydrophobic), and therefore the latter requires some modification. In the second method the presence of a larger group in a monomer may reduce and even prevent homopolymerization, making it necessary to copolymerize the transformed monomer. It is possible that some bulky chelating monomers may not polymerize or copolymerize at all.

Gregor et al.[38] suggested four conditions that the chelating group should fulfill:

1. It should be able to react with a group and become attached to the polymer matrix, or be able to polymerize for resin formation.

TABLE I

Some Chelating Groups Grafted on Polymers

CHELATING GROUP	REFERENCES
Alpha-Amino acids	13-19
Iminodiacetic and/or Iminodipropionic acids	5, 7, 14, 15, 20-34, 36
Nitrilotriacetic acid	35-37
EDTA	37-39
Hydroxamic acid	40-42
Deferoxamine B	43
8-Hydroxyquinoline	44-47
Others	3, 8, 37, 48-80

2. If the latter is the case, it should be sufficiently stable to withstand the polymerization process without alteration of its chelating function.

3. The chelating group should be compact so as not to be sterically hindered by the polymer matrix.

4. The arms of the chelate structure should be present on the same monomer unit in proper spatial configuration, for preservation of specific arrangement of the ligand.

D'Alelia et al.[28] suggest that monomeric models for the chelating polymer should be constructed, and that for an effective chelating polymer the behavior of the active group in the polymer should be similar to that of the group in the monomeric segment. They proposed the following approach for the synthesis of a chelating polymer:

1. Synthesis and determination of chelating characteristics of a model monomeric compound.

2. Synthesis and characterization of an oligomer of low molecular weight (dimer, trimer, etc.) model compound, and determination of its chelating properties.

3. When the results of the second step confirm chelation, a high molecular weight polymer should be synthesized and characterized, and its chelation capacity should be determined.

D'Alelio et al.[28,29,31-33] have followed this approach for the study of chelating polymers containing hydroxyarylaminoacetate groups as the chelating group.

A general sequence for the synthesis of grafted chelating polymers is shown in Fig. 2. In this case, polystyrene (1), usually cross-linked with divinylbenzene, is first chloromethylated, then transformed into a sulfonium salt (3), and finally reacted with sodium iminodiacetate to give the chelating polymer (4). The sulfonium salt is hydrophylic whereas the chloromethylated resin (2) is not and will not react appreciably with iminodiacetate salt.

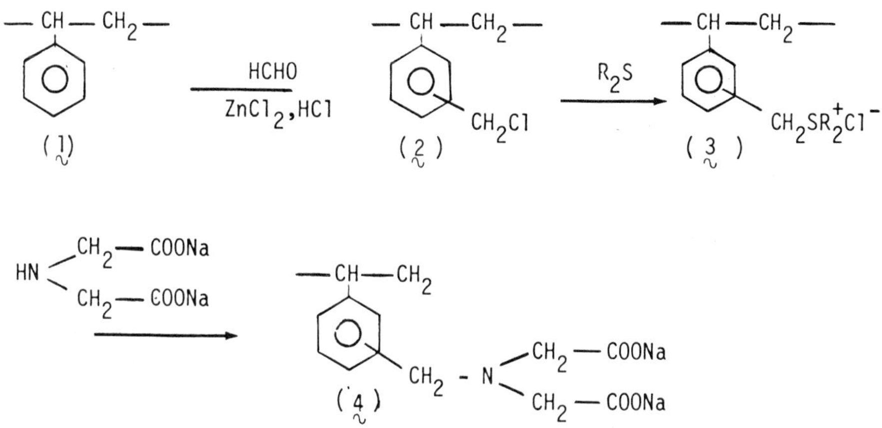

Fig. 2. Reaction sequence for the synthesis of an iminodiacetate chelating polymer [1].

Another approach to solve the problem of hydrophobicity of the chloromethylated polystyrene is to partially transform it into a quaternary salt with a tertiary amine (e.g., triethylamine), which makes the polymer more hydrophylic, and then to react it with the chelating group[81]. The reaction sequence for this process is shown in Fig. 3.

The transformation of the chelating group into a monomeric species is shown in Fig. 4[26]. In this case methacrylchloride (8) reacts with the diethyl ester of iminodiacetic acid to give amide (9), which may be homopolymerized to give (10). Monomer (9) may also be copolymerized with other vinyl monomers or graft copolymerized with radiation on preformed polymers[9] or fibers to produce chelating graft copolymers.

Distribution Coefficient and Selectivity Coefficient of a Chelating Resin

The distribution coefficient, K_D, or distribution ratio, and the selectivity coefficient, $K_{A/B}$, or concentration quotient[83],

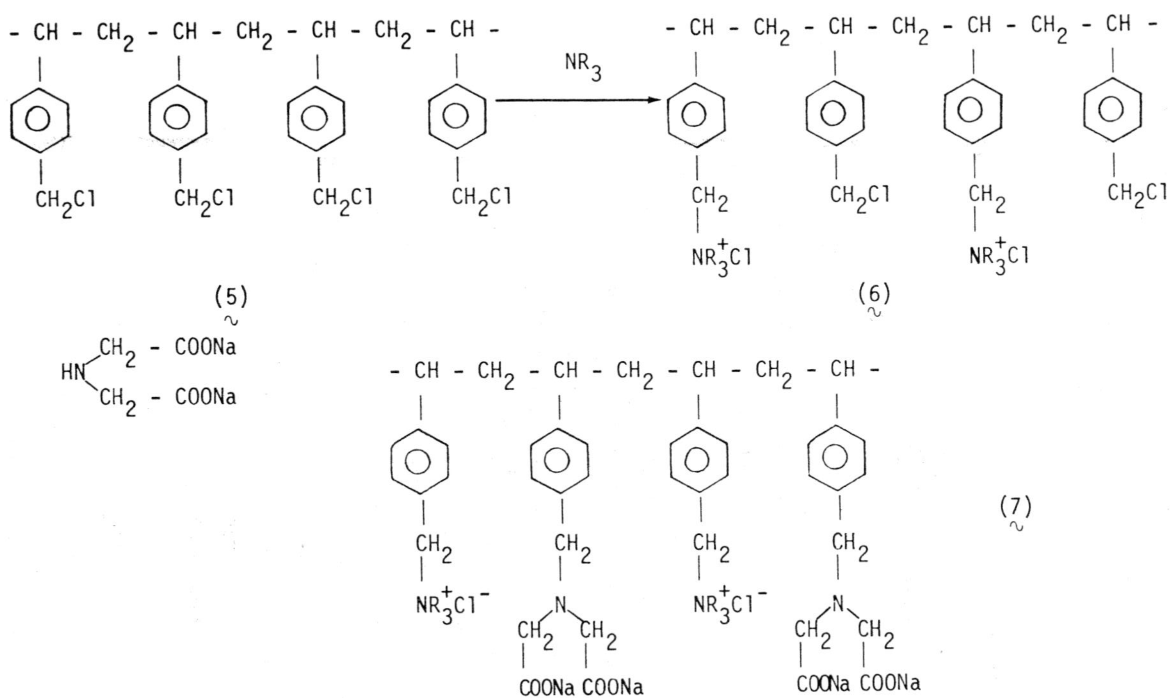

Fig. 3. Reaction sequence for the synthesis of a chelating polymer with partial substitution [1].

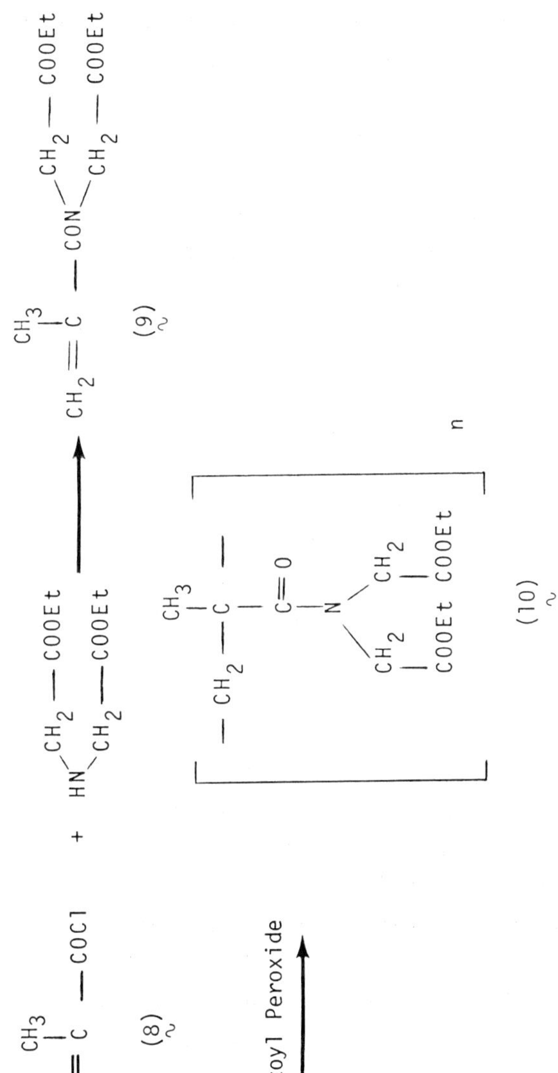

Fig. 4. Synthesis of a chelating polymer with one active group per monomer unit.

are important parameters in the characterization of ion-exchange and chelating resins.

For very dilute solutions of an ion A, the distribution coefficient for any ion exchange resin is given by

$$D = [\bar{A}]/[A] \qquad (1)$$

where $[\bar{A}]$ is the concentration of ion A in the resin phase and
$[A]$ is the concentration of ion A in the aqueous phase.

The selectivity coefficient for the exchange of two ions, A and B, of charge $|a|$ and $|b|$, respectively, is given by

$$K_{B/A} = \frac{[A]^{|b|}[\bar{B}]^{|a|}}{[\bar{A}]^{|b|}[B]^{|a|}} \qquad (2)$$

where the bars indicate the resin phase.

Equation (2) is comparable to the equilibrium constant of an ionic equilibrium reaction. Thus if one considers the exchange reaction of two ions A and B with charges a and b as being

$$|a|B^{b} + |b|\bar{A}^{a} \rightleftarrows |a|\bar{B}^{b} + |b|A^{a} \quad ; \qquad (3)$$

the equilibrium constant for this reaction is identical to $K_{B/A}$.

These two expressions, (1) and (2), have been defined for ion-exchange (nonchelating) resins; however, they can be equally applied for chelating resins[33,83].

Both the distribution and selectivity coefficients depend on the degree of cross-linking in the resin[82,84], on the specific capacity of the resin, i.e., the number of miliequivalents of fixed exchange groupings per gram of resin[84], and on the nature of the exchange (or chelating) group[84].

Detailed treatments of ion-exchange selectivity and of resin selectivity in aqueous solutions are given by Reichenberg[84] and by Diamond and Whitney[82].

CHELATING AGENTS IN MEDICINE

Chelating agents are a class of drugs for which a clear connection between action in the body and physical and chemical properties are known[85]. Several natural chelates are well known: hemoglobin (iron), chlorophyll (magnesium), vitamin B-12 (cobalt), and enzymes such as cytochrome oxidase (iron and copper). This section will be mainly concerned with synthetic chelating agents or those of natural origin that are not found in the human organism.

According to Schuber[85], chelating agents may accomplish three different functions:

1. To chelate toxic metals in blood and help the body get rid of them by excretion.
2. To deliver certain trace metals that are essential to tissues.
3. To inactivate certain bacteria and viruses by making essential metabolic metals unavailable, or by delivering metals that are harmful to them.

An effective chelating agent must form more stable complexes with the metals of interest than the biological substance holding the metals. This means that the stability constant of the metal chelate should be higher than the stability constant of the complex between the biological substance (usually proteins) and the metal. The stability constants of chelates are taken as a measure of their chelating ability. However, the stability constant determined in aqueous systems is usually different from the actual stability constants found in a living system, where interferences from proteins, enzymes, pH, and other ions may change the chelating properties of the drug[86]. Shubert[87] examined the biological implications of iron binding and described chelating agents in terms of an "effective constant", which is not actually constant but varies with the media in which the chelating agent and the metal ion are in contact.

An important property that the chelating agent should have is selectivity. It should be able to chelate the metal of interest without influencing the other essential ions present, often in larger amounts, such as Ca and Mg, as well as the trace metals, such as Zn, Mn, and Cu, whose removal could be dangerous.

Solubility of both the chelating agent and the complex in various media are important. They should be water soluble, to permeate within the intra- and extravascular spaces but this may exclude them from penetrating into the intracellular space. Water solubility also allows a more rapid excretion of the drug and its chelate.

Chelating agents used as drugs should withstand metabolic processes, and if they undergo enzymatic or other degradation, the metabolites should not have toxic side effects[86].

The administration mode of a chelating agent is also very important. For example, EDTA and DTPA (see Table 2 for formulas and chemical names) must be administered by prolonged intravenous infusions, and temporarily immobilize the patient thus restricting their use to hospitals[86]. An ideal chelating agent as a drug would be one that is taken orally, acts in the gastrointestinal tract, and is absorbed through the intestines, extending its action to the rest of the body[88].

Table 2 summarizes various chelating agents used in medicine. Some of these agents are still in the experimental stage, whereas others are presently being used for clinical treatment. Some of these chelating agents are used to remove metals from the organism (EDTA, DTPA, deferoxamine, BAL, etc.). Others are used to deliver metals into infected areas (Oxine) or to the brain to reduce fever and headaches (aspirin)[85].

One of the main uses of chelating agents in medicine has been in the treatment of acute metal poisoning, such as lead, mercury, and iron. The most success has been in the case of iron poisoning where DTPA and especially deferoxamine have proven very effective.

ACUTE IRON POISONING

Occurrence of Iron Poisoning

Young children may ingest large amounts of iron tablets (commonly ferrous sulfate) that are prescribed for their mothers. Adults less often take ferrous sulfate in suicide attempts. The estimated lethal dose of ferrous sulfate in children is 900 mg/kg of body weight[90]. Some children have survived after ingesting 15 g, and sometimes as little as 1 g has been fatal[91]. Arena[92] stated that the average human fatal dose is 200 to 250 mg of iron per kilogram of body weight. For the average 2 year-old child, that dose means the ingestion of 3 g of iron.

Crotty[93] has summarized reports from the National Clearinghouse for Poison Control Centers over a period of 32 months. Of 292,060 poisoning reports, 1194 cases were of ingestion of medicinal preparations containing more than 65 mg of iron. Of these cases, 1024 were children under 5 years of age, and ferrous sulfate ranked 6th among the medicines that caused hospitalization of children under 5. According to Crotty[93], "the reported information that 179 of 654 cases (27%) were hospitalized places iron salts in the category of the most toxic substances." This is probably because of the consideration of iron as a "vitamin" and not as a potential poison; thus parents do not take the necessary precautions.

Until 1958, Aldrich[94] estimated the mortality rate of acute iron intoxication as 45%. Whitten et al.,[95] in 1965 proposed that mortality rates may be reduced with the appropriate therapy using deferoxamine as a chelating agent for iron.

Mechanism for Iron Toxicity from Iron Salt Ingestion

Iron overdoses are accompanied by both local and systematic effects. Significant amounts of either ferric or ferrous salts lead to rapid necrosis of the mucosa of the gastrointestinal tract[96]. Absorbed iron soon exceeds the bonding capacity of iron

Table 2

Some chelating agents used in medicine[85,86]

Name	Formula and Chemical Name	Uses
Aspirin	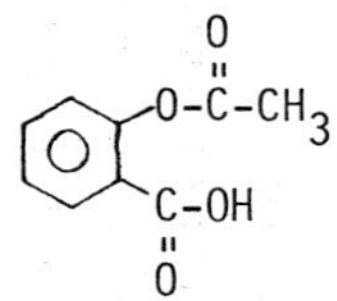 o-Acetylsalicylic acid	To reduce fever and relief of headache, muscle and joint pain. Also for symptoms of rheumatic fever.
BAL (British Anti-Lewisite)	CH_2-SH $\vert$ $CH-SH$ $\vert$ CH_2-OH 2,3-Dimercaptopropanol	In cases of arsenic, antimony, mercury, and gold poisoning. Also in lead poisoning.
Deferoxamine B (DFA)	$NH_2(CH_2)_5-N(OH)-CO(CH_2)_2-CO-NH(CH_2)_5$] $CH_3CO-N(OH)-(CH_2)_5NH-CO-(CH_2)_2CO-N(OH)$] N-{5-[3-[(5-aminopentyl)hydroxycarbamoyl]-propionamido]pentyl}-3-[[5-(N-hydroxyacetamido)-pentyl]carbamoyl]-propionohydroxamic acid	In iron poisoning cases.

Name	Formula and Chemical Name	Uses
DTPA	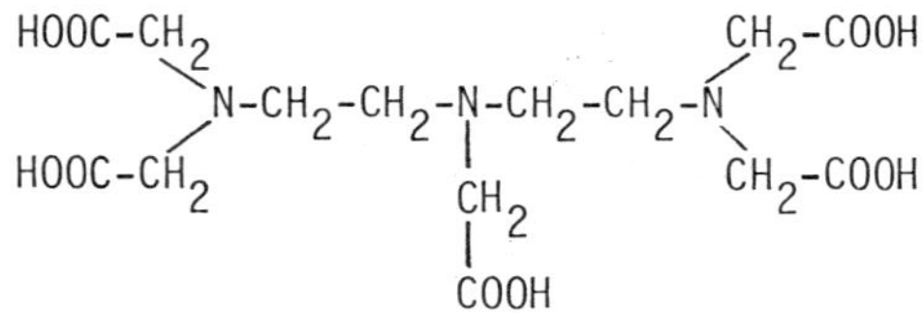 Diethylenetriaminepentaacetic acid	In poisoning by radioactive metals and in iron poisoning.
EDTA	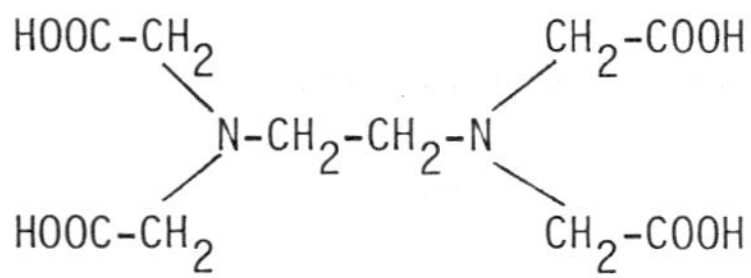 Ethylenediaminetetraacetic acid	In cases of lead poisoning. Treatment of skin irritation by metals. In treatment of atherosclerosis.
Fe-3 Specific (DHEG)	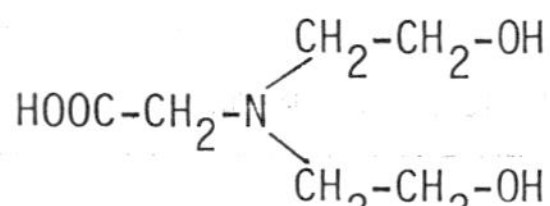 N,N- Dihydroxyethylglycine	In iron poisoning.

Table 2 (concluded)

Name	Formula and Chemical Name	Uses
Oxine	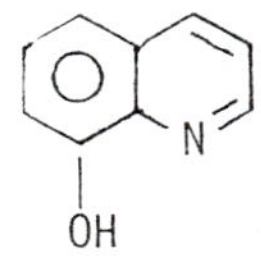 8-Hydroxyquinoline	Used for its antibacterial action.
Penicillamine	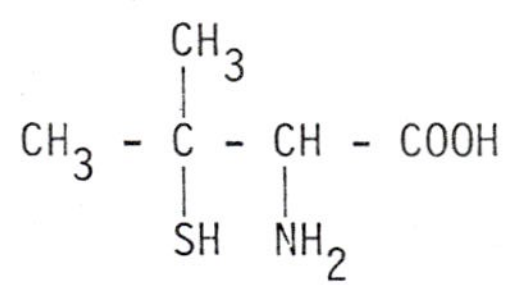 2-Amino-3-mercapto-3-methyl butanoic acid	In cases of copper overloads (Wilson disease) and in lead poisoning. Inhibits tumor growth in mice.

proteins in blood; thus serum iron concentration increases rapidly. The liver and the spleen remove sizeable amounts of iron, and this often results in hepatic necrosis[97]. Witzleben and Buck[89] proposed a hypothesis involving indirect lipid peroxidation. In the liver and elsewhere at the cellular level, iron is a potent protoplasmic poison due to its enzymatic impact[98]. However, the biochemical mechanism of iron toxicity has not yet been well established.

Symptoms of Iron Poisoning

Symptoms of iron poisoning begin with vomiting, usually accompanied by abdominal pain and diarrhea, and followed by lethargy and a state of shock[96]. Shock is attributed to hemorrhage and fluid loss from the necrotizing action of iron on the gastrointestinal tract[99].

Vasomotor collapse is explained by Smith[100] as produced by an excessive amount of ferritin formed from apoferritin in the mucosal cells. Ferritin, considered to be a vasodepresant[101], is then released into the circulation, initiating and maintaining the state of cardiovascular collapse.

Hepatic damage, although it has been found in some fatal human cases[97,100,102-105] and in some experimental studies[97,106], cannot be attributed as a causative factor in death from acute iron poisoning since no evidence of hepatic failure has been identified[104].

Other symptons include coagulation defects reported in some cases[105,107]. However, no systematic study of the various clotting factors has been reported in humans[104]. Experimental iron poisoning has shown prolongation of clotting time or complete absence of coagulation in rabbits[108]. Whitten[109] found alteration in the different stages of blood coagulation but the clotting time was not altered.

Metabolic acidosis in experimental iron poisoning is well known, and this is probably a common occurrence in lethal iron poisoning in children[104]. Metabolic acidosis may be explained by two factors: (1) The release of hydrogen ions from the conversion of iron(II) and iron(III) into the circulatory system[110] and (2) the accumulation of lactic and citric acid from anerobic metabolism characteristic of shock[110,111].

Shock and coma are characteristic of all the fatal cases of acute iron intoxication reported until 1965[95].

Serum iron levels greatly increase in iron poisoning, and they are usually a measure of the degree of intoxication. Normal serum iron levels in humans fluctuate between 60 to 160 μg/100 ml of blood. In acute iron poisoning, serum iron levels cover a wide range from 200 to 5000 μg/100 ml, although one case with serum iron of 30,494 μg has been reported[112]. Iron levels of 500 to 1000 μg/100 ml are considered dangerous, and above this range the iron level is potentially lethal.

Treatment of Acute Iron Poisoning

Emergency treatment of acute iron poisoning is described by many authors[92,96,113,114,116,117]. In general, the treatment consists of induction of vomitting with an emetic, stomach lavage with sodium bicarbonate, intravenous fluid therapy (blood or plasma) to correct acidosis and dehydration, use of a chelating agent (Deferoxamine B) to remove "free iron" and peritoneal or hemo dialysis to help the removal of chelated ironn.

Waxman and Brown[86] have an excellent review of the various chelating agents that have been employed in iron overloads. The emphasis has been on deferoxamine though DTPA has also been evaluated[118-122]. Since its introduction in 1960[123], at least 700 papers have described different aspects of the application of deferoxamine.

DEFEROXAMINE B IN IRON POISONING

The major objective in the treatment of poisoning with most substances is the removal of the toxic substance from the organism.

The human body is not provided with an effective mechanism for eliminating excess iron. Therefore, in acute iron intoxication only two ways of increasing iron excretion exist: phlebotomy and the use of a chelating agent[124].

One of the most specific and irreversible of the iron chelating agents is deferoxamine B, commercially available as the methane sulfonate salt. This derivative of deferoxamine is produced by CIBA under the brand name of Desferal. (Trademark, CIBA Pharmaceutical Co. Summit, New Jersey).

The high solubility of the iron complex of deferoxamine and its low molecular weight allow it to pass easily through the kidneys, making deferoxamine of great value in the treatment of acute iron poisoning.

Deferoxamine was initially proposed by Moeschlin and Schnider[125] as a two-fold approach for the treatment of acute iron intoxication: enternally and parenterally. Moeschlin et al.[26] showed in animal experiments that the survival rate can be greatly increased by the use of deferoxamine.

More recently, several authors[86,95,112,113,115,124,127,128] have reviewed the clinical use of deferoxamine in acute iron overloads, and the question arises whether deferoxamine should be used orally at all since it is only slightly absorbed from the gastrointestinal tract[129]. Oral deferoxamine would help to remove any iron that may still be in the intestines. However, more effective methods, such as use of emetics and lavage, exist for the evacuation of the gastrointestinal tract[128]. Westlin[112] states that intramuscular deferoxamine is preferable to avoid drug-induced hypotension and recommends that the intravenous route be used for severly intoxicated patents that are in coma or cardiovascular collapse.

Rapid infusion of deferoxamine has been reported to induce hypotension and tachycardia in dogs, cats, rabbits[130], and children[95,112]. Whitten et al.[95,114] studied the toxicity of the iron-deferoxamine complex in children and dogs. Rapid intravenous administration of 30 mg/kg of the complex produced a fall in mean arterial blood pressure in children. They gave a lethal dose of 225 mg/kg of iron-deferoxamine complex to dogs, which resulted in a fall of blood pressure, a decrease in hematocrit, a decrease of blood pH, and the development of renal shutdown.

These side effects of deferoxamine and its iron complex restricted the drug dose. An intramuscular dose[131] can be given to patients not in shock (1g initial administration followed by 0.5 g every 4 hours, with the total amount not to exceed 6 g in 24 hours. The intravenous dose[131] recommended for patients in a state of cardiovascular collapse should not exceed 15 mg/kg/hour by slow infusion, and a total dose of 6 g in 24 hours[86].

In summary, deferoxamine is very useful in the treatment of acute iron poisoning as shown by the reduction of mortality reported by several authors[95,124,128]. However, delayed therapy with deferoxamine did not reduce mortality in several cases of iron overloads[117,132,133]. Therefore, McEnery[117] recommends to "treat all cases of acute iron ingestion vigorously with deferoxamine." However, there is evidence[95,114] that deferoxamine and its iron complex are toxic when the drug is used in large amounts. The use of an extracorporeal system containing immobilized deferoxamine may prove to be useful in the treatment of very severe acute iron intoxication.

EXTRACORPOREAL TREATMENT OF ACUTE IRON POISONING

Deferoxamine may be covalently bonded to polymers containing reactive functional groups[43]. These polymers act as solid supports for the drug, which maintain its chelation ability. A number of investigators[134,135] have used an extracorporeal system based on activated carbon for the removal of certain uremic toxins and other poisons such as salicylates and barbiturates. Blood is perfused through a chamber containing the solid sorbent which removes the toxin from blood. In the case of iron poisoning, the excess iron in blood would be complexed by the immobilized deferoxamine. Then neither the drug nor its iron complex would be released into the circulatory system, avoiding the problem of toxicity of large doses of deferoxamine.

This procedure would not replace the present use of soluble deferoxamine. Instead, it would help the soluble drug in removing iron from blood in those cases of very high serum iron levels.

DEFEROXAMINE B. ORIGIN AND CHEMISTRY

Deferoxamine B is isolated from feroxamine B, an iron-bearing metabolite belonging to sideramines obtained from Streptomyces Pilosus[123,136,137]. The trivalent iron in feroxamine B is removed by chemical means[138], yielding deferoxamine.

The molecule of deferoxamine is composed of one residue of acetic acid, two residues of succinic acid, and three residues of 1-amino-5-hydroxylaminepentane, as shown in Fig. 5. The residues altogether form a trihydroxamic acid with a free amino group at one end of the molecule. This amino group gives its basic character to the molecule and allows it to react with acids (inorganic and organic) and acid derivatives (acyl chlorides and anhydrides) to form salts and amide derivatives, respectively[139].

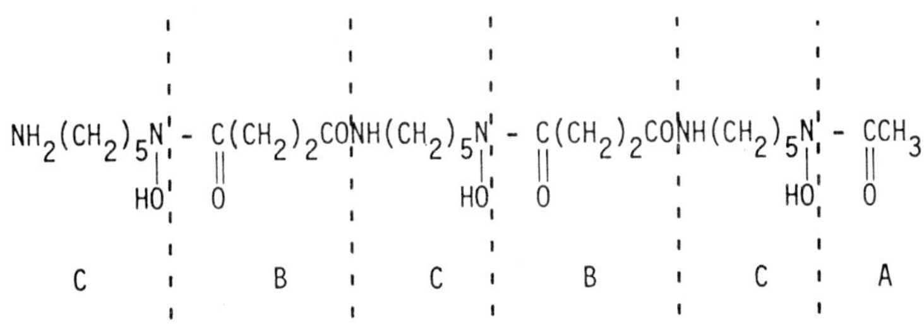

Fig. 5. Deferoxamine B, where A is acetic acid residue, B is succinic acid residue, and C is 1-amino-5-hydroxylaminopentane residue.

Deferoxamine B reacts stoichiometrically with Fe(III) to form feroxamine B, an octahedral iron complex, shown in Fig. 6. The ferric ion becomes attached to the three hydroxamic acid groups. The organic material surrounding the iron confers great stability to the complex[129].

The stability constant of the complex is of the order of 10^{31} in aqueous solutions ranging from pH 3 to pH 8. This value is at least 15 powers of 10 greater than for other complexes of deferoxamine and also greater than for other common chelating agents. Values of the stability constants for the complexes of various ions and chelating agents are given in Table 3.

Keberle[129] reported the effect of deferoxamine on the iron proteins ferritin, hemosiderin, transferrin, and hemoglobin. He found that deferoxamine removes iron from ferritin and hemosiderin until its maximum binding capacity is attained. He found, however, that deferoxamine removed only 10 to 15% of iron from saturated transferrin, and found no exchange from feroxamine to transferrin, showing that deferoxamine does not take up all the iron from transferrin under equilibrium conditions. These results are in disagreement with Bailar's conclusions[14]. He said that deferoxamine "is able to extract iron from nearly all of the tissues of the body, but not from transferrin or ferritin."

The kinetics of iron complexation of deferoxamine has been studied by Lentz et al.[142]. They found that the overall second-

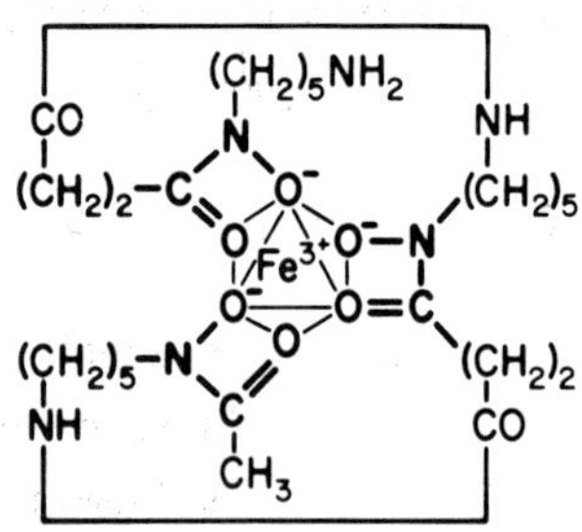

Fig. 6. Feroxamine B (also called ferrioxamine B).

TABLE III

Stability Constants for Some Metallic Complexes [86, 129, 140]

Metallic Ion	Deferoxamine B (DFA)	EDTA (a)	DTPA (b)	HEDTA (c)	EDDHA (d)	Fe-3-Specific (e)
Fe(III)	10^{31}	10^{25}	10^{29}	10^{20}	10^{32}	10^{30}
Fe(II)	10^{10}	10^{14}	10^{17}	10^{12}	-	10^{4}
Mg(II)	10^{4}	10^{9}	10^{9}	10^{5}	10^{8}	10
Ca(II)	10^{2}	10^{11}	10^{10}	10^{8}	10^{7}	-
Sr(II)	10	10^{9}	10^{11}	-	-	-
Co(II)	10^{11}	10^{16}	10^{19}	-	-	-
Ni(II)	10^{10}	10^{19}	10^{20}	-	-	-
Cu(II)	10^{14}	10^{19}	10^{21}	10^{17}	10^{16}	10^{8}
Cd(II)	10^{8}	10^{16}	-	-	-	-
Zn(II)	10^{11}	10^{16}	10^{18}	10^{14}	10^{9}	10^{5}

a Ethylenediaminetetraacetic acid

b Diethylenetriaminepentaacetic acid

c Hydroxyethylethylenediaminetriacetic acid

d Ethylenediamine-bis(o-hydroxyphenylacetic acid)

e Bis-(hydroxyethyl)-glycine

order rate constant for the complex formation of deferoxamine and iron(III) is the same in aqueous solutions of pH 3.92 as in blood plasma of pH values ranging from 4.1 to 7.1.

Whitten et al.[95] found that iron(II) also forms a complex with deferoxamine almost quantitatively at pH's of 6 or higher. For iron(III) the pH range in which complex formation is rapid and quantitative is between 4.5 to 7.0. At higher pH's it becomes increasingly incomplete, and is directly related to the pH and buffer capacity of the solution[139].

In-vivo experiments of Tripod and Keberle[143] first demonstrated that deferoxamine was also capable of removing iron from the organism. They administered ferric chloride and ferric chloride plus deferoxamine to rats and found that in the first case 0.03% of the iron was excreted in the urine and 81% of the feces with 19% absorbed. On the other hand, 0.8% of the iron administered with deferoxamine was excreted in the urine and 99.3% in feces, with no iron being absorbed.

IMMOBILIZATION OF DRUGS

Biologically Active Polymers

Immobilization of a drug by binding it to a polymer may prolong the activity of the drug. This may be achieved by grafting the drug to a reactive polymer or by converting the drug to a monomer with subsequent polymerization.

Cornell and Donamura[144] prepared monomeric drugs by reaction of methacrylyl chloride with tropolones that show biological activity. An example is

$$CH_2 = \overset{CH_3}{\overset{|}{C}} — COCl \; + \; R\text{-}OH \; \rightarrow \; CH_2 = \overset{CH_3}{\overset{|}{C}} — \underset{\underset{R}{|}}{CO}$$

R = 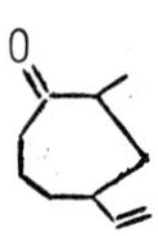

However, not all the monomers were able to polymerize, probably because of the steric effects associated with the bulky groups[145].

Donamura and co-workers have reported, in a sequence of papers, the synthesis of biologically active formaldehyde-sulfamide copolymers[146,147] and sulfonamide- and sulfone-dimethylolurea copolymers[148,149]. They found that some of the copolymers presented antimalarial activity, and that the activity depended on the comonomer present in the polymer other than the drug[152].

Phenethylamines have been bound to polymers by converting them into vinyl monomers for homopolymerization[153] or copolymerization[154]. Monomers derived from styrene and methacrylic acid were used in this work. The biological activity of the monomers, homopolymers, and copolymers was studied by Weiner et al.[154]. They propose that the biological activity depended on the hydrolysis rate of the drug, and found that the polymer suffered slower hydrolysis than the monomer.

Polymeric salicylic acid derivatives have been studied by Weiner and Zilkha[155] for their biological action. They prepared the monomer by the reaction of methacrylyl chloride and various derivatives of salicylic acid, with subsequent homopolymerization. They tested the polymeric drugs as analgesics and found the drugs had retarded analgesic action, in some cases better than aspirin. The retarded action was suggested to be due to slow release of the salicylic acid derivatives.

Several other polymer drugs have been described in the literature[156-161].

Examples where the drug is grafted onto polymers are found in the binding of noradrenaline, isonicotinic acids, and hydrazide onto oxydextran and polyacrolein[163]. Drugs have also been bound to plasma proteins, such as albumin[163-170], including acidic[163,164] and basic drugs[166-170].

In our work we grafted a basic drug, deferoxamine, on to various polymers in order to insolubilize the drug and use it as a chelating polymer for biomedical applications.

POLYMERIC SUPPORTS FOR DEFEROXAMINE

The chemical structure of deferoxamine, which contains a primary amino group, allows it to react with other functional groups, such as chloroformyl (-COCl), benzyl chloride, and aldehyde (-CHO). Polymers and copolymers containing such functional groups were studied and the rationale for the selection of the various polymers is discussed.

Polyacrolein

In a previous paper we reported the synthesis of a polyacrolein-deferoxamine copolymer for potential use in the extracorporeal treatment of iron poisoning[43]. Polyacrolein was selected because the aldehyde groups of the polymer react with amines in neutral or slightly basic solvents to form Schiff's bases[171], while they do not react with hydroxamic acid groups, also present in the deferoxamine molecule, under these conditions.

Copolymers of Acrylyl Chloride and Methacrylyl Chloride with Styrene

Our interest in the poly(acid chlorides) is because of the high reactivity of these polymers which makes it possible to transform them into many polymeric derivatives. Copolymers are similarly useful because inclusion of only a small percent of acid chloride in a polymer chain provides sites which can be used for the attachment of grafts.

The inclusion of styrene as comonomer of the acid chloride is based on the hydrophobicity of styrene groups that renders the polymeric-deferoxamide insoluble in aqueous media. If homopoly (acid chlorides) are used and reacted with deferoxamine, a completely water-soluble system is obtained[172].

Terpolymers of Methacrylyl Chloride, Vinyl Pyrrolidone, and Styrene

To allow a more open structure for easier access of the ferric ions into the chelating sites, a swellable network is necessary. We expected that by introducing vinyl pyrrolidone units into the polymer chains we would obtain a more hydrophylic, swellable (but not soluble) polymer. Styrene was included to provide the hydrophobicity necessary to prevent complete solubility.

Chloromethylated Polystyrene

Chloromethylated polystyrene is very commonly used as a support for the immobilization of enzymes[173] and for the preparation of ion-exchange resins[1]. The chlorobenzyl groups are very reactive with other groups containing active hydrogens, such as amine and hydroxyl groups[174]. Thus the low molecular weight analog, benzyl chloride, reacts with primary amines to form secondary amines, the polymer reacting in a similar way.

The bond that is formed in the reaction of chlorobenzyl and primary amine groups is very stable to hydrolysis. Therefore, if deferoxamine were reacted with chloromethylated polystyrene, the drug would not be released from the polymer even in conditions more severe than those found in blood.

Chloromethylated Amberlites

Amberlites XAD-2 and XAD-4, trademark of Rohm and Haas Company, Philadelphia, Pennsylvania 19105, are divinylbenzene cross-linked polystyrene resins available as porous beads[175,176]. The presence of styrene groups and the relatively high surface area of these beads allow them to be chloromethylated and aminated. Deferoxamine may be attached to the beads since the size of the pores is greater than the size of the deferoxamine molecule. This would allow us to prepare chelating resins without disturbing the scope of the beads and without having to fabricate the polymers into an appropriate physical form that would permit a good perfusion of blood.

EXPERIMENTAL RESULTS

SYNTHESIS OF POLYMER-DEFEROXAMINE GRAFTS

The reaction between the polymers or copolymers with deferoxamine utilized N,N-dimethylformamide (DMF) as solvent and triethylamine (TEA) as basic catalyst in order to prevent reaction of the hydroxamic acid groups with functional groups of the polymers. With the exception of the chloromethylated polystyrene-based polymers, they were soluble in DMF at the temperature of the reaction, 100-110°C.

Polyacrolein-Deferoxamine (VIII)[43]

Polyacrolein (I), prepared by free radical redox polymerization of acrolein[177], contains free aldehyde groups, hydrated aldehyde groups, and cyclic hemiacetal groups[178], all of which may react with an amine[171,179]. This is shown in Fig. 7, where after the reaction with a primary amine three types of groups may be found: azomethine, N-acetal, and tetrahydropyran. Thus, in the reaction of polyacrolein and deferoxamine all the possible structures may be found. Table 4 shows the yield and elemental analysis of the various polymer-deferoxamine grafts. The IR spectra of polyacrolein and polyacrolein-deferoxamine graft (VIII) are given in Fig. 8.

Poly(acryly chloride-co-styrene)-Deferoxamine (IX) and Poly(methacrylyl chloride-co-styrene-Deferoxamine (X)

The poly(acid chloride-co-styrene) copolymers were synthesized using a comonomer feed mole ratio of 1:1. The copolymer of acrylyl chloride and styrene (II) showed more styrene units then acid chloride units, as shown in Table 4. The reaction between the copolymers and deferoxamine proceeds with the formation of an amide linkage between the acid chloride group and the amine group of the drug, as shown in Fig. 9.

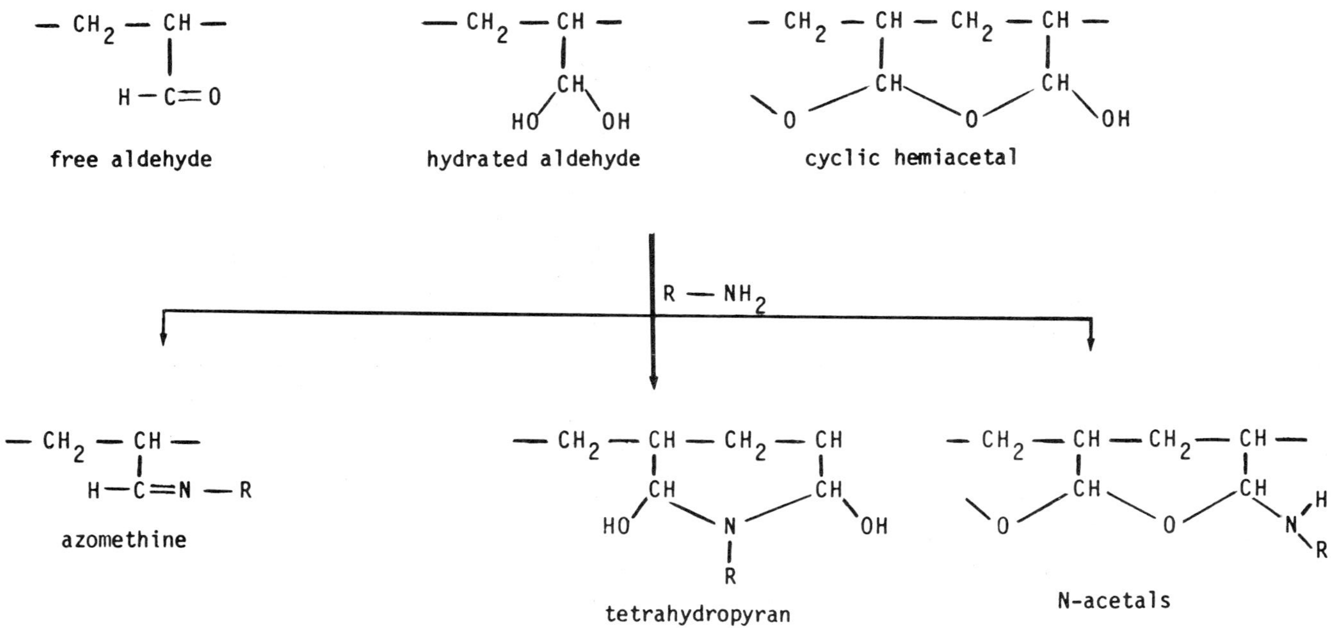

Fig. 7. Functional groups of polyacrolein and their reaction with an amine[178].

Table 4

Results of the synthesis of various polymers

Base Polymer						Graft Polymer					
No.	Composition	Elemental Analysis				No.	Yield %	Elem. Analysis			% DFA Grafted
		C	H	N	Cl			C	H	N	
I	$\left(-CH_2-CH(CHO)-\right)_n$ Polyacrolein	Calc. 64.27	7.19	-	-	VIII	49	Found 57.89	7.46	7.81	57 ratio* 1:5
		Found 61.17	6.99	-	-						
II	$\left(-CH_2-CH(C_6H_5)-\right)_3\left(-CH_2-CH(COCl)-\right)_2$ Poly(acrylyl chloride-co-styrene)	Calc. 73.02	6.13	-	14.37	IX	27	Found 75.36	7.81	3.28	30 ratio* 1:1
		Found 70.88	6.45	-	14.24						
III	$\left(-CH_2-CH(C_6H_5)-\right)\left(-CH_2-C(CH_3)(COCl)-\right)$ Poly(methacrylyl chloride-co-styrene)	Calc. 69.06	6.28	-	16.99	X	51	Found 68.76	8.08	6.05	53 ratio* 1:1
		Found 70.00	6.42	-	17.52						

Base Polymer						Graft Polymer					
No.	Composition	Elemental Analysis C	H	N	Cl	No.	Yield %	Elem. Analysis C	H	N	% DFA Grafted
IV	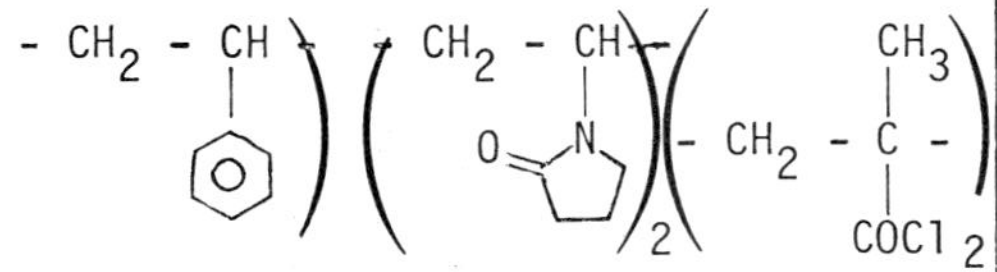 Poly(methacrylyl chloride-co-styrene co-vinyl pyrrolidone)	Calc. 62.79 Found 59.14	6.78 6.62	5.23 4.75	13.24 13.31	XI	55	Found 64.87	7.79	7.98	60 ratio* 1:1
V	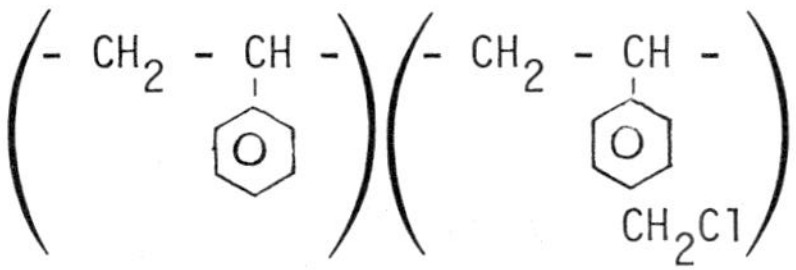 Chloromethylated polystyrene	Calc. 73.95 Found 77.23	6.20 6.07	- -	18.85 19.86	XII	63	Found 72.70	8.02	6.51	68 ratio* 1:2
VI	Chloromethylated Amberlite XAD-2 (Amberlite XAD-2 is polystyrene crosslinked with divinylbenzene)	Found -	-	-	6.77	XIII	97	Found 84.91	8.21	2.26	28 ratio* 1:1
VII	Chloromethylated(Amberlite XAD-4 Amberlite XAD-4 is polystyrene crosslinked with divinylbenzene).	Found -	-	-	11.14	XIV	96	Found 83.18	8.27	3.28	33 ratio* 1:1

* Mole ratio of reactants = polymer:DFA

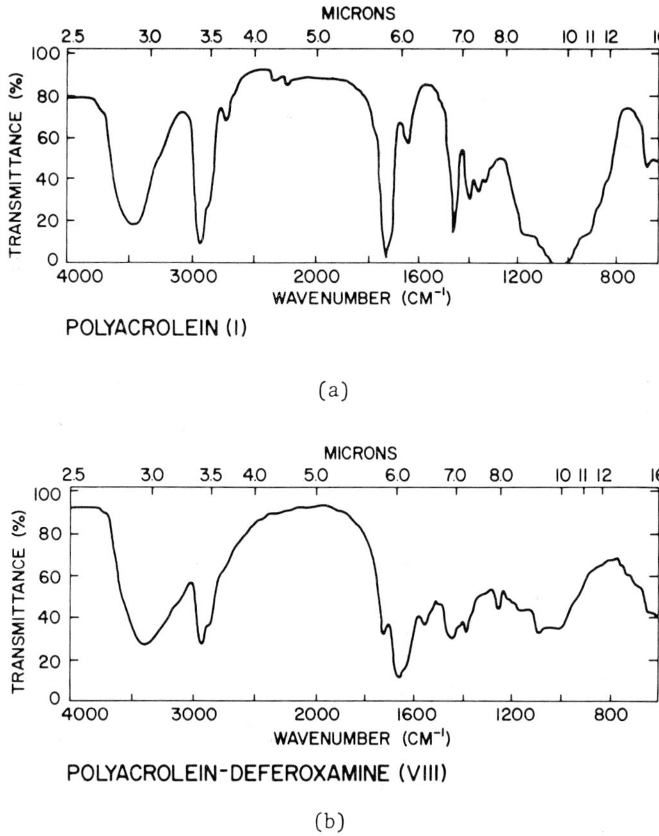

Fig. 8. IR spectra of (a) polyacrolein (I) and (b) polyacrolein-deferoxamine graft (VIII).

The reaction is not quantitative as shown by the percentage of deferoxamine grafted (Table 4). The unreacted acid chloride groups may be converted to carboxylic groups in the precipitation of the copolymer-drug graft with 10% hydrochloric acid. The IR spectra of poly(methacrylyl chloride-co-styrene) and of the copolymer-drug graft are shown in Fig. 10.

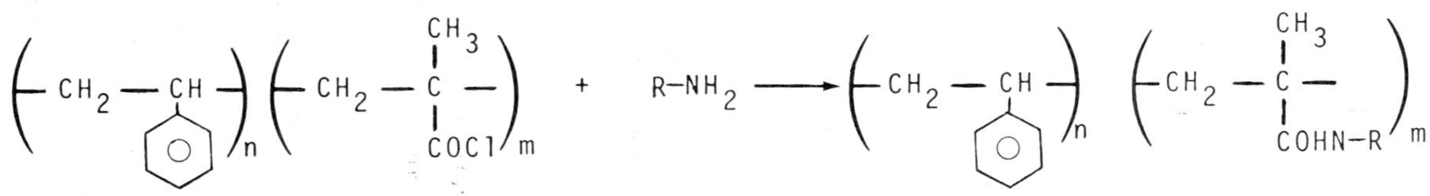

Fig. 9. Reaction between poly(methacrylyl chloride-co-styrene) (III) and deferoxamine. $R\text{-}NH_2$ is deferoxamine.

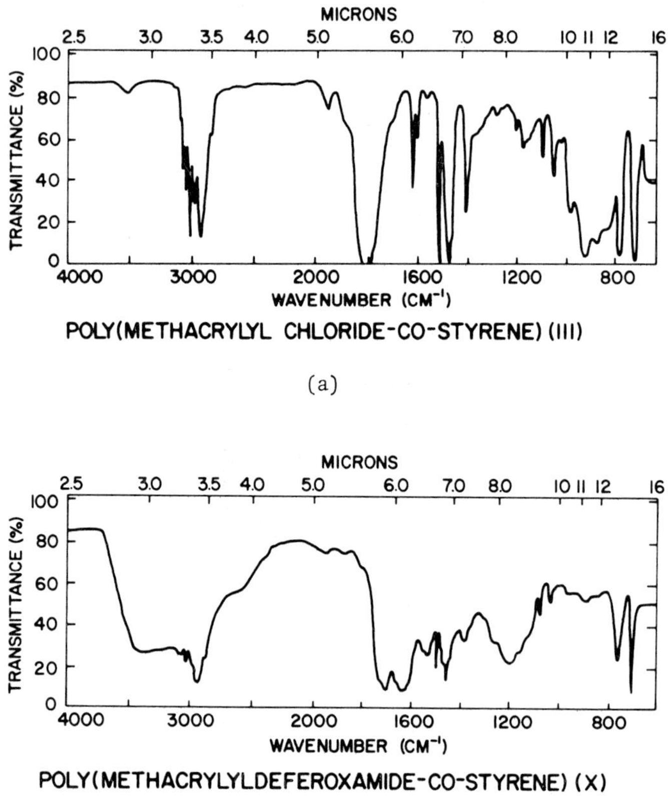

(a) (b)

Fig.10. IR spectra of (a) poly(methacrylyl-chloride-co-styrene) (III) and (b) its deferoxamine graft (X).

Poly(methacrylyl chloride-co-vinyl pyrrolidone-co-styrene)-Deferoxamine (XI)

The terpolymer (IV) was obtained by free radical co polymerization of methacryl chloride, N-vinyl pyrrolidone, and styrene in a 2:2:1 feed mole ratio. A ratio of 2 acid chloride units, 2 vinyl pyrrolidone units, and 1 styrene unit is obtained from elemental analysis (Table 4). The reaction of the terpolymer with deferoxamine proceeds with the formation of an amide linkage. The chemical equation for the reaction is similar to that of Fig. 9. Results for this reaction are summarized in Table 4, and the IR spectra of both the terpolymer (IV) and termpolymer-drug graft (XI) are shown in Fig. 11.

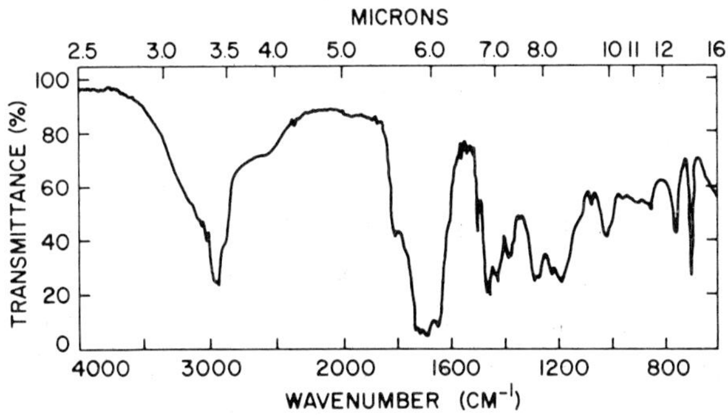

POLY(METHACRYLYL CHLORIDE-CO-VINYL PYRROLIDONE-CO-STYRENE) (IV)

(a)

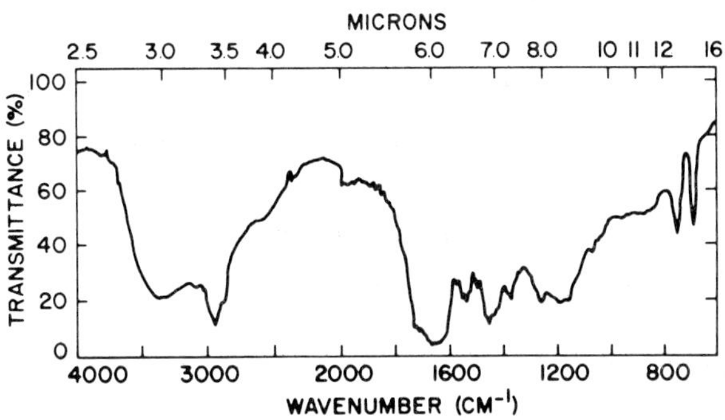

POLY(METHACRYLYL CHLORIDE-CO-VINYL PYRROLIDONE-CO-STYRENE)-DEFEROXAMINE (XI)

(b)

Fig.11. IR spectra of (a) the terpolymer (IV) and (b) its deferoxamine graft (XI).

Chloromethylated Polystyrene-Deferoxamine (XII)

Chloromethylated polystyrene (V) was prepared by the reaction of commercial polystyrene[180] with chloromethyl methyl ether and zinc chloride. To obtain a high chlorine content, the reaction time was longer than the one normally used[181,182]. The percentage of chlorine from Table 4 indicates that about 80% of the phenyl rings have been chloromethylated.

The reaction of chloromethylated polystyrene and deferoxamine is shown in Fig. 12. In this reaction a secondary amine is formed, which provides the graft with very high hydrolytic stability. The results for this reaction are shown in Table 4. In this case, there was 3.91% of unreacted chlorine, indicating that a longer reaction time is probably necessary for higher conversion. The IR spectra of chloromethylated polystyrene (V) and poly(styrene-co-vinyl benzyl)-deferoxamine (XII) are shown in Fig. 13.

Chloromethylated Amberlite XAD-2- and XAD-4-Deferoxamine (XIII) and (XIV)

Chloromethylated Amberlites were obtained by the same procedure used for the chloromethylation of polystyrene. The reaction between deferoxamine and the chlormethylated Amberlites follows the same scheme as for the synthesis of XII. Table 4 gives the results for the reactions. IR spectra could not be obtained for the chloromethylated Amberlites or the Amberlite-deferoxamine grafts due to their particulate nature.

PROPERTIES OF THE POLYMERS

Some physical properties of the polymers are given in Table 5. Viscosity measurements could be done only on those polymers that were soluble in appropriate concentrations.

Stability of the Grafts in Saline Solutions

Stability of the grafts in the present case refers to the degree of deferoxamine liberation when the polymers are placed in

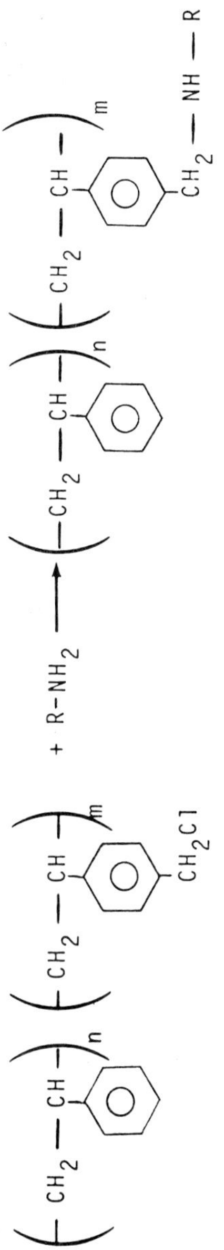

Fig.12. Reaction between chloromethylated polystyrene and deferoxamine. $R\text{-}NH_2$ is deferoxamine.

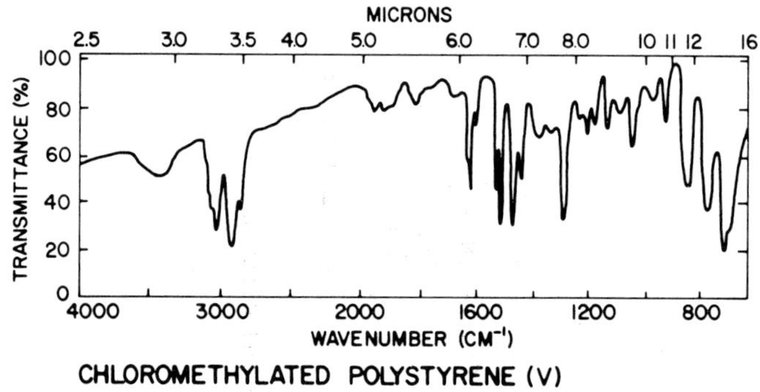

(a)

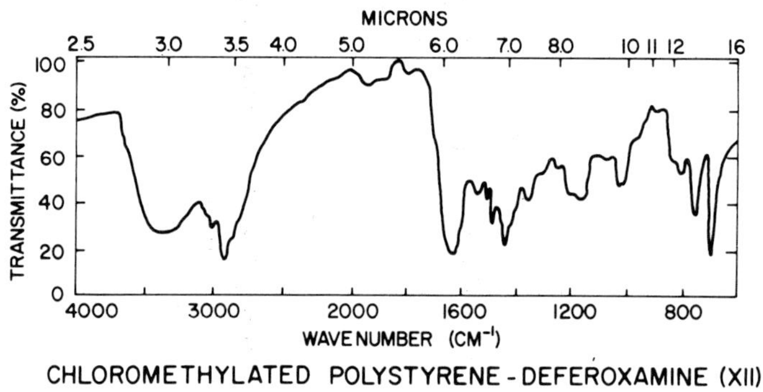

(b)

Fig.13. IR spectra of (a) chloromethylated polystyrene (V) and (b) its deferoxamine graft (XII).

contact with saline solutions for several days. The presence of deferoxamine in the solutions was determined quantitatively by UV spectrophotometry. Deferoxamine has a strong absorption with a maximum of 210 nm (epsilon max = 32,000) at pH 5.6, allowing a rapid determination of deferoxamine concentration in solution.

Fig. 14 shows the stability of the various polymer grafts in normal saline (0.9% NaCl) solution (pH 5.6). The stability of the grafts in phosphate-buffered saline solution (pH 7.4) is shown

TABLE V
Some Physical Properties of the Various Polymers

Polymer No.	[η] (Solvent)	Solubility	Transition Temperature(b) (°C)	Density(c) (g/cm^3)
I	0.38(a) (water)	Xylene, DMSO	M.T. 126(d)	1.196
II	1.05 ($CHCl_3$)	Acetone, iso-propanol, $CHCl_3$	M.T. 224	0.700
III	0.53 (DMF)	$CHCl_3$, xylene, DMF, DMSO, 1,2-dichloro-ethana, dioxane, CCl_4, acetone	M.T. 302 (some decom-position)	-
IV	0.11 (dioxane)	Swells in ethanol, xylene, DMF, DMSO	T_g 282; Dec. 358(e)	0.751
V	-	Swells in DMF(hot), xylene, DMSO, $CHCl_3$	Dec. 252	0.915
VI	-	Insoluble in all solvents tested	Dec. 335	-
VII	-	Swells in DMSO	Dec. 227	0.943
VIII	-	Xylene, DMSO, 1,2-dichlorethane	Dec. 203	1.158
IX	-	Ethanol, xylene, DMSO, Dioxane, acetone, isopropanol	M.T. 352 (with dec.)	-
X	-	Xylene, DMF	Dec. 357	1.041
XI	-	Swells in DMSO, DMF, xylene, CCl_4	T_g 261	0.964
XII	-	Swells in DMSO, DMF, xylene	Dec. 374	1.016
XIII	-	Swells in 1,2-dichloroethane	Dec. 327	0.967
XIV	-	Swells in DMSO, CCl_4 (hot)	T_g 212; Dec. 262	0.883

a Intrinsic viscosity of a polyacrolein-bisulfite adduct solution
b Determined by differential scanning calorimetry (DSC)
c Determined by pycnometry
d M.T.: Melt temperature (center of peak in DSC thermogram)
e Dec.: Temperature at which decomposition started

in Fig. 15. In both cases, the graft with the lowest stability was VIII. This is due to the easier hydrolysis of the carbon-nitrogen double bonds. In the case of graft XII, the relatively high

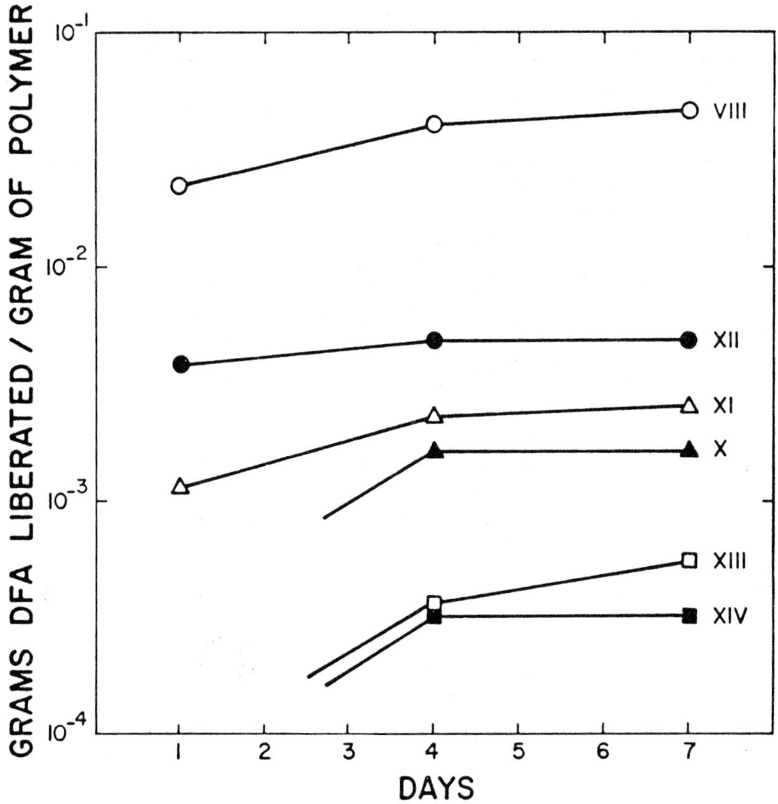

Fig.14. Stability of the various polymer-deferoxamine grafts in normal saline solution.

concentration of liberated deferoxamine is probably due to some unreacted drug entrapped in the polymer network.

The amounts of deferoxamine liberated from the polymers were low enough to allow us to expect similar behavior in biological media. Blood plasma was used to determine the stability of the grafts in the presence of other molecules, such as proteins and enzymes, which may promote or enhance degradation of the grafts.

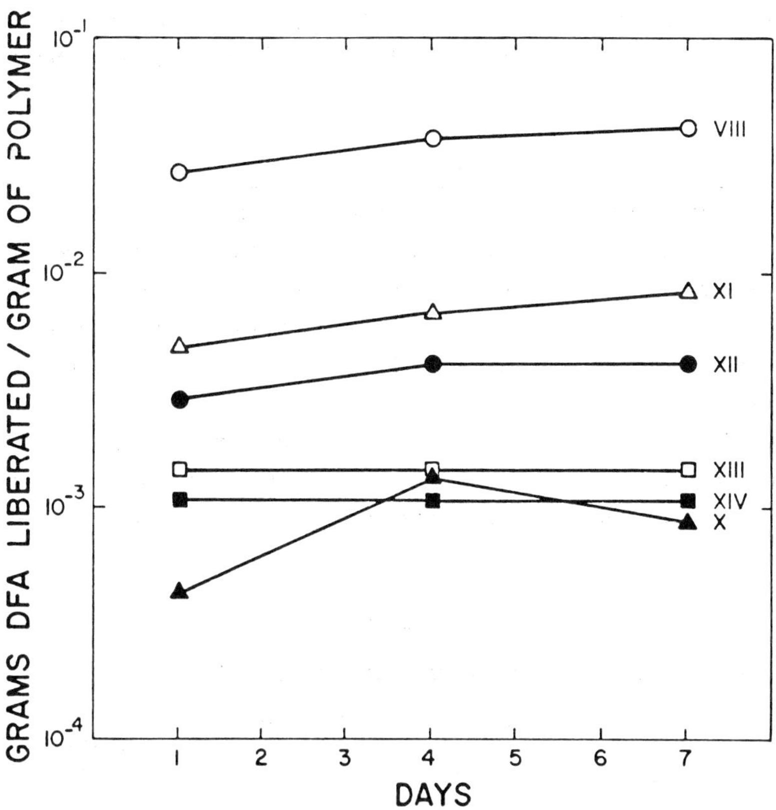

Fig.15. Stability of the various polymer-deferoxamine grafts in phosphate buffered saline solution.

Stability of the Grafts in Plasma

The same principle, deferoxamine liberation, was used to determine the stability of the polymers in plasma. In this case the liberated drug was analyzed spectrophotometrically by forming its Fe^{3+} complex and measuring absorbances at 430 nm. This method is only accurate for concentrations not lower than $9 \cdot 10^{-5}$ mole/liter as determined from a standard calibration curve. None of the polymers showed liberation of deferoxamine in concentrations higher than the accurate limit, i.e., $9 \cdot 10^{-5}$ mole/liter, after 27 hours in contact with plasma.

Iron Chelation Ability of the Polymer Grafts

Samples of the various grafts were placed in concentrated ferric chloride solutions for 3 days. All the polymers took on a brown-red color. The iron content of the polymers was determined quantitatively by chemical analysis[183] and qualitatively by energy dispersive analysis of x-rays (EDAX) in the scanning electron microscope.

A comparison of chelating capacity found for the grafts and the one calculated from the nitrogen content is given in Table 6. These results show that the capacity of grafts XIII and XIV (derived

TABLE 6

Chelating Capacity of the Various Polymer-Deferoxamine Grafts

Polymer No.	Capacity (mg Fe/g resin) Calculated	Found	Efficiency[a] (%)
VIII	49.3	18.8	38
IX	20.8	-	-
X	40.4	17.7	44
XI	30.2	10.3	34
XII	41.6	12.9	31
XIII	14.8	6.0	41
XIV	18.6	6.2	33

a Efficiency = mg Fe found/mg Fe calculated.

from Amberlite XAD-2 and XAD-4, respectively) is about half of the capacity for other polymers. However, the efficiency is comparable, indicating that the accessability of the iron into the chelating sites is approximately the same for all the grafts.

A qualitative analysis of the iron in the polymers is given in Fig. 16. The traces on the left side are controls, i.e., grafts without exposure to iron, and at the right are the results for iron-containing grafts. The peaks at 6.4 keV reveal the presence of iron[184], whereas the controls do not present any peak in that position.

Stability Constants of the Polymer Iron Chelates

When a polymer-deferoxamine iron complex (P-DFA-Fe) is placed in water, the following reaction takes place:

$$\text{P-DFA-Fe} + 3H_2O \rightleftarrows \text{P-DFA-}H_3 + Fe(OH)_3$$

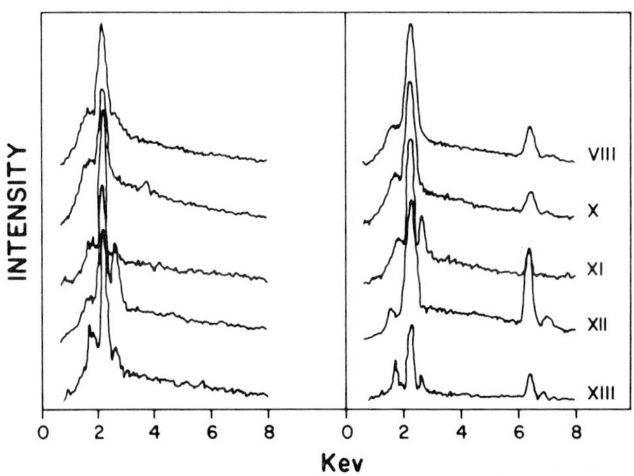

Fig.16. Energy dispersive analysis of X-rays (EDAX) of various polymer-deferoxamine-iron complexes. At the right are the traces for iron complexes, and at the left for the controls.

The equilibrium constant for this reaction is

$$K_e = \frac{[P\text{-}DFA\text{-}H_3]\,[Fe(OH)_3]}{[P\text{-}DFA\text{-}Fe]}\text{, where} \tag{4}$$

$[Fe(OH)_3] = [Fe^{3+}]sol$ = iron concentration in solution, and

$[P\text{-}DFA\text{-}Fe] = [Fe^{3+}]pol$ = iron concentration in the polymer.

Therefore

$$K_e = \frac{[Fe^{3+}]^2 sol}{[Fe^{3+}]_{pol}} \tag{5}$$

and the stability constant would be

$$K_s = \frac{1}{K_e} = \frac{[Fe^{3+}]_{pol}}{[Fe^{3+}]^2 sol} \tag{6}$$

The various polymer grafts were placed in concentrated ferric chloride solutions for 4 days to let them equilibrate with iron. Then a known amount of each polymer-chelate was placed in distilled water for 7 days. Iron and deferoxamine in solution were analyzed. The difference between total moles of iron and moles of deferoxamine gives the absolute amount of iron coming from the polymer-chelate according to the reaction given above. The iron content of the polymers was determined[183], and the stability constants (Table 7) were calculated using Eq. (6). The values in Table 7 demonstrates that a large decrease in stability of the deferoxamine-iron complex occurs when the drug is bound to a polymer backbone. This is probably due to steric hindrance from the polymer and a decrease in mobility of the drug molecule.

TABLE 7

Stability Constants of the Polymer-Deferoxamine-Iron Complexes

Polymer	K_S (liter/mole)
X	10^6
XI	10^5
XII	10^5
XIII	10^4
XIV	10^4

REMOVAL OF IRON FROM SOLUTION IN RECIRCULATING SYSTEMS

Iron in Aqueous Solutions

Preliminary experiments for iron removal were carried out in aqueous solutions. Ferric chloride solutions of relatively high concentration ($1.2 \cdot 10^{-3}$ M) compared to the concentration of serum iron in acute poisoning cases ($4.5 \cdot 10^{-5}$ to $1.8 \cdot 10^{-4}$ M) were used. Fig. 17 gives a schematic diagram of the system employed for these determinations.

The iron solution was recirculated through the chamber containing the chelating polymer, and samples were taken in the outflow. The results obtained for iron concentration in the solution at various times indicated that, with the iron concentrations used, the resins became saturated after 15 to 20 min. of initiating the experiment, showing no further decrease in iron concentration. The results for some of the polymers tested in these conditions are given in Table 8.

The polymer grafts that were obtained in powder form (VIII-XII) became agglomerated as a hard cylinder, and the solution could not penetrate into the bulk of the resin, flowing only through

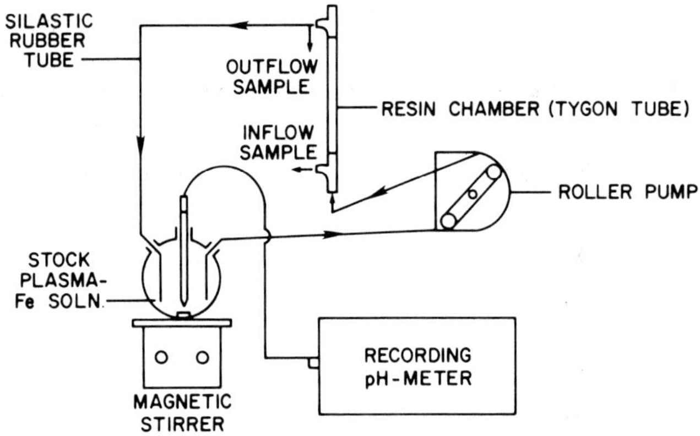

Fig.17. Diagram of the system employed in the experiments for iron removal from water and plasma.

the outer parts next to the walls of the chamber. This agglomeration effect was not present in those polymers in bead form (XIII and XIV).

Iron in Plasma Solutions

Iron was added to plasma in the form of ferric chloride in concentrations of 500 to 800 μg Fe/100 ml plasma (i.e, $8.9 \cdot 10^{-5}$

TABLE 8

Iron Removed from Aqueous Solutions by Various Polymer-Deferoxamine Grafts

Polymer	μg Fe removed/g resin
X	792
XII	2196
XIII	661
XIV	900

to $1.4 \cdot 10^{-4}$ M). Polymer X and XII gave scattered results for the rate of iron removal. Polymers VIII, IX, and XI were not tested.

Results for grafts XIII and XIV (in bead form) are shown on Figs. 18 and 19. The plot on Fig. 18 is the rate of iron removal, i.e., decrease of iron concentration in plasma with time. The two polymers show approximately the same rate. A better comparison is given on Fig. 19 where the amount of iron removed is observed during the first hours of the experiments. Then the iron removed by graft XIV starts leveling off, while the iron removed by grafts XIII continues to increase. This result agrees with the higher chelating efficiency determined for XIII (Table 6).

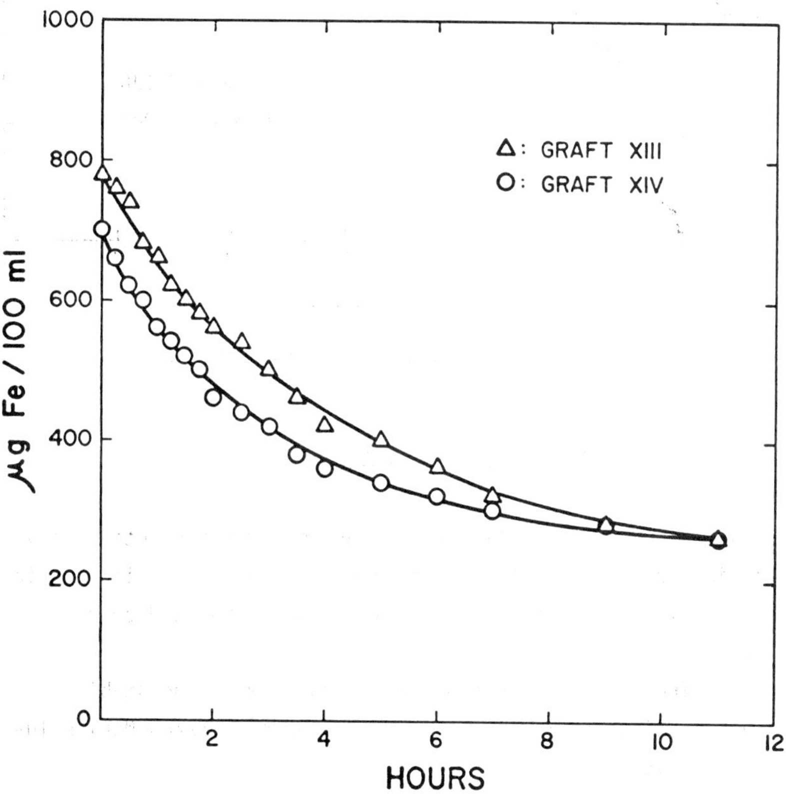

Fig.18. Rate of iron removal from plasma.

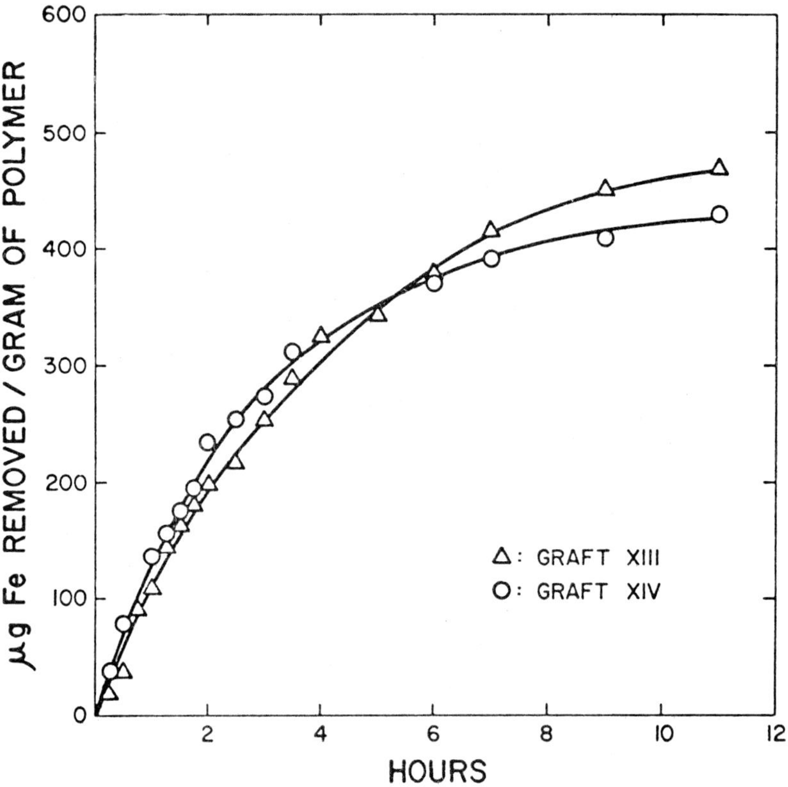

Fig.19. Iron removed from plasma.

CONCLUSIONS AND FUTURE DIRECTIONS

1. Deferoxamine B has been bound to various reactive polymers, including polyacrolein, copolymers of acrylic acid chlorides, and chloromethylated polystyrenes (uncross-linked and cross-linked).

2. The polymer-deferoxamine grafts showed good stability in saline solutions (0.9% NaCl and phosphate-buffered NaCl solutions) and in plasma.

3. The polymer-deferoxamine grafts maintained the chelation ability of the drug, although the stability of the polymer-drug-iron complex was decreased compared to the stability of feroxamine (the soluble deferoxamine-iron complex).

4. The chelation capacity of the polymer varies from 6 to 19 mg of Fe/g of resin. This value is in the range necessary for their application in acute iron poisoning where the serum iron levels vary from 0.2 to 5 mg/100 ml of blood.

5. Additional experiments should be performed in order to study the effect of the different variables that affect the chelation process in recirculating systems.

6. The ability of the polymer-deferoxamine grafts to remove iron from plasma, a protein-rich medium where iron becomes less available, opens the possibility of applying the graft copolymer to in-vivo situations. Experiments in animals will be the next logical step after complete in-vitro characterization of the grafts has been made.

ACKNOWLEDGEMENTS

We thank Dr. Anthony Temple, Poison Control Center, University of Utah, for his encouragement in doing this work. CIBA Pharmaceutical Company, Summit, New Jersey, graciously provided samples of Desferal. Dr. Donald T. Warner, Kalamazoo, Michigan, advised and aided us with many helpful suggestions.

The financial support of the Catholic University of Chile given to R. S. Ramirez is appreciated. Portions of this work were supported by NIH Grant FD 00480-01 and Biomedical Sciences Support Grant FR-07092.

REFERENCES

1. R. M. Weaton and M. J. Hatch, "Synthesis of Ion-exchange Resins," in Ion Exchange, Vol. 2, (J. A. Marinksy, ed.), Dekker, New York, 1969, p. 221.

2. Dowex A-1 Chelating Resin, Dow Chemical Co., Midland, Michigan, 1964.

3. E. M. Vasil'eva, R. K. Gavurina, and O. P. Kalomeitsev, U.S.S.R. Patent 184,451 (1966); Chem. Abstr., 66, 29770 (1966).

4. O. E. Paris (to E. I. Du Pont de Nemours & Co.) U.S. Patent 3,035,022 (1962).

5. R. Hering, J. Prakt. Chem., 32, 291 (1966).

6. W. Sterner and L. Loveless, Sci. Techn. Aerospace Rept., 4, 534 (1966).

7. R. Hering, J. Prakt. Chem., 34, 69 (1966).

8. T. R. Norton, U.S. Patent 3,362,935 (1968), to Dow Chemical Co.

9. G. F. D'Alelio, U.S. Patents 3,395,134 (1968) and 3,395,197 (1968).

10. A. Zlatkis, W. Bruening, and E. Bayer, Anal. Chim. Acta, 56 399 (1971).

11. A. M. Talati and V. N. Mistry, Mater. Sci. Eng., 10 287 (1972).

12. G. Schumckler, Talanta, 12, 281 (1965).

13. I. Shondak and A. F. Nikolaev, Vysokomol. Soedin., 7, 1835 (1965).

14. E. B. Trostyanskaya and G. F. Nefedova, Zh. Anal. Khim., 20, 775 (1965); Chem. Abstr., 872 (1966).

15. P. Hering, German (East) Patent 42,086 (1965).

16. G. Maura and F. Pochetti, Ann. Chim., 55, 1000 (1965).

17. S. S. Ivanov, L. P. Gavryuchenkova, and M. M. Koton, Vysolomol. Soedin., 8, 470 (1966).

18. Y. Matsunaga, T. Itagaki, and T. Katsuura, German Patent 2,258,535 (1973), to Mitsubishi Chemical Industries Co. Ltd., Chem. Abstr., 79, 67075 (1973).

19. S. V. Rogozhin, V. A. Davankov, and I. A. Yamskov, Vysokomol. Soedin., Ser. B., 15, 216 (1973).

20. K. W. Pepper and D. K. Hale, Ion Exchange and Its Applications, Society of Chemical Industry, London, 1955.

21. R. A. Mock, C. A. Marshall, and L. R. Morris, U.S. Patent 2,910,445 (1959).

22. R. Hering, Z. Chem., 5, 29, 194 (1965).

23. G. F. D'Alelio, U.S. Patent 3,228,920 (1966).

25. A. Gabert and B. Foigt, German (East) Patent 44,124 (1965), Chem. Abstr., 64, 19901 (1966).

26. L. O. Gunderson and H. Kerst, U. S. Patent 3,285,886 (1966), to W. R. Grace & Co.

27. R. Hering, Proc. Conf. Appl. Phys. Chem. Methods Chem. Anal., Budapest, 2, 35 (1966).

28. G. F. D'Aleio, E. T. Hofman, and J. R. Zeman, J. Macromol. Sci. Chem., A3, 959 (1969).

29. G. F. D'Alelio, E. T. Hofman, and F. J. Mevers, Ibid., A4, 361 (1970).

30. E. Dabrunz and H. Reuter, German (East) Patent 79,152 (1971); Chem. Abstr., 76, 60464 (1972).

31. G. F. D'Alelio, E. T. Hofman, and T. R. Dehner, J. Macromol. Sci. Chem., A4, 897 (1970).

32. G. F. D'Alelio, E. T. Hofman, and W. F. Strazik, Ibid., A6, 471 (1972).

33. G. F. D'Alelio, E. T. Hofman, and W. F. Strazik, Ibid., A6, 513 (1972).

34. T. Nakanishi, H. Kawamura, and H. Kakita, Japanese Patent 73 47,993 (1973), to Matsushita Electrical Industry Co. Ltd.

35. R. Hering, J. Prakt. Chem., 14, 285 (1961).

36. E. B. Trostyanskaya, I. P. Losev, A. S. Teulina, S. B. Makarova, G. Z. Nefedova, and H. J. Lu, J. Polym. Sci., 59, 379 (1962).

37. W. Langenbeck, H. W. Krause, and J. Reihsig, German (East) Patent 24,421 (1962); Chem. Abstr., 59, 8901 (1963).

38. H. P. Gregor, M. Taifer, L. Citarel, and E. I. Becker, Ind. Eng. Chem., 44, 2834 (1952).

40. J. P. Cornaz and H. Duel, Experientia, 10, 137 (1954).

41. G. Petrie, D. Locke, and C. Meloan, Anal. Chem., 37, 919 (1965).

42. G. Petrie, D. Locke, and C. Meloan, Anal. Biochem., 11, 919 (1965).

43. R. S. Ramirez and J. D. Andrade, J. Macromol. Sci. Chem., A7, 1035 (1973).

44. L. D. Pennington and M. B. Williams, Ind. Eng. Chem., 51, 759 (1959).

45. R. C. DeGeiso, L. G. Donamura, and E. A. Tomic, J. Appl. Polym. Sci., 9, 411 (1965).

46. F. Vernon and H. Eccles, Anal. Chim. Acta, 63, 403 (1973).

47. K. F. Sugawara, H. H. Weetall, and G. D. Schucker, Anal. Chem., 46, 489 (1974).

48. F. Wolf and M. Morgner, German Patent 1,056,825 (1959).

49. J. A. Mikes and L. Kovacs, J. Polym. Sci., 59, 209 (1962).

50. G. Manecke and A. Gorhmann, Makromol. Chem., 82, 146 (1965).

51. G. Manecke and H. Heller, Angew. Chem., 72, 523 (1960).

52. E. Blasius and B. Brozio, J. Chromatog., 18, 572 (1965).

53. V. M. Tolmachov, L. A. Lomako, and L. A. Gurskaya, Vysokomol. Soedin., 5, 512 (1963).

54. M. J. Hatch, U. S. Patent 3, 197,281 (1965), to Dow Chemical Co.

55. J. Krynicki, Polimery, 11, 283 (1966).

56. S. B. Savvin, O. P. Eliseeva, and Y. G. Rozovskii, Dokl. Akad. Nauk SSSR, 180(2), (1968), Chem. Abstr. 69, 36663 (1968).

57. R. M. Genik-Sas-Berezowsky and I. H. Spinner, Can. J. Chem., 48, 163 (1970).

58. A. B. Dovankov, Tr. Mosk. Khim. Tekhnol. Inst., 64, 118 (1970); Chem. Abstr., 75, 77568 (1971).

59. S. V. Vinogrova, V. A. Vasner, and V. V. Korshak, Vysokomol. Soedin. Ser. B., 13, 805 (1971); Chem. Abstr., 76, 86173 (1972).

60. V. D. Kopylova, K. M. Saldadze, and G. D. Azambadze, Zh. Prikl. Khim. (Leningrad), 44, 2512 (1971); Chem. Abstr. 76, 113823 (1972).

61. C. H. Budeanu and A. Gavriliuof, An. Stiint. Univ. "Al. I. Cuza" Iasi, Sect. 1c, 17, 197 (1971), Chem. Abstr., 76, 45901 (1972).

62. R. Hering, Z. Chem., 12, 272 (1972); Chem. Abstr, 77, 147188 (1972).

63. R. Hering and D. Haupt, German (East) Patent 88,780 (1972); Chem. Abstr, 77, 115340 (1972).

64. Y. M. Dedk, O. P. Eliseeva, A. N. Ermakov, S. B. Savvin, and M. G. Slotintseva, Zh. Anal. Khim., 27, 726 (1972); Chem. Abstr., 96530 (1972).

65. A. Erdelyi, S. Fustimolnar, K. Horvath, and M. Nyiro, Hungarian Patent 4,123 (1972); Chem. Abstr., 77, 62768 (1972).

66. B. A. Zhubanov and A. K. Zharmagambetov, Izv. Akad. Nauk Kaz. SSSR, Ser. Khim., 22, 83 (1972); Chem. Abstr., 78, 98043 (1973).

67. G. F. D'Alelio, U.S. Patent 3,700,671 (1972).

68. G. V. Nyasoedva, L. I. Bol'shakova, and S. F. Savvin, Izv. Akad. Nauk SSSR, Ser. Khim., 1972(9), 2033 (1972); Chem. Abstr. 78, 59093 (1973).

69. J. Ritz, and J. Reese, German Patent 2,164,489 (1973); Chem. Abstr., 79, 27005 (1973).

70. A. Warshawsky, German Patent 2,236,335 (1973); Chem. Abstr., 79, 116082 (1973).

71. M. Haruta, K. Kageno, and S. Harada, Makromol. Chem., 167, 95 (1973).

72. M. Haruta, K. Kageno, and S. Harada, Ibid., 170, 11 (1973).

73. Y. G. Rozovskii, N. N. Basargin, N. I. Ivanova, and O. S. Ermakova, Zh. Vses. Khim. Obsch., 18, 359 (1973); Chem. Abstr., 79, 99567 (1973).

74. W. Szcezepaniak and J. Siepak, Chem. Anal. (Warshaw), 18, 1019 (1973).

75. G. Manecke and E. Graudenz, Makromol. Chem., 170, 51 (1973).

76. H. Eccles and F. Vernon, Anal. Chim. Acta, 66, 231 (1973).

77. Y. Kobayashi, A. Tanaka, and M. Sugai, Bull. Fac. Eng. Yokohama Nat. Univ., 22, 45 (1973); Chem. Abstr., 80 30409 (1974).

78. B. Rodriguez Rios, A. Perlonso Uruena, and A. Maderos Perez, An. R. Soc. Esp. Fis. Quim., 61, 723 (1965).

79. P. Teyssie, C. Decoene, and M. T. Teyssie, Macromol Chem., 84, 51 (1965).

80. J. L. Sides and C. T. Kenner, Anal. Chem., 38, 707 (1966).

81. H. Small, Ind. Eng. Chem., 59., 147 (1967).

82. R. M. Diamond and D. C. Whitney, "Resin Selectivity in Dilute to Concentrated Aqueous Solutions," in Ion Exchange, (J. A. Markinsky, ed.), Dekker, New York, 1966, Vol. 1, p. 277.

83. R. C. DeGeiso, L. G. Donamura, and E. A. Tómic, Anal. Chem., 34, 845 (1962).

84. D. Reichenberg, "Ion Exchange Selectivity", in Ion Exchange, Vol 1 (J. A. Morinsky, ed.), Dekker, New York, 1966, p. 227.

85. J. Schubert, Sci. Am., 214, 40 (1966).

86. H. S. Waxman and E. B. Brown "Clinical Usefulness of Iron Chelating Agents," Prog. Hematol. 6 (1969).

87. J. Schubert, "The Chemical Basis of Chelations," in Iron Metabolism (F. Gross, ed.), Springer, Berlin, 1964, p. 466.

88. M. J. Seven, ed. Metal-Binding in Medicine, Lippincott, Philadelphia, 1960.

89. C. L. Witzleben and B. E. Buck, Clin. Toxicol., 4, 579 (1971).

90. B. E. Conley, J. Am. Med. Assoc., 170, 676 (1959).

91. R. D. Coli, L. A. Leone, and M. M. Albala, Rhode Island Med. J., 50, 549 (1967).

92. J. M. Arena, Poisoning, 2nd ed., Thomas, Springfield, Illinois, 1970, p. 369.

93. J. J. Crotty, Clin. Toxicol., 4, 615 (1971).

94. R. A. Aldrich, "Acute Iron Toxicity," in Iron in Clinical Medicine (R. E. Wallerstein and S. R. Mettier, eds.), University of California Press, Berkeley, 1958, p. 93.

95. C. F. Whitten, G. W. Gibson, M. H. Good, J. F. Goodwin, and J. Brough, Pediatrics, 36, 322 (1965).

96. W. O. Robertson, Mod. Treat., 8, 552 (1971).

97. M. A. Luongo and S. S. Bjornson, New Engl. J. Med., 251, 995 (1954).

98. W. O. Robertson, Ohio Med. J., 59, 1198 (1963).

99. I. O. B. Spencer, Br. Med. J., 1951-2, 1112.

100. J. P. Smith, J. Pathol. Bacteriol., 64, 467 (1952).

101. E. Short, B. W. Zweifach, and R. F. Furchgott, Science, 102, 489 (1945).

102. G. Forbes, Br. Med. J., 1947-1, 367.

103. J. H. Prain, Ibid, 1949-2, 1019.

104. C. F. Whitten and A. J. Brough, Clin. Toxicol., 4, 585 (1971).

105. E. Charney, J. Am. Med. Assoc., 178, 326 (1961).

106. K. R. Reissmann, T. J. Coleman, B. S. Berdai, and L. R. Morcarty, Blood, 10, 35 (1955).

107. R. Richards and S. E. H. Broosk, West Indian Med. J., 15, 134 (1966).

108. S. J. Wilson, H. E. Heath, P. L. Nelson, and G. G. Ens, Blood, 13, 483 (1958).

109. C. F. Whitten, Am. J. Dis. Child., 104, 538 (1962).

110. K. R. Reissmann and T. J. Coleman, Blood, 10, 46 (1955).

111. C. F. Whitten, Y. C. Chen, and G. W. Gibson, Pediatr. Res., 2, 479 (1968).

112. W. Westlin, Clin. Pediatr., 5, 531 (1966).

113. J. R. Walsh and J. B. Gillick, Ibid., 4, 633 (1965).

114. C. F. Whitten, Y. Cohen, and G. W. Gibson, Pediatrics, 38, 103 (1966).

115. J. T. McEnery and J. Greengard, J. Pediatr., 68, 773 (1966).

116. S. Moeschlin, Poisoning, Grune and Stratton, New York, 1965, p. 128.

117. J. T. McEnery, Clin. Toxicol., 4, 603 (1971).

118. P. D. Doolan, S. L. Schwartz, J. R. Hayes, J. C. Mullen, and N. B. Cummings, Toxicol. Appl. Pharmacol., 10, 481 (1967).

119. J. L. Fahey, C. E. Rath, J. V. Princioto, J. B. Brick, and M. Rubin, J. Lab. Clin. Med., 57, 436 (1961).

120. J. V. H. Kemble, Guy's Hosp. Rep., 113, 68 (1964).

121. E. F. Robertson, G. H. Maxwell, and R. B. Elliott, Med. J. Aust., 2, 705 (1963).

122. G. Whelam, V. Lazio, and J. C. Biggs, Am. J. Med., 41, 626 (1966).

123. H. Bickel, E. Gaumann, W. Keller-Schierlein, V. Prelog, E. Vischer, A. Wettstein, and H. Zahner, Experientia, 16, 129 (1960).

124. W. F. Westlin, Clin. Toxicol., 4, 597 (1971).

125. S. Moeschlin and U. Schnider, New Engl. J. Med., 269, 57 (1963).

126. S. Moeschlin and U. Schnider, Iron Metab., Intern. Symp., Aix-en-Province, France, 1963, 525.

127. S. Leikin, P. Vossough, and F. Mochir-Fatemi, J. Pediatr., 71, 425 (1967).

128. N. Movassaughi, G. G. Purugganan, and S. Leikin, Ibid., 75, 604 (1969).

129. H. Keberle, Ann. N.Y. Acad. Sci., 119, 758 (1964).

130. H. Brunner, G. Peters, and R. Jaques, Helv. Physiol. Acta, 21, C3 (1963).

131. "Desferal Mesylate," CIBA Pharmaceutical Co., Summit, New Jersey, October 1971.

132. D. G. D. Barr and D. K. B. Fraser, Br. Med. J., 1968-1, 737.

133. T. J. Covey, J. Pediat., 64, 218 (1964).

134. J. D. Andrade, K. Kunitomo, R. Van Wagenen, B. Kastagir, D. Gough, and W. J. Kolff, Trans Am. Soc. Artif. Int. Organs, 17, 222 (1971).

135. J. D. Andrade, K. Kopp, R. Van Wagenen, C. Chen, and W. J. Kolff, Proc. Eur. Dialysis Transplant Assoc., 4, 290 (1972).

136. H. Bickel, G. E. Hall, W. Keller-Schierlein, V. Prelog, E Vischer, and A. Wettstein, Helv. Chim. Acta, 43, 2129 (1960).

137. H. Bicker, R. Bosshard, E. Gaumann, P. Reusser, E. Vischer, W. Vosch, A. Wettstein, and H. Zahner, Ibid., 43, 218 (1960).

138. H. Bickel, H. Beberle, and E. Vischer, Ibid., 46, 1385 (1963).

139. G. Schwarzenbach and K. Schwarzenbach, Ibid., 46, 1390 (1963).

140. A. E. Martell, Stability Constants of Metal-Ion Complexes, Section II, Organic Ligands, Special Publication No. 17, Chemical Society, London, 1964.

141. J. C. Bailar Jr., Am. Sci., 59, 586 (1971).

142. D. J. Lentz, G. H. Henderson, and E. M. Eyring, Mol. Phamacol. 9, 514 (1973).

143. J. Tripod and H. Keberle, Helv. Physiol. Pharmacol. Acta, 20, 291 (1962).

144. R. J. Cornell and L. G. Donamura, J. Med. Chem. 8, 388 (1965).

145. R. J. Cornell land L. G. Donamura, J. Polym. Sci., 3A, 827 (1965).

146. L. G. Donamura and J. Razzano, J. Med. Chem., 9, 258 (1966).

147. J. R. Dombroski, L. G. Donamura, and J. Razzano, Ibid., 10, 963.

148. J. R. Dombroshi, L. G. Conamura, and J. Razzano, Ibid., 10, 964 (1967).

149. L. G. Donamura and J. Razzano, Ibid., 14, 244 (1971).

150. J. R. Dombroshi and L. G. Donamura, J. Appl. Polym. Sci., 15, 2119 (1971).

151. J. R. Dombroshi and L. G. Donamura, J. Med. Chem., 14, 460 (1971).

152. J. R. Dombroshi, L. G. Donamura, and J. Razzano, Ibid., 14, 993 (1971).

153. M. Tahan, N. Lender, and A. Zilkha, Israel J. Chem., 10, 835 (1972).

154. B. Z. Weiner, M. Tahan, and A. Zilkha, J. Med. Chem., 15, 410 (1972).

155. B. Z. Weiner and A. Zilkha, Israel J. Chem., 11, 567 (1973).

156. A. Heyd, J. Pharmacol. Sci., 60 1343 (1971).

157. H. G. Batz, G. Franzman, and H. Ringsdorf, Angew. Chem. Int. Ed., 11, 1103 (1972).

158. H. G. Batz, V. Hofmann, and H. Ringsdorf, Makromol. Chem., 169, 323 (1973).

159. B. Z. Weiner and A. Zilkha, J. Med. Chem., 16, 573 (1973).

160. H. G. Batz, G. Franzmann, and H. Ringsdorf, Makromol. Chem., 172, 27 (1973).

161. J. Coupek, J. Torkova, O. Haobalkova, and M. Kirvakova, German Patent 2,315,508 (1973); Chem. Abstr., 80, 30723 (1974).

162. G. Pauli, A. Gabest, G. Langhammer, and M. Gommoll, German (East) Patent 90,197 (1972); Chem. Abstr., 77, 143810 (1972).

163. A. Goldstein, Pharmacol. Rev., 1, 102 (1949).

164. M. C. Meyer and D. E. Guttman, J. Pharmaceut. Sci., 57, 895 (1968).

165. Y. Cohen, Int. Encycl. Pharmacol. Ther., 78, 241 (1970).

166. D. Borga, D. L. Azarnoff, G. P. Forshell, and F. Sjoqvist, Biochem. Pharmacol. Toxicol., 18, 2135 (1969).

167. G. Franksson and E. Angara, Acta Pharmacol Toxicol., 28, 209 (1970).

168. G. T. Tucker, R. M. Boyes, P. O. Bridenbaugh, and D. C. Moore, Anesthesiology, 33, 287 (1970).

169. L. E. Mather, G. J. Long, and J. Thomas, J. Pharm. Pharmacol., 23, 359 (1971).

170. G. D. Olsen, Science, 176, 525 (1972).

171. R. C. Schulz, R. Hollander, and W. Kern, Makromol. Chem., 40, 16 (1960).

172. R. S. Ramirez and J. D. Andrade, Unpublished Results.

173. O. Zaborsky, Immobilized Enzymes, CRC Press, Cleveland, 1973.

174. K. W. Pepper, H. M. Paisley, and M. A. Young, J. Chem. Soc., 1953, 4097.

175. Amberlite XAD-2 Technical Bulletin, Ion Exchange Department, Rohm and Haas Co., Philadelphia, July 1971.

176. Preliminary Technical Notes. Amberlite XAD-4, Rohm and Haas Co., Philadelphia, July 1971.

177. E. E. Ryder and P. Pezzaglia, J. Polym. Sci., A3 3459 (1965).

178. R. C. Schulz, "Polymerization of Acrolein," in Vinyl Polymerization, Part 1 (G. E. Ham ed.), Dekker, New York, 1967, p. 403.

179. R. C. Schulz, K. Meyersen, and W. Kern, Makromol. Chem., 54, 156 (1962); 59, 123 (1962).

180. Styron 678-U, Dow Chemical Co., Midland, Michigan.

181 D. Brown, H. Cherdon, and W. Kern, Techniques of Polymer Synthesis and Characterization, Wiley-Interscience, New York, 1972.

182. D. T. Warner, The Upjohn Company, Kalamozoo, Michigan, Person Communication.

183. Gailbraith Laboratories, Knoxville, Tennessee.

184. EDAX Peak Identification Chart, EDAX Laboratories, EDAX International Inc., Raleigh, North Carolina.